Descriptive and
Normative Approaches
to Human Behavior

ADVANCED SERIES ON MATHEMATICAL PSYCHOLOGY

Series Editors: H. Colonius (*University of Oldenburg, Germany*)
E. N. Dzhafarov (*Purdue University, USA*)

Advanced Series on Mathematical Psychology | Vol. 3

Descriptive and Normative Approaches to Human Behavior

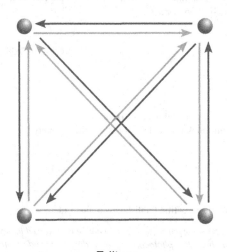

Editors

Ehtibar Dzhafarov
Purdue University, USA

Lacey Perry
Purdue University, USA

 **World Scientific**

NEW JERSEY · LONDON · SINGAPORE · BEIJING · SHANGHAI · HONG KONG · TAIPEI · CHENNAI

Published by

World Scientific Publishing Co. Pte. Ltd.

5 Toh Tuck Link, Singapore 596224

USA office: 27 Warren Street, Suite 401-402, Hackensack, NJ 07601

UK office: 57 Shelton Street, Covent Garden, London WC2H 9HE

Library of Congress Cataloging-in-Publication Data
Descriptive and normative approaches to human behavior / edited by Ehtibar Dzhafarov & Lacey Perry.
   p. cm. -- (Advanced series on mathematical psychology ; v.
   Includes bibliographical references and index.
   ISBN-13: 978-981-4368-00-1 (hbk. : alk. paper)
   ISBN-10: 981-4368-00-8 (hbk. : alk. paper)
   1. Psychology--Mathematical models. 2. Analysis (Philosophy)--Mathematical model
I. Dzhafarov, Ehtibar N. II. Perry, Lacey.
   BF39.D47 2011
   155.2--dc23

                       2011027639

British Library Cataloguing-in-Publication Data
A catalogue record for this book is available from the British Library.

Printed in Singapore.

Preface

The idea of a book of chapters some of which would be written by mathematical psychologists and some by "mathematical philosophers" (by which
term I refer to the analytic philosophers prominently using mathematical
formalisms in their work) occurred to me in 2003-2004 when I was a fellow at the Swedish Collegium for Advanced Studies in Uppsala, Sweden.
Communicating with my mathematical philosophy colleagues and listening
to their presentations I became convinced that a side-by-side display of the
two fields would be mutually beneficial. Since then I had many opportuni-
ties to witness that the factual awareness in the two fields that the methods
and styles used in them are very similar is quite low. So in 2009, when I
was a fellow at the Swedish Collegium once again, I asked my good friend
Wlodek Rabinowicz to co-organize with me a joint meeting of mathematical psychologists and mathematical philosophers, based on which we could
create a book of chapters.

We used for this purpose the 2010 Purdue Winer Memorial Lectures.
This is an annual meeting involving about 15 invited speakers who are
asked to present a representative sample of their work in 45-60 minutes
lectures. I and my colleagues at Purdue had initiated this tradition in 2002
within the framework of Benjamin Winer Fund, of which I have to say a
few words here. Ben Winer was a prominent member of the mathematical
psychology program at the Purdue Department of Psychological Sciences
in 1954-1984, perhaps best known for his influential textbook *Statistical
Principles in Experimental Design* (McGraw-Hill, 1962). At his death in
1984 he bequeathed a generous legacy to the department to be used for
purposes related to the development of "quantitative" (mathematical) psychology. For years this fund was annually contributed to by Ben Winer's
sister, Sylvia VerMeer, who considerably incremented the fund by her own
bequest at her death in 2008. I use this opportunity to express my deep
gratitude to the memory of both Ben Winer and Sylvia VerMeer.

The Purdue Winer Memorial Lectures in question took place in October

2010, and this book, as planned, is a collection of chapters written by the invited lecturers. I thank all of them for doing this work (and most of them for being prompt in doing it). My co-editor Lacey Perry, a graduate student at the Purdue mathematical psychology program, was a co-organizer of the meeting with me and Wlodek Rabinowicz. She has done the considerable work of copy-editing the chapters, converting them to Latex, and putting them together in the right format.

The chapters are arranged in alphabetic order of their authors' names.

Ehtibar Dzhafarov, West Lafayette, June 2011

Contents

Chapter 1

The Impossibility of a Satisfactory Population Ethics

Gustaf Arrhenius

Stockholm University, Swedish Collegium for Advanced Studies,
L'institut d'études avancées-Paris

Population axiology concerns how to evaluate populations in regard to their goodness, that is, how to order populations by the relations "is better than" and "is as good as". This field has been riddled with paradoxes and impossibility results which seem to show that our considered beliefs are inconsistent in cases where the number of people and their welfare varies. All of these results have one thing in common, however. They all involve an adequacy condition that rules out Derek Parfit's *Repugnant Conclusion*. Moreover, some theorists have argued that we should accept the Repugnant Conclusion and hence that avoidance of this conclusion is not a convincing adequacy condition for a population axiology. As I shall show in this chapter, however, one can replace avoidance of the Repugnant Conclusion with a logically weaker and intuitively more convincing condition. The resulting theorem involves, to the best of my knowledge, logically weaker and intuitively more compelling conditions than the other theorems presented in the literature. As such, it challenges the very existence of a satisfactory population ethics.

1.1. Introduction

Population axiology concerns how to evaluate populations in regard to their goodness, that is, how to order populations by the relations "is better than" and "is as good as". This field has been riddled with impossibility results which seem to show that our considered beliefs are inconsistent in cases where the number of people and their welfare varies.[1] All of these results

---

[1] The informal Mere Addition Paradox in Parfit (1984), pp. 419ff is the *locus classicus*. For an informal proof of a similar result with stronger assumptions, see Ng (1989), p. 240. A formal proof with slightly stronger assumptions than Ng's can be found in Blackorby and Donaldson (1991). For theorems with much weaker assumptions, see my (1999), (2000b), and especially (2000a), (2001), and (2011).

have one thing in common, however. They all involve an adequacy condition that rules out Derek Parfit's Repugnant Conclusion:

> *The Repugnant Conclusion*: For any perfectly equal population with very high positive welfare, there is a population with very low positive welfare which is better, other things being equal.[2]

A few theorists have argued that we should accept the Repugnant Conclusion and hence that avoidance of this conclusion is not a convincing adequacy condition for a population axiology.[3] As I showed in Arrhenius (2003), however, one can replace avoidance of the Repugnant Conclusion in a version of Parfit's Mere Addition Paradox with a weaker condition, namely avoidance of the following conclusion:

> *The Very Repugnant Conclusion*: For any perfectly equal population with very high positive welfare, and for any number of lives with very negative welfare, there is a population consisting of the lives with negative welfare and lives with very low positive welfare which is better than the high welfare population, other things being equal.

This conclusion seems much harder to accept than the Repugnant Conclusions. Here we are comparing one population where everybody enjoys very high quality of lives with another population where people either have very low positive welfare or very negative welfare. Even if we were to accept the Repugnant Conclusion, we are not forced to accept the Very Repugnant Conclusion. We might, for example, accept the Repugnant Conclusion but not the Very Repugnant Conclusion because we give greater moral weight to suffering than to positive welfare.

In my 2003 paper, I made use of one controversial principle, namely a version of the Mere Addition Principle. I claimed that this principle could be replaced with other conditions that are intuitively much more compelling. To properly show this is the aim of this chapter. The theorem presented here involves, to the best of my knowledge, logically weaker and intuitively more compelling conditions than all the other impossibility

---

[2]See Parfit (1984), p. 388. My formulation is more general than Parfit's apart from that he does not demand that the people with very high welfare are equally well off. Expressions such as "a population with very high positive welfare", "a population with very low positive welfare", etc., are elliptical for the more cumbersome phrases "a population consisting only of lives with very high positive welfare", "a population consisting only of lives with very low positive welfare", etc.

[3]See e.g, Mackie (1985), Hare (1988), Tännsjö (1991, 1998, 2002), Ryberg (1996).

theorems presented in the literature.

In the present theorem we shall use a condition that is slightly logically stronger than avoidance of the Very Repugnant Conclusion but still intuitively very compelling:

> *The Weak Quality Addition Condition*: For any population X, there is a perfectly equal population with very high positive welfare, and a very negative welfare level, and a number of lives at this level, such that the addition of the high welfare population to X is at least as good as the addition of any population consisting of the lives with negative welfare and any number of lives with very low positive welfare to X, other things being equal.

Consider some arbitrary population X. Roughly, according to the above condition there is at least some number of people suffering horribly, and some number of people enjoying excellent lives, such that it is better to add the people with the excellent lives to X rather than the suffering lives and any number of lives barely worth living.

The Weak Quality Addition Condition implies avoidance of the Very Repugnant Conclusion. An example of a principle that violates this condition is Total Utilitarianism according to which a population is better than another if and only if it has greater total welfare. Consider the following populations:

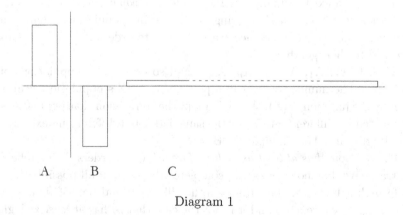

A   B   C

Diagram 1

The blocks in the above diagram represent three populations, A, B and C. The width of each block represents the number of people in the corresponding population, the height represents their welfare. Dashes indicate that the block in question should intuitively be much wider than shown, that is, the population size is intuitively much larger than shown (in this case population C).

The A-people have very high positive welfare, the B-people have very negative welfare, and the C-people have very low positive welfare. However, if there is just sufficiently many C-people, the total well-being in B∪C will be higher than in A. Thus, Total Utilitarianism ranks B∪C as better than A. This holds irrespective of how much people suffer in B and of how many they are.

1.2. The Basic Structure

For the purpose of proving the theorem, it will be useful to state some definitions and assumptions, and introduce some notational conventions. A *life* is individuated by the person whose life it is and the kind of life it is. A *population* is a finite set of lives in a possible world.[4] We shall assume that for any natural number n and any welfare level **X**, there is a possible population of n people with welfare **X**. Two populations are identical if and only if they consist of the same lives. Since the same person can exist (be instantiated) and lead the same kind of life in many different possible worlds, the same life can exist in many possible worlds. Moreover, since two populations are identical exactly if they consist of the same lives, the same population can exist in many possible worlds. A *population axiology* is an "at least as good as" quasi-ordering of all possible populations, that is, a reflexive, transitive, but not necessarily complete ordering of populations in regard to their goodness.

A, B, C,..., A_1, A_2,..., A_n, A∪B, and so on, denote populations of finite size. The number of lives in a population X (X's population size) is given by the function $N(X)$. We shall adopt the convention that populations represented by different letters, or the same letter but different indexes, are pairwise disjoint. For example, A∩B = $A_1 \cap A_2$ = ∅.

The relation *"has at least as high welfare as"* quasi-orders (being reflexive, transitive, but not necessarily complete) the set **L** of all possible lives. A life p_1 has higher welfare than another life p_2 if and only if p_1 has at least as high welfare as p_2 and it is not the case that p_2 has at least as high welfare as p_1. A life p_1 has the same welfare as p_2 if and only if p_1 has at

---

[4]For some possible constraints on possible populations, see Arrhenius (2000a; 2011, Chapter 2).

least as high welfare as p_2 and p_2 has at least as high welfare as p_1.

We shall also assume that there are possible lives with positive or negative welfare. We shall say that a life has *neutral welfare* if and only if it is equally good for the person living it as a neutral welfare component (that is, a component that neither makes her life better, nor worse for her), and that a life has *positive* (*negative*) welfare if and only if it has higher (lower) welfare than a life with neutral welfare.[5]

By a *welfare level* **A** we shall mean a set such that if a life a is in **A**, then a life b is in **A** if and only if b has the same welfare as a. In other words, a welfare level is an equivalence class on **L**. Let a^* be a life which is representative of the welfare level **A**. We shall say that a welfare level **A** is higher (lower, the same) than (as) a level **B** if and only if a^* has higher (lower, the same) welfare than (as) b^*; that a welfare level **A** is positive (negative, neutral) if and only if a^* has positive (negative, neutral) welfare; and that a life b has welfare above (below, at) **A** if and only if b has higher (lower, the same) welfare than (as) a^*.

We shall assume that *Discreteness* is true of the set of all possible lives **L** or some subset of **L**:

Discreteness: For any pair of welfare levels **X** and **Y**, **X** higher than **Y**, the set consisting of all welfare levels **Z** such that **X** is higher than **Z**, and **Z** is higher than **Y**, has a finite number of members.

The statement of the informal version of some of the adequacy conditions below, for example the Non-Elitism Condition, involve the not so exact relation "slightly higher welfare than". In the exact statements of those adequacy conditions, we shall instead make use of two consecutive welfare levels, that is, two welfare levels such that there is no welfare level in between them. Discreteness ensures that there are such welfare levels. Intuitively speaking, if **A** and **B** are two consecutive welfare levels, **A** higher than **B**, then **A** is just slightly higher than **B**. More importantly, the in-

---

[5]A welfare component is neutral relative to a certain life x iff x with this component has the same welfare as x without this component. A hedonist, for example, would typically say that an experience which is neither pleasurable nor painful is neutral in value for a person and as such does not increase or decrease the person's welfare. The above definition can of course be combined with other welfarist axiologies, such as desire and objective list theories. For a discussion of alternative definitions of a neutral life, many of which would also work fine in the present context, see Arrhenius (2011), Chapter 2. Notice that we actually do not need an analysis of a neutral welfare in the present context but rather just a criterion, and the criterion can vary with different theories of welfare.

tuitive plausibility of the adequacy conditions is preserved. Of course, this presupposes that the order of welfare levels is fine-grained, which is exactly what is suggested by expressions such as "Marc is slightly better off than Vito" and the like. Notice that Discreteness does not exclude the view that for any welfare level, there is a higher and a lower welfare level (compare with the integers).

Discreteness can be contrasted with Denseness:

Denseness: There is a welfare level in between any pair of distinct welfare levels.

My own inclination is that Discreteness rather than Denseness is true. If the latter is true, then for any two lives p_1 and p_2, p_1 with higher welfare than p_2, there is a life p_3 with welfare in between p_1 and p_2, and a life p_4 with welfare in between p_3 and p_2, and so on *ad infinitum*. It is improbable, I think, that there are such fine discrimination between the welfare of lives, even in principle. Rather, what we will find at the end of such a sequence of lives is a pair of lives in between which we cannot find any life or only lives with roughly the same welfare as both of them.

One might think otherwise, and a complete treatment of this topic would involve a detailed examination of the features of different welfarist axiologies. We shall not engage in such a discussion here. The important question is whether the validity and plausibility of the theorem below depend on whether Denseness or Discreteness is true. But that is not the case (indeed, it would have been an interesting result if the existence of a plausible axiology hinged on whether Denseness or Discreteness is true). If Denseness is true of the set of all possible lives $\mathbf{L}$, then we can form a subset $\mathbf{L}_1$ of $\mathbf{L}$ such that Discreteness is true of $\mathbf{L}_1$, and such that all the conditions which are intuitively plausible in regard to populations which are subsets of $\mathbf{L}$ are also intuitively plausible in regard to populations which are subsets of $\mathbf{L}_1$. Given that Denseness is true of $\mathbf{L}$, one cannot plausibly deny that there is such a subset $\mathbf{L}_1$ since the order of the welfare levels in $\mathbf{L}_1$ could be arbitrarily fine-grained even though Discreteness is true of $\mathbf{L}_1$. Now, since all the populations which are subsets of $\mathbf{L}_1$ also are subsets of $\mathbf{L}$, if we can show that there is no population axiology satisfying the adequacy conditions in regard to the populations which are subsets of $\mathbf{L}_1$, then it follows that there is no population axiology satisfying the adequacy conditions in regard to the populations which are subsets of $\mathbf{L}$.

Given Discreteness, we can index welfare levels with integers in a natural manner. Discreteness in conjunction with the existence of a neutral welfare level and a quasi-ordering of lives implies that there is at least one positive welfare level in **L** such that there is no lower positive welfare level.[6] Let $\mathbf{W}_1$, $\mathbf{W}_2$, $\mathbf{W}_3$,... and so forth represent positive welfare levels, starting with one of the positive welfare level for which there is no lower positive one, such that for any pair of welfare levels $\mathbf{W}_n$ and $\mathbf{W}_{n+1}$, $\mathbf{W}_{n+1}$ is higher than $\mathbf{W}_n$, and there is no welfare level **X** such that $\mathbf{W}_{n+1}$ is higher than **X**, and **X** is higher than $\mathbf{W}_n$. Analogously, let $\mathbf{W}_{-1}$, $\mathbf{W}_{-2}$, $\mathbf{W}_{-3}$,... and so on represent negative welfare levels.[7] The neutral welfare level is represented by $\mathbf{W}_0$.

A *welfare range* $\mathbf{R}(x, y)$ is a union of at least *three* welfare levels defined by two welfare levels $\mathbf{W}_x$ and $\mathbf{W}_y$, $x < y$, such that for any welfare level $\mathbf{W}_z$, $\mathbf{W}_z$ is a subset of $\mathbf{R}(x, y)$ if and only if $x \leq z \leq y$.[8] We shall say that a welfare range $\mathbf{R}(x, y)$ is higher (lower) than another range $\mathbf{R}(z, w)$ if and only if $x > w$ $(y < z)$; that a welfare range $\mathbf{R}(x, y)$ is positive (negative) if and only if $x > 0$ $(y < 0)$; and that a life p has welfare above (below, in) $\mathbf{R}(x, y)$ if and only if p is in some $\mathbf{W}_z$ such that $z > y$ $(z < x, y \geq z \geq x)$.

1.3. Adequacy Conditions

We shall make use of the following five adequacy conditions:

The Egalitarian Dominance Condition: If population A is a perfectly equal population of the same size as population B, and every person in A has higher welfare than every person in B, then A is better than B, other things being equal.

---

[6] There might be more than one since we only have a quasi-ordering of lives, that is, there might be lives and thus welfare levels which are incomparable in regard to welfare.

[7] Another way to put it is that we have a division of welfare levels into threads, each of which is completely ordered. If we assume that all lives with neutral welfare are comparable and have the same welfare level, which seems natural given our definition of a neutral welfare level, then the neutral welfare level is comparable with all other welfare levels, irrespective of the thread to which the latter level belongs. Moreover, any negative welfare level in any thread can be compared to any positive level in any thread. Both of these implications are desirable. For example, it would be odd to claim that p_1 enjoys positive welfare and p_2 suffers negative welfare but p_1 and p_2 are incomparable in regard to welfare. Still, two positive welfare levels might be incomparable, and two negative welfare levels might be incomparable. I am grateful to Kaj Børge Hansen for pressing this issue.

[8] The reason for restricting welfare ranges to unions of at least three welfare levels, as opposed to at least two welfare levels, is that this restriction allows us to simplify the exact statements of the adequacy conditions.

The Egalitarian Dominance Condition (exact formulation): For any populations A and B, $N(A)=N(B)$, and any welfare level $\mathbf{W}_x$, *if* all members of B have welfare below $\mathbf{W}_x$, and $A \subset \mathbf{W}_x$, *then* A is better than B, other things being equal.

The General Non-Extreme Priority Condition: There is a number n of lives such that for any population X, and any welfare level $\mathbf{A}$, a population consisting of the X-lives, n lives with very high welfare, and one life with welfare $\mathbf{A}$, is at least as good as a population consisting of the X-lives, n lives with very low positive welfare, and one life with welfare slightly above $\mathbf{A}$, other things being equal.

The General Non-Extreme Priority Condition (exact formulation): For any $\mathbf{W}_z$, there is a positive welfare level $\mathbf{W}_u$, and a positive welfare range $\mathbf{R}(1, y)$, $u > y$, and a number of lives $n > 0$ such that *if* $A \subset \mathbf{W}_x$, $x \geq u$, $B \subset \mathbf{R}(1,y)$, $N(A)=N(B)=n$, $C \subset \mathbf{W}_z$, $D \subset \mathbf{W}_{z+1}$, $N(C)=N(D)=1$, *then,* for any E, $A \cup C \cup E$ is at least as good as $B \cup D \cup E$, other things being equal.

The Non-Elitism Condition: For any triplet of welfare levels $\mathbf{A}$, $\mathbf{B}$, and $\mathbf{C}$, $\mathbf{A}$ slightly higher than $\mathbf{B}$, and $\mathbf{B}$ higher than $\mathbf{C}$, and for any one-life population A with welfare $\mathbf{A}$, there is a population C with welfare $\mathbf{C}$, and a population B of the same size as $A \cup C$ and with welfare $\mathbf{B}$, such that for any population X consisting of lives with welfare ranging from $\mathbf{C}$ to $\mathbf{A}$, $B \cup X$ is at least as good as $A \cup C \cup X$, other things being equal.

The Non-Elitism Condition (exact formulation): For any welfare levels $\mathbf{W}_x$, $\mathbf{W}_y$, $x - 1 > y$, there is a number of lives $n > 0$ such that *if* $A \subset \mathbf{W}_x$, $N(A)=1$, $B \subset \mathbf{W}_y$, $N(B)=n$, and $C \subset \mathbf{W}_{x-1}$, $N(C)=n+1$, *then,* for any $D \subset \mathbf{R}(y,x)$, $C \cup D$ is at least as good as $A \cup B \cup D$, other things being equal.

The Weak Non-Sadism Condition: There is a negative welfare level and a number of lives at this level such that an addition of any number of people with positive welfare is at least as good as an addition of the lives with negative welfare, other things being equal.

The Weak Non-Sadism Condition (exact formulation): There is a welfare level $\mathbf{W}_x$, $x < 0$, and a number of lives n, such that *if* $A \subset \mathbf{W}_x$,

$N(A)=n$, $B \subset \mathbf{W}_y$, $y > 0$, *then*, for any population C, $B \cup C$ is at least as good as $A \cup C$, other things being equal.

The Weak Quality Addition Condition: For any population X, there is a perfectly equal population with very high positive welfare, and a very negative welfare level, and a number of lives at this level, such that the addition of the high welfare population to X is at least as good as the addition of any population consisting of the lives with negative welfare and any number of lives with very low positive welfare to X, other things being equal.

The Weak Quality Addition Condition (exact formulation): For any population X, there is a negative welfare level $\mathbf{W}_x$, $x < 0$, two positive welfare ranges $\mathbf{R}(u,v)$ and $\mathbf{R}(1,y)$, $u > y$, and two population sizes $n > 0$, $m > 0$, such that *if* $A \subset \mathbf{W}_z$, $z \geq u$, $N(A)=n$, $B \subset \mathbf{R}(1,y)$, $C \subset \mathbf{W}_x$, $N(C)=m$ *then* $A \cup X$ is at least as good as $B \cup C \cup X$, other things being equal.

Notice that in the exact formulation of the adequacy conditions, we have eliminated concepts such as "very high positive welfare", "very low positive welfare", "very negative welfare", and the like. Hence, such concepts are not essential for our discussion and results. For example, in the exact formulation of the Weak Quality Addition Condition, we have eliminated the concepts "very low positive welfare" and "very high positive welfare" and replaced them with two non-fixed positive welfare ranges, one starting at the lowest positive welfare level, and the other one starting anywhere above the first range.

1.4. The Impossibility Theorem

The Impossibility Theorem: There is no population axiology which satisfies the Egalitarian Dominance, the General Non-Extreme Priority, the Non-Elitism, the Weak Non-Sadism, and the Weak Quality Addition Condition.

Proof. We shall show that the contrary assumption leads to a contradiction. We shall first prove two lemmas to the effect that the Non-Elitism and the General Non-Extreme Priority Conditions each imply another condition, Condition β and Condition δ respectively. We shall then prove a lemma to the effect that Weak Quality Addition Condition and Condition δ imply what we shall call the Restricted Quality Addition Condition. Finally, we shall then show that there is no population axiology which satisfies this condition in conjunction with Conditions β and δ, the Egalitarian Dominance, and the Weak Non-Sadism Condition.

1.4.1. *Lemma 1.1*

Lemma 1.1: The Non-Elitism Condition implies Condition β.

Condition β: For any triplet $\mathbf{W}_x$, $\mathbf{W}_y$, $\mathbf{W}_z$ of welfare levels, $x > y > z$, and any number of lives $n > 0$, there is a number of lives $m > n$ such that if $A \subset \mathbf{W}_x$, $N(A) = n$, $B \subset \mathbf{W}_z$, $N(B) = m$, and $C \subset \mathbf{W}_y$, $N(C) = m + n$, then, for any $D \subset \mathbf{R}(z, y + 1)$, $C \cup D$ is at least as good as $A \cup B \cup D$, other things being equal.

We shall prove Lemma 1.1 by first proving

Lemma 1.1.1: The Non-Elitism Condition entails Condition α.

Condition α: For any welfare levels $\mathbf{W}_x$, $\mathbf{W}_y$, $x - 1 > y$, and for any number of lives $n > 0$, there is a number of lives $m \geq n$ such that *if* $A \subset \mathbf{W}_x$, $N(A) = n$, $B \subset \mathbf{W}_y$, $N(B) = m$, $C \subset \mathbf{W}_{x-1}$, $N(C) = m + n$, then, for any $D \subset \mathbf{R}(y, x)$, $C \cup D$ is at least as good as $A \cup B \cup D$, other things being equal.

Proof: Let

(1) $\mathbf{W}_x$ and $\mathbf{W}_y$ be any welfare levels such that $x - 1 > y$;

(2) n be any number of lives such that $n > 0$;

(3) $p > 0$ be a number which satisfies the Non-Elitism Condition for $\mathbf{W}_x$ and $\mathbf{W}_y$.

Let $A_1, \ldots, A_{n+1}$, $B_1, \ldots, B_{n+1}$, and $C_0, \ldots, C_n$, be any three sequences of populations satisfying

(4) $A_i \subset \mathbf{W}_x$; $N(A_i) = 1$ for all i, $1 \leq i \leq n$; $A_{n+1} = \emptyset$;

(5) $B_i \subset \mathbf{W}_y$; $N(B_i) = p$, for all i, $1 \leq i \leq n$; $B_{n+1} = \emptyset$;

(6) $C_i \subset \mathbf{W}_{x-1}$; $N(C_i) = p + 1$, for all i, $1 \leq i \leq n$; $C_0 = \emptyset$.

Finally, let

(7) D be any population such that $D \subset \mathbf{R}(y, x)$.

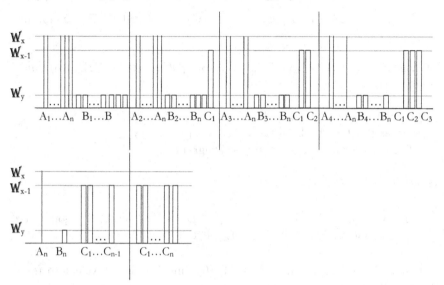

Diagram 2

The above diagram shows a selection of the involved populations in a case where $n \geq 6$. Dots in between two blocks indicate that there is a number of same sized blocks which have been omitted from the diagram. Population D is omitted throughout.

Recall that we have adopted the convention that populations represented by different letters, or the same letter but different indexes, are pairwise disjoint. Hence, for example, $A_1 \cap A_2 = \emptyset$.

Since $\mathbf{W}_x$ and $\mathbf{W}_y$ can be any pair of welfare levels separated by at least one welfare level, and D can be any population consisting of lives with welfare ranging from $\mathbf{W}_y$ to $\mathbf{W}_x$, and $N(A_1 \cup \ldots \cup A_n) = n$ (by (4)) can be any number of lives greater than zero, and $N(B_1 \cup \ldots \cup B_n) = np \geq n$ (by (5)), we can show that Lemma 1.1.1 is true by showing that $C_1 \cup \ldots \cup C_n \cup D$ is at least as good as $A_1 \cup \ldots \cup A_n \cup B_1 \cup \ldots \cup B_n \cup D$. This suffices since $A_1, \ldots,$ $A_{n+1}, B_1, \ldots, B_{n+1}, C_1, \ldots, C_n$, and D are arbitrary populations satisfying (4)-(7).

It follows from (3)-(6) and the Non-Elitism Condition that

(8) $C_i \cup E$ is at least as good as $A_i \cup B_i \cup E$ for all i, $1 \leq i \leq n$ and any

$E \subset \mathbf{R}(y, x)$

and from (4)-(7) that

(9) $A_{i+1} \cup \ldots \cup A_{n+1} \cup B_{i+1} \cup \ldots \cup B_{n+1} \cup C_0 \cup \ldots \cup C_{i-1} \cup D \subset \mathbf{R}(y, x)$ for all i, $1 \leq i \leq n$.

Letting $E = A_{i+1} \cup \ldots \cup A_{n+1} \cup B_{i+1} \cup \ldots \cup B_{n+1} \cup C_0 \cup \ldots \cup C_{i-1} \cup D$, (8) and (9) imply that

(10) $C_i \cup [A_{i+1} \cup \ldots \cup A_{n+1} \cup B_{i+1} \cup \ldots \cup B_{n+1} \cup C_0 \cup \ldots \cup C_{i-1} \cup D]$ is at least as good as $A_i \cup B_i \cup [A_{i+1} \cup \ldots \cup A_{n+1} \cup B_{i+1} \cup \ldots \cup B_{n+1} \cup C_0 \cup \ldots \cup C_{i-1} \cup D]$ for all i, $1 \leq i \leq n$ (see Diagram 2).

Transitivity and (10) yield

(11) $C_n \cup A_{n+1} \cup B_{n+1} \cup C_0 \cup \ldots \cup C_{n-1} \cup D$ is at least as good as $A_1 \cup B_1 \cup A_2 \cup \ldots \cup A_{n+1} \cup B_2 \cup \ldots \cup B_{n+1} \cup C_0 \cup D$

and since $A_{n+1} = B_{n+1} = C_0 = \emptyset$ by (4)-(6), line (11) is equivalent to (see Diagram 2)

(12) $C_1 \cup \ldots \cup C_n \cup D$ is at least as good as $A_1 \cup \ldots \cup A_n \cup B_1 \cup \ldots \cup B_n \cup D$.

$\square$

To show that Lemma 1.1 is true, we now need to prove

Lemma 1.1.2: Condition α entails Condition β.

Proof. Let

(1) $\mathbf{W}_x$, $\mathbf{W}_y$, $\mathbf{W}_z$ be any three welfare levels such that $x > y > z$;
(2) $r = x - y$.

Let $A_1, \ldots, A_{r+1}$ and $B_1, \ldots, B_{r+1}$ be any two sequences of populations, $m_0, \ldots, m_r$ any sequence of integers, and f a function satisfying

(3) $m_0 > 0$;
(4) $f(m_i) = m_0 + m_1 + \ldots + m_i$, for all i, $0 \leq i \leq r$;
(5) $m_i \geq f(m_{i-1})$ satisfies Condition α for $\mathbf{W}_{x-(i-1)}$, $\mathbf{W}_z$, and $f(m_{i-1})$ for all i, $1 \leq i \leq r$;

(6) $A_i \subset \mathbf{W}_{x-(i-1)}$, $N(A_i) = f(m_{i-1})$ for all i, $1 \le i \le r+1$;

(7) $B_i \subset \mathbf{W}_z$, $N(B_i) = m_i$, for all i, $1 \le i \le r$; $B_{r+1} = \emptyset$.

Finally, let

(8) D be any population such that $D \subset \mathbf{R}(z, y+1)$.

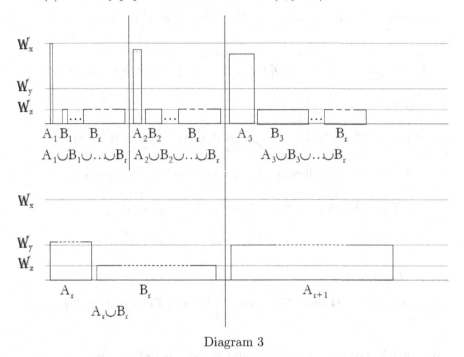

Diagram 3

The above diagram shows a selection of the involved populations in a case where $r \ge 4$. Population D is omitted throughout.

We can conclude from (3)-(7) that $N(B_1 \cup \ldots \cup B_r) > m_0 = N(A_1)$. Consequently, since $\mathbf{W}_x$, $\mathbf{W}_y$, and $\mathbf{W}_z$ can be any welfare levels such that $x > y > z$, and D can be any population consisting of lives with welfare ranging from $\mathbf{W}_z$ to $\mathbf{W}_{y+1}$, we can show that Condition α implies Condition β by showing that $A_{r+1} \cup D$ is at least as good as $A_1 \cup B_1 \cup \ldots \cup B_r \cup D$. This suffices since $A_1, \ldots, A_{r+1}$, $B_1, \ldots, B_r$, and D are arbitrary populations satisfying (6)-(8).

From (3)-(7) and Condition α, it follows that

(9) $A_{i+1} \cup E$ is at least as good as $A_i \cup B_i \cup E$ for all i, $1 \le i \le r$ and any

$E \subset \mathbf{R}(z, y+1)$,

and from (7) and (8) that

(10) $B_{i+1} \cup \ldots \cup B_{r+1} \cup D \subset \mathbf{R}(z, y+1)$ for all i, $1 \leq i \leq r$.

Consequently, letting $E = B_{i+1} \cup \ldots \cup B_{r+1} \cup D$, (9) and (10) imply that

(11) $A_{i+1} \cup [B_{i+1} \cup \ldots \cup B_{r+1} \cup D]$ is at least as good as
$A_i \cup B_i \cup [B_{i+1} \cup \ldots \cup B_{r+1} \cup D]$ for all i, $1 \leq i \leq r$ (see Diagram 3).

Transitivity and (11) yield

(12) $A_{r+1} \cup B_{r+1} \cup D$ is at least as good as $A_1 \cup B_1 \cup \ldots \cup B_{r+1} \cup D$

and since $B_{r+1} = \emptyset$ (7), line (12) is equivalent to (see Diagram 3)

(13) $A_{r+1} \cup D$ is at least as good as $A_1 \cup B_1 \cup \ldots \cup B_r \cup D$. $\square$

It follows trivially from Lemmas 1.1.1 and 1.1.2 that Lemma 1.1 is true. $\square$

1.4.2. *Lemma 1.2*

Lemma 1.2: The General Non-Extreme Priority Condition implies Condition δ.

Condition δ: For any $\mathbf{W}_z$, $z < 0$, and any number of lives $m > 0$, there is a positive welfare level $\mathbf{W}_u$, and a positive welfare range $\mathbf{R}(1, y)$, $u > y$, and a number of lives $n > 0$ such that *if* $A \subset \mathbf{W}_x$, $x \geq u$, $B \subset \mathbf{R}(1, y)$, $N(A) = N(B) = n$, $C \subset \mathbf{W}_z$, $D \subset \mathbf{W}_3$, $N(C) = N(D) = m$, *then*, for any E, $A \cup C \cup E$ is at least as good as $B \cup D \cup E$, other things being equal.

We shall prove Lemma 1.2 by first proving

Lemma 1.2.1: The General Non-Extreme Priority Condition implies *Condition χ*.

Condition χ: For any $\mathbf{W}_z$, $z < 0$, there is a positive welfare level $\mathbf{W}_u$, and a positive welfare range $\mathbf{R}(1, y)$, $u > y$, and a number of lives $n > 0$ such that *if* $A \subset \mathbf{W}_x$, $x \geq u$, $B \subset \mathbf{R}(1, y)$, $N(A) = N(B) = n$, $C \subset \mathbf{W}_z$, $D \subset \mathbf{W}_3$,

$N(C)=N(D)=1$, *then*, for any E, AUCUE is at least as good as BUDUE, other things being equal.

Proof: Let

(1) $\mathbf{W}_z$ be any welfare level such that $z < 0$;

(2) $r = 3 - z$;

(3) $\mathbf{W}_{u_i}$ be a positive welfare level, $\mathbf{R}(1, v_i)$ be a positive welfare range, and n_i a number of lives which satisfy the General Non-Extreme Priority Condition for $\mathbf{W}_{z+(i-1)}$ for all i, $1 \leq i \leq r$;

(4) $\mathbf{W}_u$ be a welfare level such that u equals the maximal element in $\{u_i: 1 \leq i \leq r\}$;

(5) $\mathbf{W}_x$ be a welfare level such that $x \geq u$;

(6) y be a number such that y equals the minimal element in $\{v_i: 1 \leq i \leq r\}$.

Let $A_1, \ldots, A_{r+1}$, $B_0, \ldots, B_r$, and $C_1, \ldots, C_{r+1}$, be any three sequences of populations satisfying

(7) $A_i \subset \mathbf{W}_x$, $N(A_i) = n_i$, for all i, $1 \leq i \leq r$; $A_{r+1} = \emptyset$;

(8) $B_i \subset \mathbf{R}(1, y)$, $N(B_i) = n_i$, for all i, $1 \leq i \leq r$; $B_0 = \emptyset$;

(9) $C_i \subset \mathbf{W}_{z+(i-1)}$, $N(C_i)=1$, for all i, $1 \leq i \leq r+1$.

Finally, let

(10) E be any population.

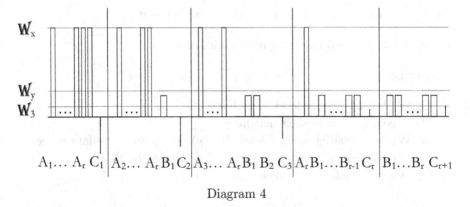

Diagram 4

The above diagram shows a selection of the involved populations in a case where $r \geq 4$. Population E is omitted throughout.

Since $\mathbf{W}_z$ can be any negative welfare level (by (1)), and $\mathbf{W}_x$ can be any welfare level at least as high as $\mathbf{W}_u$ (by (5)), and since it follows from (3) and (6) that $\mathbf{R}(1,y)$ is a welfare range such that $u > y$, we can show that Lemma 1.2.1 is true by showing that $A_1 \cup \ldots \cup A_r \cup C_1 \cup E$ is at least as good as $B_1 \cup \ldots \cup B_r \cup C_{r+1} \cup E$. This suffices since $A_1, \ldots, A_r, B_1, \ldots, B_r$, $C_1, \ldots, C_{r+1}$, and E are arbitrary populations satisfying (7)-(10).

The General Non-Extreme Priority Condition and (3)-(9) imply that

(11) $A_i \cup C_i \cup F$ is at least as good as $B_i \cup C_{i+1} \cup F$ for all i, $1 \leq i \leq r$ and any population F.

Letting $F = A_{i+1} \cup \ldots \cup A_{r+1} \cup B_0 \cup \ldots \cup B_{i-1} \cup E$, it follows from (11) that

(12) $A_i \cup C_i \cup [A_{i+1} \cup \ldots \cup A_{r+1} \cup B_0 \cup \ldots \cup B_{i-1} \cup E]$ is at least as good as $B_i \cup C_{i+1} \cup [A_{i+1} \cup \ldots \cup A_{r+1} \cup B_0 \cup \ldots \cup B_{i-1} \cup E]$ for all i, $1 \leq i \leq r$ (see Diagram 4).

Transitivity and (12) yield

(13) $A_1 \cup C_1 \cup A_2 \cup \ldots \cup A_{r+1} \cup B_0 \cup E$ is at least as good as $B_r \cup C_{r+1} \cup A_{r+1} \cup B_0 \cup \ldots \cup B_{r-1} \cup E$,

and since $A_{r+1} = B_0 = \emptyset$ by (7)-(8), line (13) is equivalent to (see Diagram 4)

(14) $A_1 \cup \ldots \cup A_r \cup C_1 \cup E$ is at least as good as $B_1 \cup \ldots \cup B_r \cup C_{r+1} \cup E$. $\square$

To show that Lemma 1.2 is true, we now need to prove

Lemma 1.2.2: Condition χ implies Condition δ.

Proof: Let

(1) $\mathbf{W}_z$ be any welfare level such that $z < 0$;
(2) m be any number such that $m > 0$;
(3) $\mathbf{W}_u$ be a positive welfare level, $\mathbf{R}(1,y)$ be a positive welfare range, and n a number of lives which satisfy Condition χ for $\mathbf{W}_z$;
(4) $\mathbf{W}_x$ be a welfare level such that $x \geq u$.

Let $A_1, \ldots, A_{m+1}$, $B_0, \ldots, B_m$, $C_1, \ldots, C_{m+1}$, and $D_0, \ldots, C_m$, be any four sequences of populations satisfying

(5) $A_i \subset \mathbf{W}_x$, $N(A_i) = n$, for all i, $1 \le i \le m$; $A_{m+1} = \emptyset$;

(6) $B_i \subset \mathbf{R}(1, y)$, $N(B_i) = n$, for all i, $1 \le i \le m$; $B_0 = \emptyset$;

(7) $C_i \subset \mathbf{W}_z$, $N(C_i) = 1$, for all i, $1 \le i \le m$; $C_{m+1} = \emptyset$;

(8) $D_i \subset \mathbf{W}_3$, $N(D_i) = 1$, for all i, $1 \le i \le m$; $D_0 = \emptyset$.

Finally, let

(9) E be any population.

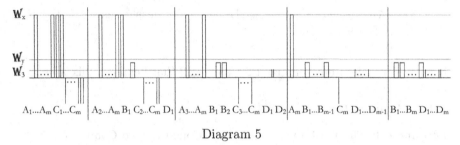

Diagram 5

The above diagram shows a selection of the involved populations in a case where $m \ge 4$. As before, population E is omitted throughout.

Since $\mathbf{W}_z$ can be any negative welfare level (by (1)), and $\mathbf{W}_x$ can be any welfare level at least as high as $\mathbf{W}_u$ (by (5)), and m can be any number of lives greater than zero, and $\mathbf{R}(1, y)$ is a welfare range such that $u > y$, and n is a number greater than zero (by (3)), we can show that Lemma 1.2.2 is true by showing that $A_1 \cup \ldots \cup A_m \cup C_1 \cup \ldots \cup C_m \cup E$ is at least as good as $B_1 \cup \ldots \cup B_m \cup D_1 \cup \ldots \cup D_m \cup E$. This suffices since $A_1, \ldots, A_m$, $B_1, \ldots, B_m$, $C_1, \ldots, C_m$, $D_1, \ldots, D_m$, and E are arbitrary populations satisfying (5)-(9).

It follows from (3)-(8) and Condition χ that

(10) $A_i \cup C_i \cup F$ is at least as good as $B_i \cup D_i \cup F$ for all i, $1 \le i \le m$, and any population F

which, for $F = A_{i+1} \cup \ldots \cup A_{m+1} \cup C_{i+1} \ldots \cup C_{m+1} \cup B_0 \cup \ldots \cup B_{i-1} \cup D_0 \cup \ldots \cup D_{i-1} \cup E$, in turn implies

(11) $A_i \cup C_i \cup [A_{i+1} \cup \ldots \cup A_{m+1} \cup C_{i+1} \ldots \cup C_{m+1} \cup B_0 \cup \ldots \cup B_{i-1} \cup D_0 \cup \ldots \cup D_{i-1} \cup E]$ is at least as good as $B_i \cup D_i \cup [A_{i+1} \cup \ldots \cup A_{m+1} \cup C_{i+1} \cup \ldots \cup C_{m+1} \cup B_0 \cup \ldots \cup B_{i-1} \cup D_0 \cup \ldots \cup D_{i-1} \cup E]$ for all i, $1 \le i \le m$

(see Diagram 5).

Transitivity and (11) yield

(12) $A_1 \cup C_1 \cup A_2 \cup \ldots \cup A_{m+1} \cup C_2 \ldots \cup C_{m+1} \cup B_0 \cup D_0 \cup E$ is at least as good as $B_m \cup D_m \cup A_{m+1} \cup C_{m+1} \cup B_0 \cup \ldots \cup B_{m-1} \cup D_0 \ldots \cup D_{m-1} \cup E$

and since $A_{m+1} = B_0 = C_{m+1} = D_0 = \emptyset$ (by (5)-(8)), line (12) is equivalent to (see Diagram 5)

(13) $A_1 \cup \ldots \cup A_m \cup C_1 \ldots \cup C_m \cup E$ is at least as good as $B_1 \cup \ldots \cup B_m \cup D_1 \ldots \cup D_m \cup E$.   □

It follows trivially from Lemmas 1.2.1 and 1.2.2 that Lemma 1.2 is true. □

1.4.3. *Lemma 1.3*

Lemma 1.3: The Weak Quality Addition Condition and Condition δ imply the Restricted Quality Addition Condition.

The Restricted Quality Addition Condition (exact formulation): For any population X, there is a positive welfare level $\mathbf{W}_x$ and a positive welfare range $\mathbf{R}(1, y)$, $x > y$, and a population size n and m such that *if* $A \subset \mathbf{W}_z$, $z \geq x$, $N(A){=}n$, $B \subset \mathbf{R}(1, y)$, $N(B){=}p$, $p \geq m$, *then* $A \cup X$ is at least as good as $B \cup X$, other things being equal.

Proof. Let

(1) X be any population;
(2) $\mathbf{R}(w, t)$ and $\mathbf{R}(1, v)$,$w > v$, be two welfare ranges, $\mathbf{W}_z$ a negative welfare level, and p and m two population sizes, which satisfy the Weak Quality Addition Condition for X;
(3) $\mathbf{W}_u$ be a positive welfare level, $\mathbf{R}(1, y)$ a welfare range, and n a number of lives, which satisfy Condition δ for $\mathbf{W}_z$ and m;
(4) Let $\mathbf{W}_x$ be a welfare level such that $x{=}\max(w, u)$;[9]
(5) $A_1 \subset \mathbf{W}_x$, $N(A_1) = n$;
(6) $A_2 \subset \mathbf{W}_x$, $N(A_2) = p$;
(7) $C \subset \mathbf{W}_z$, $N(C){=}m$;
(8) $B_1 \subset \mathbf{R}(1, v) \cap \mathbf{R}(1, y)$, $N(B_1) = n$;

---

[9] We define the function $\max(x, y)$ in the ordinary way: $\max(x, y) = x$ if $x \geq y$, otherwise $\max(x, y) = y$.

(9) $B_3 \subset \mathbf{W}_3$, $N(B_3) = m$;

(10) $q \geq 0$ be any population size;

(11) $B_2 \subset \mathbf{R}(1, v) \cap \mathbf{R}(1, y)$, $N(B_2) = q$.

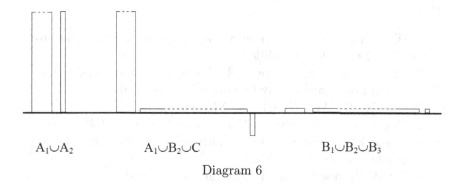

$A_1 \cup A_2$ $\qquad$ $A_1 \cup B_2 \cup C$ $\qquad$ $B_1 \cup B_2 \cup B_3$

Diagram 6

Population X is omitted throughout in the above diagram.

We can conclude from (8)-(11) that $B_1 \cup B_2 \cup B_3$ can be of any size greater than or equal to $p + m$ since B_2 can be of any size. Moreover, since $A_1 \cup A_2 \subset \mathbf{W}_x$, $x > v$ (by (2), (4)-(6)) and $B_1 \cup B_2 \cup B_3 \subset \mathbf{R}(1, v) \cap \mathbf{R}(1, y)$ (by (8), (9), (11)), we can show that Lemma 1.3 is true by showing that $A_1 \cup A_2 \cup X$ is at least as good as $B_1 \cup B_2 \cup B_3 \cup X$. This suffices since X can be any population (by (1)).

It follows from (2), (4), (6), (11), and the Weak Quality Addition Condition that

(12) $A_1 \cup A_2 \cup X$ is at least as good as $A_1 \cup B_2 \cup C \cup X$ (see Diagram 6).

It follows from (3)-(5), (7)-(9), and Condition δ that

(13) $A_1 \cup B_2 \cup C \cup X$ is at least as good as $B_1 \cup B_2 \cup B_3 \cup X$ (see Diagram 6).

By transitivity, it follows from (12) and (13) that

(14) $A_1 \cup A_2 \cup X$ is at least as good as $B_1 \cup B_2 \cup B_3 \cup X$. $\qquad$ □

1.4.4. *Lemma 1.4*

We shall show that the theorem is true by proving

Lemma 1.4: There is no population axiology which satisfies Condition β

and δ, the Egalitarian Dominance, the Restricted Quality Addition, and the Weak Non-Sadism Condition.

Proof. We show that the contrary assumption leads to a contradiction. Let

(1) $\mathbf{W}_z$ be a negative welfare level and m a population size which satisfy the Weak Non-Sadism Condition;

(2) $\mathbf{W}_u$ be a positive welfare level, $\mathbf{R}(1,y)$ a welfare range, and n a number of lives, which satisfy Condition δ for $\mathbf{W}_z$ and m;

(3) $B_1 \subset \mathbf{W}_3$, $B_2 \subset \mathbf{W}_3$, $N(B_1) = n$, $N(B_2) = m$;

(4) $\mathbf{W}_w$ be a welfare level, and $\mathbf{R}(1,v)$, $w > v$, be a welfare range, and p and k two population sizes, which satisfy the Restricted Quality Addition Condition for $B_1 \cup B_2$;

(5) Let $\mathbf{W}_x$ be a welfare level such that $x = \max(w, u)$;

(6) $A \subset \mathbf{W}_x$, $N(A) = p$;

(7) $H \subset \mathbf{W}_x$, $N(H) = n$;

(8) $E \subset \mathbf{W}_z$, $N(E) = m$.

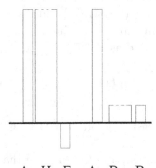

A∪H∪E   A∪B₁∪B₂

Diagram 7

It follows from the definition of a welfare range that $\mathbf{W}_3 \subset \mathbf{R}(1,y)$. Accordingly, from (3) we know that $B_1 \subset \mathbf{R}(1,y)$. Consequently, from (2), (3), (7), (8), and Condition δ we get that

(9) A∪H∪E is at least as good as A∪B₁∪B₂ (see Diagram 7).

Let

(10) $r > n + p$ be a number of lives which satisfies Condition β for the three welfare levels $\mathbf{W}_x$, $\mathbf{W}_2$, and $\mathbf{W}_1$ and for $n + p$ lives at $\mathbf{W}_x$;
(11) q be any number of lives such that $q \geq m + k$ and $q \geq r$;
(12) $G \subset \mathbf{W}_2$, $N(G) = n + p + r$;
(13) $I \subset \mathbf{W}_1$, $N(I) = q - r$;
(14) $F \subset \mathbf{W}_1$, $N(F) = r$.

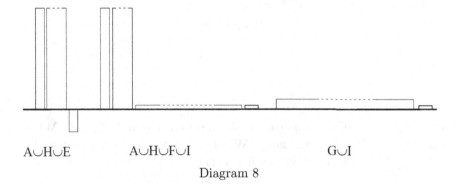

AᵁHᵁE   AᵁHᵁFᵁI   GᵁI

Diagram 8

Since AᵁH $\subset \mathbf{W}_x$, and $N($AᵁH$) = n + p$ (by (6) and (7)), and I $\subset \mathbf{R}(1, 3)$ (by the definition of a welfare range), it follows from (10)-(14) and Condition β that

(15) GᵁI is at least as good as AᵁHᵁFᵁI (see Diagram 8).

Since the F- and the I-lives have positive welfare (by (13) and (14)), it follows from (1), (8) and the Weak Non-Sadism Condition that

(16) AᵁHᵁFᵁI is at least as good as AᵁHᵁE (see Diagram 8).

By transitivity, it follows from (15) and (16) that

(17) GᵁI is at least as good as AᵁHᵁE.

Let

(18) $C \subset \mathbf{W}_3$, $N(C) = p + q - m$.

Since $\mathbf{W}_3 \subset \mathbf{R}(1, v)$, we can conclude that $C \subset \mathbf{R}(1, v)$ and since $q \geq m + k$

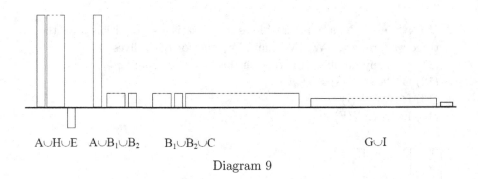

$$A \cup H \cup E \quad A \cup B_1 \cup B_2 \qquad B_1 \cup B_2 \cup C \qquad\qquad\qquad G \cup I$$

Diagram 9

(by (11)) and $N(C)=p+q-m$ (by (18)), we know that $N(C) \geq k$. Moreover, since $x \geq w$ (by (5)), and $A \subset \mathbf{W}_x$ (by (6)), it follows from (4) and the Restricted Quality Addition Condition that

(19) $A \cup B_1 \cup B_2$ is at least as good as $B_1 \cup B_2 \cup C$ (see Diagram 9).

Since $B_1 \cup B_2 \cup C \subset \mathbf{W}_3$ (by (3) and (18)) and $G \cup I \subset \mathbf{W}_1 \cup \mathbf{W}_2$, (by (12) and (13)) and $N(B_1 \cup B_2 \cup C)=N(G \cup I)$, the Egalitarian Dominance Condition implies that

(20) $G \cup I$ is worse than $B_1 \cup B_2 \cup C$ (see Diagram 9).

By transitivity, it follows from (19) and (20) that

(21) $G \cup I$ is worse than $A \cup B_1 \cup B_2$

and from (9) and (21) that

(22) $G \cup I$ is worse than $A \cup H \cup E$

which contradicts (17).   □

It follows trivially from Lemmas 1.1-1.4 that the impossibility theorem is true.   □

1.5. Discussion

The above theorem shows that our considered moral beliefs are mutually inconsistent, that is, necessarily at least one of our considered moral beliefs is false. Since consistency is, arguably, a necessary condition for moral justification, we would thus seem to be forced to conclude that there is no moral theory which can be justified. In other words, the cases in population ethics involving future generations of different sizes constitute a serious challenge to the existence of a satisfactory moral theory.

The theorem presupposes that the relation "is at least as good as" is transitive. Some theorists find this a matter of logic, claiming that it is part of the meaning of "better than" and "equally as good as" (Broome, 1991, p. 11). Although we are inclined to agree, one might think otherwise, and argue that the impossibility theorem actually demonstrates that these relations are non-transitive.[10] What is attractive with this move is that given non-transitivity of "at least as good as" and "better than", we can stick to our axiological evaluations without any contradiction.[11] However, as I have shown elsewhere, the axiological population theorems, including the above one, can be reconstructed on the normative level, in terms of what one ought to choose, without any appeal to transitivity (see Arrhenius 2004, 2011). Instead of a non-transitive ordering of populations, one gets a situation in which all of the available actions are forbidden, i.e., a moral dilemma. Hence, it does not look like we can exorcise the paradoxes of population ethics by giving up some formal condition like the transitivity of "better than".

In our discussion we have assumed that welfare is at least sometimes interpersonally comparable. Without this assumption, claims such as "Iwao is better off than Ben" would not be meaningful. In other words, conditions such as the Egalitarian Dominance and the Non-Elitism Condition, in

---

[10]Among others, Larry Temkin and Stuart Rachels have suggested something like this, see Rachels (1998, 2001) and Temkin (1987, 1996). See Broome (2004), Section 4.1, for a thorough discussion of various arguments against the transitivity of betterness, including Temkin's and Rachel's arguments.

[11]However, as many seem to fear (including Temkin), non-transitivity of "better than" might spell the end for any axiology-based morality and practical reason in general. Temkin suggests that arguments to the effect that "better than" is non-transitive "are [perhaps] best interpreted as a frontal assault on the intelligibility of consequentialist reasoning about morality and rationality. Such reasoning may need to be severely limited, if not jettisoned altogether." (Temkin 1987, p. 186, fn. 49). Elsewhere, he considers that non-transitivity "... opens the possibility that there would be no rational basis for choosing between virtually any alternatives" (Temkin 1996, p. 209). This needs to be shown, however. I argue the contrary in Arrhenius (2011), Chapter 12.

their normative or axiological guise, would not make sense. The adequacy conditions and the theorems are quite undemanding, however, in regard to measurement of welfare. It does not matter whether welfare is measurable on an ordinal, interval or ratio scale, for example. The conditions and theorems only presuppose that lives are quasi-ordered by the relation "has at least as high welfare as".

It is interesting to compare the information demands of the present theorems with that of Arrow's famous impossibility theorem (see Arrow, 1963). Without interpersonal comparability of welfare, one gets Arrowian impossibility results already in a fixed population size setting.[12] Not surprisingly then, the standard remedy for such impossibility results is to introduce some kind of interpersonal comparability of welfare.[13] But with interpersonal comparability of welfare, and some minimal demands on the orderings of lives, we come up against the impossibility theorem presented here. Moreover, and more worryingly, this result holds even if we have complete interpersonal comparability of welfare on a ratio scale, that is, access to all the possible information about people's welfare for which we could wish.

Acknowledgments

This chapter is a significantly revised and expanded version of Arrhenius (2009). I would like to thank Kaj Børge Hansen for his exceptionally helpful and detailed comments on an earlier version of this theorem. Thanks also to CERSES, CNRS, and IEA-Paris for being such a generous host during some of the time when this work was done. Financial support from the Bank of Sweden Tercentenary Foundation, IEA-Paris, and the Swedish Collegium for Advanced Study is gratefully acknowledged.

References

Arrhenius, G. (1999). An impossibility theorem in population axiology with weak ordering assumptions. In R. Sliwinski (Ed.), _Philosophical Crumbs. Uppsala Philosophical Studies_, vol. 49. Uppsala: Department of Philosophy, Uppsala University.

Arrhenius, G. (2000a). Future generations: A challenge for moral theory (Doctoral dissertation). Uppsala University.

---

[12]See Sen (1970), pp. 123-125, 128-130, and Roemer (1996), pp. 26-36.

[13]Roemer (1996), p. 36, among many others, suggests this.

Arrhenius, G. (2000b). An impossibility theorem for welfarist axiologies. *Economics and Philosophy, 16*, 247–266.

Arrhenius, G. (2001). What Österberg's population theory has in common with Plato's. In E. Carlson and R. Sliwinski (Eds.), *Omnium-gatherum. Uppsala Philosophical Studies*, vol. 50. Uppsala: Department of Philosophy, Uppsala University.

Arrhenius, G. (2003). The Very Repugnant Conclusion. In K. Segerberg and R. Sliwinski (Eds.), *Logic, Law, Morality: Thirteen Essays in Practical Philosophy. Uppsala Philosophical Studies*, vol. 51. Uppsala: Department of Philosophy, Uppsala University.

Arrhenius, G. (2009). One more axiological impossibility theorem. In R. Sliwinski, L. Johansson, and J. Österberg, (Eds.), *Logic, Ethics, and All That Jazz. Uppsala Philosophical Studies*, vol. 57. Uppsala: Department of Philosophy, Uppsala University.

Arrhenius, G. (2011). *Population Ethics.* Oxford: Oxford University Press.

Arrow, K. J. (1963). *Social Choice and Individual Values* (2nd ed.). New Haven and London: Yale University Press.

Blackorby, C., & Donaldson, D. (1991). Normative population theory: A comment. *Social Choice and Welfare, 8*, 261–267.

Broome, J. (1991). *Weighing Goods: Equality, Uncertainty, and Time.* Oxford: Basil Blackwell.

Broome, J. (2004). *Weighing Lives.* Oxford: Oxford University Press.

Hare, R. M. (1988). Possible people. *Bioethics, 2*(4) (reprinted in *Essays on Bioethics*, Oxford: Clarendon Press, 1993).

Mackie, J. L. (1985). Parfit's population paradox. In J. Mackie and P. Mackie (Eds.), *Persons and Values.* Oxford: Oxford University Press.

Ng, Y. K. (1989). What should we do about future generations? Impossibility of Parfit's theory X. *Economics and Philosophy, 5*, 235–253.

Rachels, S. (1998). Counterexamples to the transitivity of 'better than'. *Australasian Journal of Philosophy, 76*, 71–83.

Rachels, S. (2001). A set of solutions to Parfit's problems. *Noûs, 35*, 214–238.

Parfit, D. (1984). *Reasons and Persons.* Oxford: Oxford University Press.

Roemer, J. E. (1996). *Theories of Distributive Justice.* Cambridge, MA: Harvard University Press.

Ryberg, J. (1996). Topics on Population Ethics. Doctoral dissertation, University of Copenhagen, Copenhagen, Denmark.

Sen, A. (1970). *Collective Choice and Social Welfare. Mathematical Economics Texts, 5.*

Tännsjö, T. (1991). *Göra Barn.* Borås: Sesam förlag.

Tännsjö, T. (1998). *Hedonistic Utilitarianism.* Edinburgh: Edinburgh University Press.

Tännsjö, T. (2002). Why we ought to accept the Repugnant Conclusion. *Utilitas, 14* (3) (reprinted in J. Ryberg and T. Tännsjö (Eds.), *The Repugnant Conclusion: Essays on Population Ethics,* Dordrecht: Kluwer Academic Publisher, 2005).

Temkin, L. S. (1987). Intransitivity and the Mere Addition Paradox. *Philosophy and Public Affairs, 16,* 138–187.

Temkin, L. S. (1996). A continuum argument for intransitivity. *Philosophy and Public Affairs, 25,* 175–210.

Chapter 2

Explaining Interference Effects Using Quantum Probability Theory

Jerome R. Busemeyer and Jennifer S. Trueblood

Indiana University

This chapter examines the empirical evidence for interference effects in psychological experiments. It also reviews the competing interpretations of these effects with respect to traditional cognitive models and new quantum cognition models.

2.1. Introduction

Human behavior is not deterministic. If the same person is asked the same question on two different occasions, then there is some reasonable chance that the answers are inconsistent, even when there is no *known* intervention to account for this change. Behavior seems to vary across people, and within a person, it varies across occasions in ways that are far from perfectly predictable. Consequently psychologists need to use some type of probability theory to account for this indeterministic behavior. What kind of probability theory is best for modeling human behavior?

Almost all previous theoretical work in psychology has been developed along the lines of Kolmogorov's theory (Kolmogorov, 1933/1950). According to this theory, probabilities are assigned to events represented as sets from a sample space that form a Boolean algebra. While this has certainly proved to be a useful theory, the Boolean logic that lies at its foundation may be too restrictive to fully account for human behavior. Another probability theory is based on the intuitionistic logic of Brouwer (Narens, 2007). According to this theory, probabilities are assigned to events represented by open sets from a topology. This theory relaxes the complementation axiom of Boolean algebra, and provides one way to generalize Kolmogorov's theory. A third probability theory is based on quantum principles (von Neumann, 1932/1955). According to this theory, probabilities are assigned to events represented as subspaces of a vector space. This theory relaxes the distributive axiom of Boolean algebra, which provides another way to generalize Kolmogorov's theory. This chapter explores the latter approach.

Why consider quantum probability for human behavior? One good reason is the pervasive finding of interference effects in psychology (Khrennikov, 2010). An interference effect is a violation of the law of total probability, which is a theorem of Kolmogorov theory derived from the distributive axiom. In other words, interference effects suggest that the distributive axiom may be violated in psychology. The finding of interference effects in particle physics was the primary reason for the construction of quantum theory by physicists from the beginning. What is an interference effect? What is the empirical evidence for these effects? What is the best explanation for these effects?

The purpose of this chapter is to answer these three questions, but before we do, we should note that there are two different lines of research using quantum theory in psychology. One is to develop a quantum physical model of the brain (Hammeroff, 1998), and the other is to develop models that are called 'quantum like' (Khrennikov, 2010) or generalized quantum (Atmanspacher & Filk, 2010) or quantum structural (Aerts, 2009) models. The latter are not quantum physics models of the brain, but instead they are mathematical models of human behavior derived from principles abstracted and extrapolated from quantum theory. This chapter is only concerned with the latter type of theory.

2.2. What Is an Interference Effect?

Suppose we have two different judgment tasks: task A with J different levels of a response variable (e.g., $J = 2$ binary forced choice); and task B with K levels of a response measure (e.g., $K = 7$ point confidence rating). Participants are randomly assigned to two groups: group A receives only task A, but group BA receives task B followed immediately by task A. (Other variations are of course possible.) We obtain estimates of the response probabilities for (a) $p(A = j) :=$ the probability of choosing level j from the response to task A from group A-alone, (b) $p(B = k) :=$ the probability of first responding with level k from task B, and (c) $p(A = j|B = k) :=$ the probability of responding with level j from task A given that the person responded with level k on the earlier task B. From the latter two probability distributions we can compute the *total probability* for response to task

A as

$$p_T(A = j) = \sum_{k=1}^{K} p(B = k) \cdot p(A = j | B = k).$$

The interference effect for level j of the response to task A (produced by responding to task B) equals by definition

$$\delta_A(j) = p(A = j) - p_T(A = j).$$

Note that $\sum_{j=1}^{J} p(A = j) = 1 = \sum_{j=1}^{J} p_T(A = j)$ so that $\sum_j \delta_A(j) = 0$. Given these definitions, we can write

$$p(A = j) = p_T(A = j) + \delta_A(j).$$

Do these interference effects, $\delta_A(j)$, occur in human psychology experiments? If they do, how do we explain them?

2.3. What Is the Evidence for Interference Effects?

Below we summarize several lines of research that provide evidence for interference effects.

2.3.1. *Perception of ambiguous figures*

Interference effects were first investigated in the perceptual domain by Elio Conte (Conte et al., 2009). Approximately 100 students randomly were divided into two groups: One was given 3 seconds to make a single binary choice (plus vs. minus) concerning an ambiguous figure A, and the other group was given 3 seconds to make a single binary choice for an ambiguous figure B followed 800 msec later by a 3-second presentation requesting another single binary choice (plus vs. minus) for figure A. The results produced significant interference effects. For example, for one type of testing stimuli, when test B preceded test A, the following results were obtained ($p(B+) :=$ probability of plus to figure B, $p(A + |B+) :=$ probability of plus to figure A given plus to figure B, $p_T(A+) :=$ total probability of plus to figure A, $p(A+) :=$ probability of plus to figure A alone):

$p(B+)$	$p(A + |B+)$	$p(A + |B-)$	$p_T(A+)$	$p(A+)$
.62	.78	.54	.69	.55

The interference effect equals $\delta_A(+) = p(A = +) - p_T(A = +) = .55 - .69 = -.14$, and $\delta_A(-) = +.14$.

2.3.2. Categorization - decision making

Townsend (Townsend, Silva, Spencer-Smith, & Wenger, 2000) and later Wang (Busemeyer, Wang, & Lambert-Mogiliansky, 2009) use a paradigm to study the interactions between categorization and decision making, which is highly suitable for investigating interference effects. On each trial, participants are shown pictures of faces, which vary along two dimensions (face width and lip thickness). Two different distribution of faces were used: on average a 'narrow' face distribution had a narrow width and thick lips; on average a 'wide' face distribution had a wide width and thin lips. The participants were asked to categorize the faces as belonging to either a 'good' guy (G) or 'bad' guy (B) group, and/or they were asked to decide whether to take an 'attack' (A) or 'withdraw' (W) action. The participants were informed that 'narrow' faces had a .60 probability to come from the 'bad guy' (B) population, and 'wide' faces had a .60 chance to come from the 'good guy' (G) population. The participants were usually (.70 chance) rewarded for attacking 'bad guys' and they were usually (.70 chance) rewarded for withdrawing from 'good guys.' The primary manipulation was produced by using the following two test conditions, presented across a series of trials, to each participant. In the C-then-D condition, participants made a categorization followed by an action decision; in the D-Alone condition, participants only made an action decision.

The categorization-decision paradigm provides a simple test of the law of total probability. In particular, this paradigm allows one to compare the probability of taking an 'attack' (A) action obtained from the D-Alone condition with the total probability computed from the C-then-D condition. Townsend et al. (2000) reported significant deviations from the law of total probability using chi square tests, but they did not examine the direction of these effects. The later study by Wang found a significant interference effect for the narrow faces and a smaller effect in the same direction for the wide faces. For example, using the narrow face data, when categorization preceded decisions, the following results were obtained ($p(G)$:= probability categorize face as good guy, $p(A|G)$:= probability attack given face categorized as good guy, $p(A|B)$:= probability of attack given face categorized as a bad guy, $p_T(A)$:= total probability to attack, $p(A)$:= probability of attack when making only a decision):

$p(G)$	$p(A|G)$	$p(A|B)$	$p_T(A)$	$p(A)$
.17	.41	.63	.59	.69

Surprisingly, the probability of attacking without categorization was even higher than the probability of attacking after categorizing the face as a bad guy. The interference effect equals $\delta(A) = p(A) - p_T(A) = .69 - .59 = +.10$, and $\delta_A(W) = -.10$.

2.3.3. *Disjunction effect in decision making*

Perhaps the earliest report of interference effects is the disjunction effect (Tversky & Sharif, 1992). The original studies were designed to test a rational axiom of decision theory called the sure thing principle (Savage, 1954). According to the sure thing principle, if under state of the world X you prefer action A over B, and if under the complementary state of the world ˜X you also prefer action A over B, then you should prefer action A over B even when you do not know the state of the world. Tversky and Shafir experimentally tested this principle by presenting students with a two-stage gamble, that is a gamble which can be played twice. At each stage the decision was whether or not to play a gamble that has an equal chance of winning \$200 or losing \$100 (the real amount won or lost was actually \$2.00 and \$1.00 respectively). The key result is based on the decision for the second play, after finishing the first play. The experiment included three conditions: one in which the students imagined that they already won the first gamble, a second condition in which they imagined that they lost the first gamble, and a third in which they did not know the outcome of the first gamble. If they thought they won the first gamble, the majority (69%) chose to play again; if they thought they lost the first gamble, then again the majority (59%) chose to play again; but if they did not know whether they won or lost, then the majority chose not to play (only 36% wanted to play again). Tversky and Shafir replicated this experiment using both a within-subject design (the same person made choices under all conditions separated by a week) as well as with a between subject design (different groups of participants received known and unknown conditions).

This disjunction effect can be interpreted as an interference effect for the following reason. Define G as the event of playing the second gamble, W is observing a win on the first gamble, and L is observing a loss on the first gamble. The player can choose G or $\bar{G}$ alone (without observing the outcome of the first game); or the player can *observe* the outcome (W, L) first and then choose G or $\bar{G}$. Then $p(G)$ is the probability of gambling under the unknown condition, and $p_T(G) = p(W) \cdot p(G|W) + p(L) \cdot p(G|L)$ is the probability of choosing to gamble after observing the first play outcome. The total probability is a weighted average of the two known conditions, and so it requires that the probability of playing under the unknown condition must lie in between the two known probabilities. The results show that the probability for the unknown condition is below the smaller probability for the known condition. Therefore we have $p(G) < p(G|L) < p_T(G)$, which implies a negative interference effect.

This result is cited quite frequently, but the result remains controversial. Note that the gamble is actually quite attractive and it has a very positive expected value. Barkan and Busemeyer conducted a very similar study using the same gamble and under conditions in which participants chose to play the gamble a second time under three conditions: planning for a win or planning for a loss or without planning at all for the outcome of the first gamble; but the participants always preferred to play the gamble about 70% of the time and no disjunction effect was found (Barkan & Busemeyer, 1999). Another study by Kuhberger et al. attempted a direct replication of Tversky and Shafir's gambling study, but they failed to find a disjunction effect (Kuhberger, Komunska, & Perner, 2001): if told that they won the first gamble, 60% chose to play again; if told that they lost the first gamble, 47% chose to play again; and if the first play was unknown, then 47% again chose to play again. There was no difference between the known loss condition and the unknown condition.

Another paradigm, using a prisoner dilemma (PD) game, was used by Shafir and Tversky (Shafir & Tversky, 1992) to test the sure thing principle. In all PD's, the two players need to decide independently whether to cooperate with the other player or to defect against the other. The player who stands to gain the most is the one who defects against a cooperating player. Mutual cooperation yields the second-highest payoff for each player. Mutual defection gives the players a payoff lower than that gained from mutual cooperation. Finally, the player who cooperates with a defective player gains the least. No matter what the other player does, an individual player always gains more when he defects; this makes defection the dominant option when the game is played only once against a given opponent (one-shot PD). A total of 80 participants were involved and each person played 6 PD games. Shafir and Tversky found that when a player was informed that the opponent defected, then 97% of the time they defected; if the player was informed that the other opponent cooperated, then 84% of the time they defected; but if they did not know the opponent's choice then only 63% chose to defect.

Several other studies were conducted to replicate and extend the disjunction effect using the prisoner dilemma game. The first was done by Rachel

Croson who used 80 participants, each playing 2 PD's, and half were required to predict or guess what the opponent would do and half were not asked to make this prediction (Croson, 1999). In the first of Croson's conditions, the following results were obtained ($p(GD) :=$ probability guess opponent defected, $p(D|GD) :=$ probability player defects given opponent predicted to defect, $p(D|GC) :=$ probability player defects given opponent predicted to cooperate, $p_T(D) :=$ total probability to defect, $p(D) :=$ probability to defect when opponent's action is unknown):

$p(GD)$	$p(D|GD)$	$p(D|GC)$	$p_T(D)$	$p(D)$
.54	.68	.17	.45	.23

The interference effect equals $\delta(D) = p(D) - p_T(D) = .23 - .45 = -.22$, and $\delta(C) = +.22$. The cooperation rates were much higher in this study as compared to the Shafir and Tversky (1992) study.

A later study by Li and Taplan also found evidence for disjunction effects but much weaker than Shafir and Tversky's original study (Li & Taplin, 2002). Most recently, however, a very robust disjunction effect and replication of Shafir and Tversky (1992) was obtained by Matthew (Busemeyer, Matthews, & Wang, 2006). A total of 88 students played 6 PD games for real money against a computer agent. When told that the agent defected, then 92.4% defected; when told that the agent cooperated, then 83.6% defected; but when the agent's action was unknown, only 64.9% defected.

2.4. What Are the Explanations for These Effects?

Interference effects are empirical results that need a scientific explanation. One cannot immediately jump to the conclusion that they are evidence for quantum mechanisms. Nor can one jump to the conclusion that interference effects are explained psychologically without quantum theory. The same 'psychological' explanation can be formulated probabilistically as either a Kolmogorov or a quantum model, and so it does not discriminate between these two theoretical competitors. The scientific way to determine which is best is to derive formal predictions from each theory and then compare the predictions with the data. The model that best predicts the experimental results is taken as the best explanation. Below we compare some competing explanations for these interference effects. We initially focus on the categorization - decision experiment, but the same models also apply to all of the findings summarized earlier.

2.4.1. *Markov model*

Markov models are commonly used in cognitive psychology. They provide the basis for random walk and diffusion models of decision making (Ratcliff & Smith, 2004), stochastic models of information processing (Townsend & Ashby, 1983), and they are also the basis for multinomial processing tree models of memory retrieval (Batchelder & Reiffer, 1999) and memory recognition (Jacoby, 1991).

Let us first consider a very simple Markov model for the categorization - decision making experiment proposed by Townsend et al. (2000). The person has to infer whether the face comes from the 'good' or 'bad' category (represented by two mutually exclusive Markov states $|B\rangle$ and $|G\rangle$, respectively), and given this inference, the person can intend to take an attack or withdraw action (represented by two mutually exclusive Markov states $|A\rangle$ and $|W\rangle$, respectively). The person starts in a state $|S\rangle$ determined by the face. Then $\phi(B|S)$ is the probability of transiting to inference state $|B\rangle$, and $\phi(G|S) = 1 - \phi(B|S)$ is the probability of starting in inference state $|G\rangle$. These probabilities form an initial distribution

$$\phi_0 = \begin{bmatrix} \phi(B|S) \\ \phi(G|S) \end{bmatrix}.$$

The distribution ϕ_0 deserves some additional comment and interpretation. According to the Markov model, at the beginning of a trial, the person enters exactly one of two states, the person either enters state $|B\rangle$ or enters state $|G\rangle$ and the person does not enter both states. The probabilities in ϕ_0 represent the theorist's uncertainty about the person's specific unknown state, and ϕ_0 is used to make predictions about the exact state at that moment.

If the player starts out inferring the face is a 'bad guy' $|B\rangle$, then the player can transit to the attack action state $|A\rangle$ with probability $\phi(A|B)$ or transit to the withdraw action state $|W\rangle$ with probability $\phi(W|B)$. If the player starts out inferring the face is a 'good guy' $|G\rangle$, then player can transit to the attack action $|A\rangle$ with probability $\phi(A|G)$ or transit to the withdraw action $|W\rangle$ with probability $\phi(W|G)$. These conditional probabilities form a 2×2 transition probability matrix

$$T = \begin{bmatrix} \phi(A|B) & \phi(A|G) \\ \phi(W|B) & \phi(W|G) \end{bmatrix}.$$

The sum across rows within each column must equal one for a transition matrix. This is required to guarantee that the probabilities sum to one after making a transition. After the transition from inference to action, the probability distribution across action states equals

$$\phi_1 = T \cdot \phi_0 \tag{2.1}$$

$$\begin{bmatrix} \phi(A|S) \\ \phi(W|S) \end{bmatrix} = \begin{bmatrix} \phi(A|B) \cdot \phi(B|S) + \phi(A|G) \cdot \phi(G|S) \\ \phi(W|B) \cdot \phi(B|S) + \phi(W|G) \cdot \phi(G|S) \end{bmatrix}.$$

This is the Chapman-Kolmogorov equation for Markov models. The Chapman-Kolmogorov equation is simply a restatement of the law of total probability expressed in terms of the Markov states.

The probabilities in ϕ_1 deserve some further comment. As (2.1) shows, there are two paths starting from the initial state $|S\rangle$ traveling to the final state $|A\rangle$. One is the path $|S\rangle \rightarrow |B\rangle \rightarrow |A\rangle$ that passes through the 'bad guy' inference, and the other is the path $|S\rangle \rightarrow |G\rangle \rightarrow |A\rangle$ that passes through the 'good guy' inference. The person can travel along one or the other but not both of these paths, because they are mutually exclusive. The person ends up either in the $|A\rangle$ state or the $|W\rangle$ state and not both. The particular action state the person enters determines the choice response for an action. So the final probability distribution ϕ_1 represents the theorist's uncertainty about the person's final action state, and these probabilities are used by the theorist to predict the person's choices.

The preceding Markov model assumed that the states were directly observable. Now we explore the possibility that the states are mapped into responses by some 'noisy' process that allows measurement 'errors.' When measurements are noisy, it becomes important to introduce a distinction between states and observed responses. To do this, the categorization response is denoted by a variable C that can take on labels b or g for choosing the 'bad' or 'good' category respectively; and the choice response for an action is denoted by a variable D that can take on labels a or w for the choice of attack and withdraw actions, respectively.

The noisy measurement is achieved by introducing a probabilistic response map from states to responses. This is done by employing probabilistic state to response mappings. The following two matrices map inference states to category responses

$$C_b = \begin{bmatrix} C(b|B) & 0 \\ 0 & C(b|G) \end{bmatrix}, \ C_g = \begin{bmatrix} C(g|B) & 0 \\ 0 & C(g|G) \end{bmatrix}.$$

For example, if the person enters the inference state $|B\rangle$, there is a probability $C(b|B)$ of categorizing the face as 'bad' and $C(g|B)$ that it is categorized as good. The next two matrices map action states to choices of each action

$$D_a = \begin{bmatrix} D(a|A) & 0 \\ 0 & D(a|W) \end{bmatrix}, \quad D_w = \begin{bmatrix} D(w|A) & 0 \\ 0 & D(w|W) \end{bmatrix}.$$

For example, if the person is in the state $|A\rangle$, then the person may actually choose to attack with probability $D(a|A)$, but the person may instead choose to withdraw with a probability $D(w|A)$. To guarantee that the categorization response probabilities sum to unity, we require that C_i contains probabilities such that $C_b + C_g = I$, where I is the identity matrix; to require the action response probabilities to sum to unity, we require that D_j contains probabilities such that $D_a + D_w = I$. This model reduces to the original Markov model without noise when we set $C(b|B) = 1 = C(g|G)$ and $D(a|A) = 1 = D(w|W)$. It is convenient to define a row vector $L = \begin{bmatrix} 1 & 1 \end{bmatrix}$ which is used to sum across states.

Using these definitions, we can compute the following response probabilities. The probability that a face is categorized $C = i$ equals

$$p(C = i) = L \cdot C_i \cdot \phi_0.$$

If we observe the $C = i$ category response, then the probability distribution across inference states is revised by Bayes's rule to become

$$\phi_i = \begin{bmatrix} \phi(B|C = i) \\ \phi(G|C = i) \end{bmatrix} = \frac{1}{p(C = i)} \cdot C_i \cdot \phi_0$$

$$= \begin{bmatrix} \frac{\phi(B) \cdot C(i|B)}{p(C=i)} \\ \frac{\phi(G) \cdot C(i|G)}{p(C=i)} \end{bmatrix}.$$

As the above equation shows, the categorization response changes the distribution across inference states. The probability of choosing action $D = j$ given that we observe a categorization response $C = i$ equals

$$p(D = j|C = i) = L \cdot D_j \cdot T \cdot \phi_i.$$

Therefore, the probability of choosing category $C = i$ and then choosing an action $D = j$ equals the matrix product

$$p(C = i, D = j) = L \cdot D_j \cdot T \cdot C_i \cdot \phi_0. \tag{2.2}$$

The probability that the face is first categorized as a 'bad guy' and then the person attacks equals

$$p(C = b, D = a) = L \cdot D_a \cdot T \cdot C_b \cdot \phi_0.$$

The probability that the face is first categorized as a 'good guy' and then the person attacks equals

$$p(C = g, D = a) = L \cdot D_a \cdot T \cdot C_g \cdot \phi_0.$$

The probability of attacking under the decision alone condition equals

$$
\begin{aligned}
p(D = a) &= L \cdot D_a \cdot T \cdot \phi_0 \qquad\qquad\qquad\qquad (2.3) \\
&= L \cdot D_a \cdot T \cdot I \cdot \phi_0 \\
&= L \cdot D_a \cdot T \cdot (C_b + C_g) \cdot \phi_0 \\
&= L \cdot D_a \cdot T \cdot C_b \cdot \phi_0 + L \cdot D_a \cdot T \cdot C_g \cdot \phi_0 \\
&= p(C = b, D = a) + p(C = g, D = a) \\
&= p_T(D = a).
\end{aligned}
$$

In summary, this model satisfies the law of total probability, which of course fails to explain the interference effects found with the categorization-decision paradigm making task.

This Markov model also applies to all of the other findings as follows. For the ambiguous figure results, we use states $|B\rangle$ and $|G\rangle$ to represent the plus or minus perceptions to figure B, and we use $|A\rangle$ and $|W\rangle$ to represent the plus or minus perceptions to figure A. For the two-stage gambling game, we use states $|G\rangle$ and $|B\rangle$ to represent the 'win' or 'loss' inference about the first play of the gamble, and we use $|A\rangle$ and $|W\rangle$ to represent the 'play' or 'do not play' actions. For the PD game, we use states $|B\rangle$ and $|G\rangle$ to represent the 'defect' or 'cooperate' inference about the opponent, and we use $|A\rangle$ and $|W\rangle$ to represent the 'defect' or 'cooperate' strategy for the player. But this Markov model fails to explain any of the interference effects found in these other paradigms.

In fact, the matrix equations (2.2) and (2.3) hold for any finite hidden Markov system. We could assume n inference states and m actions states for arbitrary n and m numbers of states. So these equations are not limited to a model with only two inference states and two action states with which we began. As long as the *same* initial state ϕ_0 and the same transition matrix T is applied to both conditions (whether the inference state is measured or not), the Markov model fails to account for the interference effects for all of these experiments.

2.4.2. *Quantum model*

The original explanation for the disjunction effect was a psychological explanation based on the failure of consequential reasoning under the unknown conditions. Shafir and Tversky (1992) explained the finding in terms of choice based on reasons as follows. Consider for example, the two-stage gambling problem. If the person knew they won, then they had extra house money with which to play and for this reason they chose to play again; if the person knew they had lost, then they needed to recover their losses and for this other reason they chose to play again; but if they did not know the outcome of the game, then these two reasons did not emerge into their minds. Why not? If the first play is unknown, it must definitely be either a win or a loss, and it cannot be anything else. So the mystery is why these reasons do not emerge for the unknown condition. If choice is based on reasons, then the unknown condition has two good reasons. Somehow these two good reasons cancel out to produce no reasons at all! This sounds a lot like wave interference where one wave is rising and the other is falling. From this it follows that there is an interest in quantum models.

The psychological explanation given by Shafir and Tversky (1992) is quite consistent with a formal quantum mechanism for the effect. Busemeyer, Wang, and Townsend (2006) originally suggested a quantum interference interpretation for the disjunction effect, and since that time, various quantum models for this effect have been proposed, each one ultimately explaining the effects by interference terms, which includes Pothos and Busemeyer (2009), Khrennikov and Haven (2009), Aerts (2009), Yukalov and Sornette (2009), and Accardi, Khrennikov and Ohya (2009).

Busemeyer et al. (2006) started with the following simple quantum model.[1] Consider the category - decision making experiment once again. As in the Markov model, the person has to infer whether the face is a 'bad' or 'good' guy (represented by two mutually exclusive quantum states $|B\rangle$ and $|G\rangle$, respectively), and given this inference, the person may intend to attack or withdraw (represented by two mutually exclusive quantum states $|A\rangle$ and $|W\rangle$, respectively). Quantum theory replaces transition probabilities such as $\phi(A|B)$ with transition amplitudes such as $\langle A|B\rangle$ and $\phi(A|B) = |\langle A|B\rangle|^2$.

The person starts in a state $|S\rangle$. Then there is an amplitude $\langle B|S\rangle$ of transiting to the 'bad guy' inference and another amplitude $\langle G|S\rangle$ of transiting to the 'good guy' inference, $|\langle B|S\rangle|^2 + |\langle G|S\rangle|^2 = 1$. These amplitudes form an amplitude distribution (wave function) across inference

---

[1] In quantum terminology, this model treats the inference as an observable operating within a two-dimensional Hilbert space, and the action is another incompatible observable operating within the same Hilbert space.

states

$$\psi_0 = \begin{bmatrix} \langle B|S \rangle \\ \langle G|S \rangle \end{bmatrix}.$$

This amplitude distribution ψ_0 deserves more interpretation. If the face is known to come from the 'bad guy' population, then $\langle B|S \rangle = 1$, and the initial state corresponds exactly to state $|B\rangle$; if the face is known to come from the 'good guy' population, then $\langle G|S \rangle = 1$, and the initial state corresponds exactly to state $|G\rangle$. But if $1 > |\langle B|S \rangle| > 0$ and $1 > |\langle G|S \rangle| > 0$, then the person is *not* exactly in state $|B\rangle$, and the person is *not* exactly in state $|G\rangle$ either. Furthermore, the person is *not* in both states at the same time. The person is exactly in an indefinite or superposition state represented by the wave function ψ_0. In the latter case, at a single moment in time, there is some *potential* to generate either one of the two mutually exclusive categorization responses. But only one of these potentials can become actualized to create an observed response.

On the one hand, if the state starts in the 'bad guy' inference, then there is an amplitude $\langle A|B \rangle$ of transiting to the attack action and another amplitude $\langle W|B \rangle$ of transiting to the withdraw action, $|\langle A|B \rangle|^2 + |\langle W|B \rangle|^2 = 1$. On the other hand, if the state starts in the 'good guy' inference, then there is an amplitude $\langle A|G \rangle$ of transiting to the attack action and another amplitude $\langle W|G \rangle$ of transiting to the withdraw action, $|\langle A|G \rangle|^2 + |\langle W|G \rangle|^2 = 1$. These transition amplitudes form a unitary matrix

$$U = \begin{bmatrix} \langle A|B \rangle & \langle A|G \rangle \\ \langle W|B \rangle & \langle W|G \rangle \end{bmatrix}.$$

Normally in quantum theory, the matrix U is required to be unitary: $U^\dagger U = I = UU^\dagger$. This is required to guarantee that the transformed amplitude distribution remains unit length, which is needed to guarantee that the final probabilities sum to one. The orthogonality restriction of the unitary matrix implies the equality

$$\langle A|B \rangle^* \langle A|G \rangle = -\langle W|B \rangle^* \langle W|G \rangle. \tag{2.4}$$

If we square the magnitudes of the entries of the unitary matrix, we obtain the transition probability matrix

$$T = \begin{bmatrix} |\langle A|B \rangle|^2 & |\langle A|G \rangle|^2 \\ |\langle W|B \rangle|^2 & |\langle W|G \rangle|^2 \end{bmatrix}.$$

In order to satisfy the unitary property, this transition matrix must be doubly stochastic, that is both rows and columns sum to one. Double stochasticity implies that

$$|\langle W|B\rangle|^2 = 1 - |\langle A|B\rangle|^2 = |\langle A|G\rangle|^2 \tag{2.5}$$
$$|\langle A|G\rangle|^2 = |\langle W|B\rangle|^2$$
$$|\langle A|B\rangle|^2 = |\langle W|G\rangle|^2 .$$

Equations (2.4) and (2.5) are very strong constraints on this simple quantum model.

The final amplitude distribution across the action states is equal to

$$\psi_1 = U \cdot \psi_0$$
$$\begin{bmatrix} \langle A|S\rangle \\ \langle W|S\rangle \end{bmatrix} = \begin{bmatrix} \langle A|B\rangle\langle B|S\rangle + \langle A|G\rangle\langle G|S\rangle \\ \langle W|B\rangle\langle B|S\rangle + \langle W|G\rangle\langle G|S\rangle \end{bmatrix} .$$

The amplitude distribution ψ_1 across action states requires some comment. The amplitude $\langle A|S\rangle$ represents the direct path $|S\rangle \rightarrow |A\rangle$ from the initial state to the attack state. This path amplitude can be broken down by the theorist as the sum of two other path amplitudes. One is the path $|S\rangle \rightarrow |B\rangle \rightarrow |A\rangle$ from the initial state to the 'bad guy' inference and then to the attack; and the other is the path $|S\rangle \rightarrow |G\rangle \rightarrow |A\rangle$ from the initial state to the 'good guy' inference and then to the attack. But we cannot conclude from this mathematical decomposition that the person passes through one or the other and not both of these two paths to get to the attack conclusion. Also one cannot conclude that the person travels both paths. Instead the person can travel directly from the initial state to the attack conclusion. Also the final amplitude distribution across action states is an indefinite or superposition state. One cannot conclude that the person ends up definitely in the $|A\rangle$ state or the $|W\rangle$ state and not both states immediately before the choice is made. Nor is the person exactly in both states. At that moment, both actions have some potential to be expressed, and the choice actualizes one of these potentials to produce the observed response.

If the person is exactly in the 'bad guy' state, then $\langle B|S\rangle = 1$ and the probability that the person attacks equals $|\langle A|B\rangle|^2$; if the person is exactly in the 'good guy' state, then $\langle G|S\rangle = 1$ and the probability that the person attacks equals $|\langle A|G\rangle|^2$. For the indefinite or superposed state, the action probabilities equal

$$|\langle A|S\rangle|^2 = |\langle A|B\rangle\langle B|S\rangle + \langle A|G\rangle\langle G|S\rangle|^2$$
$$= |\langle A|B\rangle|^2|\langle B|S\rangle|^2 + |\langle A|G\rangle|^2|\langle G|S\rangle|^2 + \delta_1,$$
$$\delta_1 = 2 \cdot \text{Re}[\langle A|B\rangle\langle B|S\rangle\langle A|G\rangle\langle G|S\rangle],$$
$$|\langle W|S\rangle|^2 = |\langle W|B\rangle\langle B|S\rangle + \langle W|G\rangle\langle G|S\rangle|^2$$
$$= |\langle W|B\rangle|^2|\langle B|S\rangle|^2 + |\langle W|G\rangle|^2|\langle G|S\rangle|^2 + \delta_2,$$
$$\delta_2 = 2 \cdot \text{Re}[\langle W|B\rangle\langle B|S\rangle\langle W|G\rangle\langle G|S\rangle].$$

The probabilities from this quantum model can violate the law of total probability because of the cross product interference terms, δ_1 and δ_2, generated by squaring the sum. The orthogonality restriction from the unitary matrix (see (2.4)) implies that

$$\delta_1 = 2 \cdot \text{Re}[\langle A|B\rangle\langle A|G\rangle\langle B|S\rangle\langle G|S\rangle]$$
$$= -2 \cdot \text{Re}[\langle W|B\rangle\langle W|G\rangle\langle B|S\rangle\langle G|S\rangle] = -\delta_2.$$

It is useful to express the complex numbers in complex exponential form:

$$\langle A|B\rangle^*\langle A|G\rangle = |\langle A|B\rangle\langle A|G\rangle| \cdot e^{i\cdot\theta},$$
$$\langle W|B\rangle^*\langle W|G\rangle = -|\langle A|B\rangle\langle A|G\rangle| \cdot e^{i\cdot\theta},$$
$$\langle B|S\rangle^*\langle G|S\rangle = |\langle B|S\rangle\langle G|S\rangle| \cdot e^{i\omega}.$$

Then we obtain the well-known formula for the quantum interference

$$\delta_1 = 2 \cdot |\langle A|B\rangle\langle A|G\rangle\langle B|S\rangle\langle G|S\rangle| \cdot \cos(\theta + \omega) = -\delta_2.$$

To account for the categorization-decision paradigm results obtained with the narrow faces, we can set $\langle B|S\rangle = \sqrt{.8}$ to approximate the observed initial probability of categorizing the narrow face as a bad guy (the actual value was .83 for the narrow faces), and we can set $\langle A|B\rangle = \sqrt{.60} = \langle W|G\rangle$, and $\langle W|B\rangle = \sqrt{.40}\cdot e^{-i\cdot 1.2313}$ and $\langle A|G\rangle = \sqrt{.40}\cdot e^{i\cdot 1.2313}$ to closely approximate the probabilities for actions conditioned on each categorization, while at the same time satisfying the requirements for a unitary matrix. Then the predicted probability of attacking under the decision alone condition equals $|\langle A|S\rangle|^2 = .69$, which closely approximates the results for the narrow face condition of the categorization-decision paradigm making experiment.

The problem that we run into when we apply this model to the PD game results of Shafir and Tversky is that the observed transition probabilities violate double stochasticity, and therefore they cannot be generated from a unitary matrix. For the PD game, define $C = b$ as the observation that the opponent has defected, $C = g$ as the observation that the opponent has cooperated, and $D = a$ as the defect response by the player. Then the results show that $p(D = a|C = b) = .97$ and $p(D = a|C = g) = .84$ but the unitary property requires the latter to be equal to $1 - p(D = a|C = b) = .03$ rather than $.84$. The unitary property is also violated by the two-stage gambling game results of Tversky and Shafir. The perception results from Conte also violate the unitary property. The unitary property holds pretty well for the narrow face data obtained in the categorization-decision paradigm experiment by Wang, but it was violated by the wide face data. In short, the unitary property implied by the two-dimensional model does not hold up well for this two-dimensional model. To address this problem, Pothos and Busemeyer (2009) developed a four-dimensional quantum model. But here we present a newer and much simpler two-dimensional model.

2.4.3. *Quantum noise model*

On the one hand, the two-dimensional Markov model fails because the interference effects violate the law of total probability. On the other hand, the two-dimensional quantum model fails because the observed transition matrices violate double stochasticity. An interesting idea is to combine the two classes of models and form a quantum Markov model (Accardi, Khrennikov, & Ohya, 2009). The following is a new model inspired by – but much simpler than – the Accardi et al. (2009) model. It assumes that the noisy measurements are used to assess the hidden quantum states.

Once again consider the analysis of the categorization-decision paradigm experiment. As before, $|S\rangle$ is the person's initial state, and we assume that there are two mutually exclusive states of inference: infer that the face belongs to the 'bad guy' category $|B\rangle$, or infer that the face belongs to the good guy category $|G\rangle$. The initial amplitude distribution is again represented by

$$\psi_0 = \begin{bmatrix} \langle B|S\rangle \\ \langle G|S\rangle \end{bmatrix}.$$

The initial state represents the amplitude distribution at the very beginning of the choice process, immediately after instructions. From these states the person can transition to two different intended actions $|A\rangle$ and $|W\rangle$ representing attack and withdraw. The initial amplitude distribution over inferences evolves for some period of time to produce a final amplitude distribution ψ_1 over actions, which is used to make a choice. The final amplitude distribution is a unitary transformation of the initial distribution

$$\psi_1 = \begin{bmatrix} \langle A|S \rangle \\ \langle W|S \rangle \end{bmatrix}$$

$$= U \cdot \psi_0.$$

The unitary transformation is defined as

$$U = \begin{bmatrix} \langle A|B \rangle & \langle A|G \rangle \\ \langle W|B \rangle & \langle W|G \rangle \end{bmatrix}$$

$$= \begin{bmatrix} \sqrt{u} & \sqrt{1-u} \cdot e^{i \cdot \theta} \\ -\sqrt{1-u} \cdot e^{-i \cdot \theta} & \sqrt{u} \end{bmatrix}$$

for $0 \leq u \leq 1$, which satisfies the unitary property $U^\dagger U = UU^\dagger = I$ required to retain unit length following transformation.

According to this model, if the person starts out in state $|B\rangle$, then the person passes through one line of thought (with amplitude $\sqrt{u}$) that leads to one reason for attacking; if the person starts out in state $|W\rangle$, then the person passes through a different line of thought (with amplitude $\sqrt{1-u}$) that leads to a different reason for attacking; but if the person starts out in a superposition of these two states, then a direct path from $|S\rangle \rightarrow |A\rangle$ is taken in which the two lines of thought can constructively or destructively interfere (depending on the parameter θ).

The present model assumes that the quantum states are not directly observable because of measurement 'errors' or 'noise.' As before, it is important to distinguish between states and observed responses. Once again, the categorization response is denoted by a variable C that can take on label b or g for choosing the 'bad' or 'good' category respectively; and the choice response for an action is denoted by a variable D that can take on label a or w for the choice of attack and withdraw actions, respectively.

The choices are represented by measurement operators (Gardiner, 1991) that map states in observed responses. The two noisy measurement operators for categorizing as 'bad guy' or 'good guy' are defined by

$$C_b = \begin{bmatrix} \sqrt{C(b|B)} & 0 \\ 0 & \sqrt{C(b|G)} \end{bmatrix}, \ C_g = \begin{bmatrix} \sqrt{C(g|B)} & 0 \\ 0 & \sqrt{C(g|G)} \end{bmatrix}.$$

For example, if the person is in the inference state $|B\rangle$, then the person may actually categorize the face as 'bad' with probability $C(b|B)$, but the person may instead categorize the face as good with probability $C(g|B)$. The two noisy measurement operators for choosing to attack or withdraw actions are defined by

$$D_a = \begin{bmatrix} \sqrt{D(a|A)} & 0 \\ 0 & \sqrt{D(a|W)} \end{bmatrix}, \quad D_w = \begin{bmatrix} \sqrt{D(w|A)} & 0 \\ 0 & \sqrt{D(w|W)} \end{bmatrix}.$$

For example, if the person is in action state $|A\rangle$, then the person may choose to attack with probability $C(a|A)$, but instead the person may choose to withdraw with probability $C(w|A)$. This model reduces to the original quantum model without noise when we set $C(b|B) = 1 = C(g|G)$ and $D(a|A) = 1 = D(w|W)$. These two measurement operators form a complete set because they satisfy the completeness property $C_b^\dagger C_b + C_g^\dagger M_g = I$ and $D_a^\dagger D_a + D_w^\dagger D_w = I$ needed to guarantee that the choice probabilities sum to one across actions.

Using these definitions, we can compute the following response probabilities. The probability that a face is categorized $C = i$ equals

$$p(C = i) = ||C_i \cdot \psi_0||^2.$$

If we observe the $C = i$ category response, then the amplitude distribution across inference states is revised by a quantum version of Bayes's rule to become

$$\psi_i = \begin{bmatrix} \psi(B|C = i) \\ \psi(G|C = i) \end{bmatrix} = \frac{1}{\sqrt{p(C = i)}} \cdot C_i \cdot \psi_0$$

$$= \begin{bmatrix} \dfrac{\psi(B) \cdot \sqrt{C(i|B)}}{\sqrt{p(C=i)}} \\ \dfrac{\psi(G) \cdot \sqrt{C(i|G)}}{\sqrt{p(C=i)}} \end{bmatrix}.$$

As the above equation shows, the categorization response changes the amplitude distribution across inference states. The probability of choosing action $D = j$ given that we observe a categorization response $C = i$ equals

$$p(D = j|C = i) = ||D_j \cdot U \cdot \psi_i||^2.$$

Therefore, the probability of choosing category $C = i$ and then choosing an action $D = j$ equals the matrix product

$$p(C = i, D = j) = ||D_j \cdot U \cdot C_i \cdot \psi_0||^2. \tag{2.6}$$

The probability that the face is first categorized as a 'bad guy' and then the person attacks equals

$$p(C = b, D = a) = ||D_a \cdot U \cdot C_b \cdot \psi_0||^2.$$

The probability that the face is first categorized as a 'good guy' and then the person attacks equals

$$p(C = g, D = a) = ||D_a \cdot U \cdot C_g \cdot \psi_0||^2.$$

The probability of attacking under the decision alone condition equals

$$
\begin{aligned}
p(D = a) &= ||D_a \cdot U \cdot \psi_0||^2 \\
&= ||D_a \cdot U \cdot I \cdot \psi_0||^2 \\
&= ||D_a \cdot U \cdot (C_b + C_g) \cdot \psi_0||^2 \\
&= ||D_a \cdot U \cdot C_b \cdot \psi_0 + D_a \cdot U \cdot C_g \cdot \psi_0||^2 \\
&= p(C = b, D = a) + p(C = g, D = a) + \delta_A \\
&\neq p_T(D = a).
\end{aligned}
\tag{2.7}
$$

In summary, this model can violate the law of total probability, and it can explain the interference effects found with the categorization-decision paradigm making task.

To see how this works let us consider the two example applications where the original quantum model failed. First consider the two-stage gambling game. In this case, we define $|B\rangle :=$ inferring a loss on the first play, $|G\rangle :=$ inferring a win on the first play, $|A\rangle$ choosing to play the second gamble, $|W\rangle$ choosing not to play the second gamble, $C = g$ represents being told that you won the first round, $C = b$ represents being told that you lost the first round, and $D = a$ represents choosing to play the gamble again on the second round. Setting $C(b|B) = 1 = C(g|G)$, $D(a|A) = 1$, $D(a|W) = .28$, $u = .57$, and $\theta = .79 \cdot \pi$ exactly reproduces all of the results for the two-stage gambling game reported by Tversky and Shafir (1992).

Next consider the results for the PD game. In this case, we define $|B\rangle :=$ inferring opponent defects, $|G\rangle :=$ inferring opponent cooperates, $|A\rangle$ player chooses to defect, $|W\rangle$ player chooses to cooperate, $C = g$ represents being told that opponent chose to cooperate, $C = b$ represents being told that opponent chose to defect, and $D = a$ represents player choosing to defect. Because the player is informed exactly about the opponent's decision, we simply set $C(b|B) = 1 = C(g|G)$. If we also set $D(a|A) = 1$, $D(a|W) = .68$, $u = .61$, and $\theta = \pi$, then this model produces the following results. If the opponent is known to defect, then the probability that the player defects equals $p(D = a|C = b) = .88$; if the opponent is known to cooperate, then the probability that the player defects $p(D = a|C = g) = .81$; and if the opponent's action is unknown, we set $\langle I_1|S\rangle = \langle I_2|S\rangle = \frac{1}{\sqrt{2}}$ to produce a probability of defection equals to $p(A_1) = .69$. This approximates the results obtained in the PD game. This model can also exactly fit both the wide and narrow face data from the categorization-decision paradigm task as well as the results obtained with the ambiguous figures (details not shown here). In short, this model can perfectly fit many of the results demonstrating interference effects. But it does not provide a simple way to test double stochasticity. Also it has too many parameters relative to the number of data points produced by these experiments and so it is difficult to empirically test. New experiments are needed that generate more conditions and data points to test the model.

2.5. What Next?

Quantum explanations for interference effects found in psychology have already made one important contribution. Quantum theory has provided a common way to understand a number of paradoxical findings that have never been connected before, nor even mentioned together in the same articles. By examining interference effects and providing a common quantum account of these effects, quantum theorists have organized a new general and uniform way to think about all these seemingly unrelated problems. This is a step forward. We think that these initial promising steps made toward understanding all of the various interference effects are encouraging other researchers to begin examining quantum models in other applications in psychology.

What is needed next is stronger tests of these models. The applications reviewed above involve too many free parameters and too little data. For the simplest models – the two-state Markov model and the two-state quantum model – it was still possible to test the key properties (law of total probability and double stochasticity, respectively). Unfortunately, when these properties were tested, they failed for the simple models. The more complex models can account for the findings in a post hoc way, but they do not provide a strong empirical test with such small data sets. New experiments on interference effects are needed with many more conditions to provide more data points for testing these models.

Quantum probability is considered by many to be a very specialized probability theory that is only useful in physics. We believe there are useful applications of this theory outside of physics (Bruza & Gabora, 2009). One of the founding fathers of quantum theory, Neils Bohr (Bohr, 1958), speculated on this possibility and so did David Bohm (Bohm, 1979). In fact, quantum and Markov probability theories are highly similar. Both

Markov and quantum theories are defined by states and transition operators, and the only difference is the final way that the probabilities are computed. There are two key differences. Markov theory operates directly on probabilities and therefore it obeys the law of total probability but it does not have to obey the doubly stochastic law. Quantum theory operates on amplitudes and probabilities obtained by squaring the amplitudes, consequently it does not have to obey the law of total probability but instead it must obey the law of double stochasticity.

References

Accardi, L., Khrennikov, A. Y., & Ohya, M. (2009). Quantum markov model for data from Shafir-Tversky experiments in cognitive psychology. *Open Systems and Information Dynamics, 16*, 371–385.

Aerts, D. (2009). Quantum structure in cognition. *Journal of Mathematical Psychology, 53*, 314–348.

Atmanspacher, H., & Filk, T. (2010). A proposed test of temporal nonlocality in bistable perception. *Journal of Mathematical Psychology, 54*, 314–321.

Barkan, R., & Busemeyer, J. R. (1999). Changing plans: Dynamic inconsistency and the effect of experience on the reference point. *Psychological Bulletin and Review, 10*, 353–359.

Batchelder, W. H., & Reiffer, D. M. (1999). Theoretical and empirical review of multinomial process tree modeling. *Psychonomic Bulletin and Review, 6*, 57–86.

Bohm, D. (1979). *Quantum Theory*. Dover.

Bohr, N. (1958). *Atomic Theory and Human Knowledge*. NY: John Wiley & Sons.

Bruza, P., Busemeyer, J. R., & Gabora, L. (2009). Introduction to the special issue on quantum cognition. *Journal of Mathematical Psychology, 53*, 303–305.

Busemeyer, J. R., Matthews, M., & Wang, Z. (2006). A quantum information processing explanation of disjunction effects. In R. Sun and N. Myake (Eds.), *Proceedings of the 29th Annual Conference of the Cognitive Science Society and the 5th International Conference of Cognitive Science* (pp. 131–135). Erlbaum.

Busemeyer, J. R., Wang, Z., & Lambert-Mogiliansky, A. (2009). Comparison of markov and quantum models of decision making. *Journal of Mathematical Psychology, 53*, 423–433.

Busemeyer, J. R., Wang, Z., & Townsend, J. T. (2006). Quantum dynamics of human decision making. *Journal of Mathematical Psychology, 50,* 220–241.

Conte, E., Khrennikov, A. Y., Todarello, O., Federici, A., Mendolicchio, L., & Zbilut, J. P. (2009). Mental states follow quantum mechanics during perception and cognition of ambiguous figures. *Open Systems and Information Dynamics, 16,* 1–17.

Croson, R. (1999). The disjunction effect and reason-based choice in games. *Organizational Behavior and Human Decision Processes, 80,* 118–133.

Gardiner, C. W. (1991). *Quantum Noise.* Springer-Verlag.

Hammeroff, S. R. (1998). Quantum computation in brain microtubles? The Penrose–Hammeroff "Orch OR" model of consiousness. *Philosophical Transactions Royal Society London A, 356,* 1869–1896.

Jacoby, L. L. (1991). A process dissociation framework: Separating automatic from intentional uses of memory. *Journal of Memory and Language, 30,* 513–541.

Khrennikov, A. Y. (2010). *Ubiquitous Quantum Structure: From Psychology to Finance.* Springer.

Khrennikov, A. Y., & Haven, E. (2009). Quantum mechanics and violations of the sure thing principle: The use of probability interference and other concepts. *Journal of Mathematical Psychology, 53,* 378–388.

Kolmogorov, A. N. (1933/1950). *Foundations of the Theory of Probability.* NY: Chelsea Publishing Co.

Kuhberger, A., Komunska, D., & Perner, J. (2001). The disjunction effect: Does it exist for two-step gambles? *Organizational Behavior and Human Decision Processes, 85,* 250–264.

Li, S., & Taplin, J. (2002). Examining whether there is a disjunction effect in prisoner's dilemma games. *Chinese Journal of Psychology, 44,* 25–46.

Narens, L. (2007). *Theories of Probability: An Examination of Logical and Qualitative Foundations.* World Scientific Publishing Company.

Pothos, E. M., & Busemeyer, J. R. (2009). A quantum probability model explanation for violations of 'rational' decision making. *Proceedings of the Royal Society B, 276,* 2171–2178.

Ratcliff, R., & Smith, P. (2004). A comparison of sequential sampling models for two-choice reaction time. *Psychological Review, 111,* 333–367.

Savage, L. J. (1954). *The Foundations of Statistics.* NY: John Wiley & Sons.

Shafir, E., & Tversky, A. (1992). Thinking through uncertainty: Nonconsequential reasoning and choice. *Cognitive Psychology, 24*, 449–474.

Townsend, J. T., & Ashby, G. F. (1983). *Stochastic Modeling of Elementary Psychological Processes.* Cambridge, MA: Cambridge University Press.

Townsend, J. T., Silva, K. M., Spencer-Smith, J., & Wenger, M. (2000). Exploring the relations between categorization and decision making with regard to realistic face stimuli. *Pragmatics and Cognition, 8*, 83–105.

Tversky, A., & Shafir, E. (1992). The disjunction effect in choice under uncertainty. *Psychological Science, 3*, 305–309.

Von Neumann, J. (1932/1955). *Mathematical Foundations of Quantum Theory.* Princeton University Press.

Yukalov, V., & Sornette, D. (2010). Decision theory with prospect inter-ference and entanglement. *Theory and Decision, 70*, 283–328.

Chapter 3

Defining Goodness and Badness in Terms of Betterness Without Negation

Erik Carlson

Uppsala University

There is a long tradition of attempts to define the monadic value properties of intrinsic or final goodness and badness in terms of the dyadic betterness relation. Such definitions, if possible, would seem desirable for reasons of theoretical simplicity. It appears that every extant proposal of this kind relies on the concept of *negation*, presupposing that the value bearers are proposition-like entities, such as states of affairs, facts, or propositions. In this chapter, we shall investigate the possibility of defining the monadic value properties in terms of betterness, without assuming that negation or other logical connectives can be applied to the value bearers. Many value theorists believe that, for example, physical objects can have intrinsic or final value. It is therefore worthwhile to explore whether the monadic value properties can be defined in terms of betterness, within a framework that puts no restrictions on what kinds of entities that can be bearers of value. The conclusion will be that such definitions are possible if, and probably only if, certain admittedly disputable axiological principles are accepted.

3.1. Introduction

There is a long tradition of attempting to define the monadic value properties of intrinsic or final goodness and badness in terms of the dyadic betterness relation.[1] Such definitions, if possible, would arguably be desirable for reasons of theoretical simplicity. To the best of my knowledge, every extant proposal of this kind relies on the concept of *negation*, presupposing that the value bearers are proposition-like entities, such as states of affairs,

---

[1] I shall make no attempt here to define the concepts of intrinsic or final value. The "Introduction" to Rønnow-Rasmussen and Zimmerman, 2005, pp. xiii-xxxv, gives a helpful overview of the large literature on this topic. For the sake of ease of expression, I will generally leave it implicit that we are dealing with intrinsic or final value, rather than with any kind of value.

facts, or propositions. The simplest suggestion in this vein, first espoused by Albert Brogan in 1919, and accepted by many later authors, is to define p as *good* iff p is better than $\neg p$, and define p as *bad* iff $\neg p$ is better than p (Brogan, 1919).[2]

In 1966 Roderick Chisholm and Ernest Sosa published a very influential paper, in which they rejected Brogan's definitions (Chisholm & Sosa, 1966). Letting $p =$ "There are no unhappy egrets", they claimed that p is not intrinsically good, although better than $\neg p$. The reason why p is not intrinsically good is that the mere absence of unhappiness does not "rate any possible universe a plus". Similarly, $p =$ "There are no happy egrets" is not intrinsically bad, although worse than $\neg p$, since the mere absence of happiness does not "rate any possible universe a minus". As a remedy, Chisholm and Sosa suggested somewhat more complex definitions of intrinsic goodness and badness. They defined p as *indifferent* iff p is not better than $\neg p$, and $\neg p$ is not better than p. Then, p is *good* iff p is better than some indifferent q. Further, p is *bad* iff some indifferent q is better than p. Finally, p is *neutral* iff p is equal in value to some indifferent q.

Brogan, Chisholm, and Sosa's definitions exemplify two slightly different formats for defining goodness and badness in terms of betterness. The first format defines the monadic properties directly in terms of betterness and the concept of negation. The second format first defines a class of indifferent value bearers, in terms of betterness and negation, and then defines goodness and badness with the help of betterness and members of this class. Both formats thus assume that the value bearers can be negated.

In this chapter, we shall investigate the possibility of defining the monadic value properties in terms of betterness, without assuming that negation or other logical connectives can be applied to the value bearers. During most of the last century the orthodox view among philosophers in the analytic tradition was that the bearers of intrinsic value are proposition-like entities. In recent years, this orthodoxy has often been questioned, and many value theorists now believe that other kinds of entities, for example persons or physical objects, can have intrinsic or final value (see, e.g., Rabinowicz & Rønnow-Rasmussen, 2005; Kagan, 2005; Korsgaard, 2005; O'Neill, 1992). It is therefore worthwhile to explore whether the monadic value properties can be defined in terms of betterness, within a framework that puts no restrictions on what kinds of entities can be bearers of value. The conclusion will be that such definitions are possible if, and probably only if, certain admittedly disputable axiological principles are accepted.

---

[2]See Hansson, 1990, p. 136, for a list of writers who have embraced these definitions.

3.2. Assumptions

The key assumption in the definition format we shall discuss is that the value bearers can be put together or concatenated, so as to form composite value bearers. The structures we shall consider are thus of the form $(X, \succsim, \circ)$, where X is a set of value bearers, $\succsim$ stands for the relation "at least as good as", and $\circ$ denotes a closed binary operation of concatenation, mapping $X \times X$ into X. We adopt the standard definitions of "equally good as", denoted $\sim$, and "better than", denoted $\succ$, in terms of $\succsim$. Thus, for all $a, b \in X$: $a \sim b$ iff $a \succsim b$ & $b \succsim a$, and a iff $a \succsim b$ & $\neg(b \succsim a)$.

Further, the following three axioms will be presumed to hold, for all $a, b, c \in X$:

Axiom 3.1. *Quasi-order:* $(X, \succsim)$ is a quasi-order; i.e., reflexive and transitive.

Axiom 3.2. *Weak associativity:* $a \circ (b \circ c) \sim (a \circ b) \circ c$.

Axiom 3.3. *Weak commutativity:* $a \circ b \sim b \circ a$.

The reason for the qualification "weak" in Axioms 3.2 and 3.3 is that only value equality, rather than identity, is assumed.

The plausibility of the assumptions that the value bearers can be concatenated, in general, and that $\circ$ is a closed operation, in particular, might seem to depend on what kind of entities are value bearers. However, if any two value bearers have a mereological sum or fusion, we may identify concatenation with mereological summation. Since mereology puts no ontological restrictions on the parts making up a sum, this implies that concatenation is ontologically neutral. Mereological summation is usually taken to be associative, commutative, and idempotent (see Varzi, 2009, pp. 20f). The first two properties thus validate Axioms 3.2 and 3.3. Idempotency, however, creates a problem. If $a \circ a = a$, then $a \circ a \sim a$. If Axioms 3.2 and 3.4 (see Section 3.3) hold, the assumption that $a \circ a \sim a$, for all $a \in X$, implies that all value bearers are equally good. The natural solution is to define $a \circ a$ as the mereological sum of a and a numerically different but qualitatively identical entity (or an entity that is identical at least in the value-relevant aspects).

If concrete entities, such as physical objects, are value bearers, this interpretation of self-concatenation implies that X contains merely possible, in addition to actually existing value bearers. At least, this is so unless the actual world contains an infinite number of qualitatively identical copies of any given concrete object. It may be noted that self-concatenation is defined by means of qualitatively identical copies also in the context of extensive measurement, such as the measurement of length or mass (see Krantz et al., 1971, pp. 3f). Hence, the relevant domain contains merely possible objects in the case of extensive measurement, as well.

Admittedly, such a distinct but qualitatively identical entity may be difficult to find for some putative value bearers, such as the state of affairs that there is happiness in the world.[3] The value of this state, however, appears to be derived from the values of more "basic" value bearers, for example states to the effect that a certain person is happy to a certain degree at a certain time. Our primary aim is to provide definitions of the monadic value properties that apply to basic value bearers.[4]

It is, of course, a controversial assumption that there is, for any two value bearers, a third entity that is their mereological sum, and also a value bearer. This is especially controversial if there are entities of different ontological kinds among the value bearers. Suppose, for example, that the King of Sweden's tie and the state of affairs p = "There are happy elks" both are value bearers. If ∘ is closed, this implies that there is a value bearer consisting of p and the King's tie. Such proliferation of entities and value bearers might seem ontologically extravagant.[5] However, we need not assume the set X to contain *all* value bearers. It could be restricted to value bearers of a certain ontological kind. The scope of the definitions of the monadic value properties to be proposed will then be similarly restricted.

The standard definition of "better than" in terms of "at least as good as" has been questioned by Sven Danielsson (Danielsson, 2005, p. 377). Assuming states of affairs as value bearers, Danielsson suggests that if p is better than q, the disjunctive state $p \vee q$ is at least as good as q, while q is not at least as good as $p \vee q$. The standard definition then implies that $p \vee q$ is better than q, a judgement Danielsson is inclined to deny. Furthermore, q and $p \vee q$ are not equally good, since this would imply that q is at least as good as $p \vee q$. This means that "at least as good as" is not equivalent to "better than or equally good as".

In my view, the plausibility of the claim that $p \vee q$ is at least as good

---

[3]I owe this example to Jens Johansson.

[4]The notion of basic value bearers, or basic value, is often employed in attempts to solve the general problem of how the value of a composite value bearer is related to the values of its parts. See, e.g., Zimmerman, 2001, Chapter 5; Feldman, 2005; Carlson 2001, 2005; Danielsson, 1997, 2005.

[5]For a defense of such "unrestricted composition", see Lewis, 1991, pp. 79-81.

as q, but not vice versa, is evidence that $p \lor q$ is, in fact, better than q. For the purposes of this chapter, however, nothing hangs on this. We could simply assign "at least as good as" the stipulative meaning "better than or equally good as", and allow that it may have a somewhat different meaning in ordinary language.

As regards Axiom 3.1, reflexivity of $\succsim$ seems to be a safe assumption. Surely, any value bearer is at least as good as itself.[6] Transitivity has, on the other hand, been questioned by several philosophers (see, e.g., Temkin, 1996; Rachels, 1998). Their arguments will not be discussed here. We shall simply proceed on the assumption that "at least as good as" is a transitive relation, at least as regards intrinsic or final value. We noted above that Axioms 3.2 and 3.3 are satisfied if ○ is identified with mereological summation, and the latter operation is associative and commutative.

If concatenation is defined as involving physical interaction between objects, on the other hand, Axioms 3.2 and 3.3 may be less plausible. As Fred Roberts notes in a discussion of the stronger associativity assumption that $a \circ (b \circ c) = (a \circ b) \circ c$, "combining a with b first and then bringing in c might create a different object from that obtained when b and c are combined first. To give an example, if a is a flame, b is some cloth, and c is a fire retardant, then combining a and b first and then combining with c is quite different from combining b and c first and then combining with a." (Roberts, 2009, p. 125). Clearly, this difference may be evaluatively relevant. Similar remarks apply to Axiom 3.3. If $a \circ b$ is defined as "a put on top of b", $a \circ b$ may be very different from $b \circ a$, if a is a porcelain vase, and b is a heavy slab of marble.

3.3. Defining the Monadic Value Predicates

Let us define an object $a \in X$ as *positive* iff $a \circ a \succ a$, as *negative* iff $a \succ a \circ a$, and as *null* iff $a \circ a \sim a$. The following definitions of goodness, badness and neutrality suggest themselves:

Definitions 3.1:

1G. An object in X is *good* iff it is positive.

---

[6]Ruth Chang, 2005, p. 341, n. 12, reports, however, that one philosopher, Joshua Gert, has expressed doubts as to whether "equally good as" is reflexive!

1*B*. An object in X is *bad* iff it is negative.

1*N*. An object in X is *neutral* iff it is null.

Clearly, these definitions do not presuppose any concept of negation. However, Axioms 3.1-3.3 are not sufficient to vindicate Definitions 3.1. The following four claims are arguably conceptual or necessary truths about the monadic value properties:

First conceptual claim: Any value bearer has at most one of the properties of goodness, badness, and neutrality.

Second conceptual claim: If a is good and b is neutral or bad, or a is neutral and b is bad, then a is better than b.

Third conceptual claim: If a is good, and b is at least as good as a, then b is good. Similarly, if a is bad, and at least as good as b, then b is bad. Lastly, if a is neutral and has equal value as b, then b is neutral.

Fourth conceptual claim: If a and b are both neutral, then they have equal value.

As regards the first claim, it should be stressed that it concerns intrinsic or final value "overall", or "all things considered". It may be possible for an object to be good in one respect, and bad in another respect.

Given Definitions 3.1, the first conceptual claim corresponds to the following proposition:

Proposition 3.1. Any object in X has at most one of the properties of positivity, negativity, and nullity.

This proposition follows immediately from the definitions of $\sim$ and $\succ$ in terms of $\succsim$. The remaining three conceptual claims correspond, respectively, to the following propositions:

Proposition 3.2. If a is positive (null) and b is null or negative (negative), then $a \succ b$.

Proposition 3.3. If a is positive and $b \succsim a$, then b is positive. Also, if a is negative and $a \gtrsim b$, then b is negative. Finally, if a is null and $a \sim b$, then b is null.

Proposition 3.4. If a and b are both null, then $a \sim b$.

None of these propositions is implied by Axioms 3.1-3.3. These axioms do not rule out, for example, that $a \succ b$ although a is negative and b is positive. This violates Propositions 3.2 and 3.3. In violation of Proposition 3.4, it may be the case that $a \succ b$ although a and b are both null. Given Definitions 3.1, therefore, only the first of the four conceptual claims is guaranteed to hold.

However, Propositions 3.2, 3.3 and 3.4 all follow if we, in addition to Axioms 3.1-3.3, assume the following axiom to hold, for all a, b, and $c \in X$:

Axiom 3.4. *Monotonicity:* $a \gtrsim b$ iff $a \circ c \gtrsim b \circ c$.[7]

Given Axioms 3.1-3.4, therefore, Definitions 3.1 validate all four conceptual claims. The plausibility of Axiom 3.4 will be discussed in Section 3.5.

3.4. Incomparability and Indeterminacy

It may seem that we should add the claim that any value bearer is either good, bad, or neutral to the list of conceptual truths about the monadic value properties. This claim would immediately follow from Definitions 3.1, if we were to assume completeness; i.e., that $a \gtrsim b$ or $b \gtrsim a$ holds for all a, $b \in X$. Obviously, completeness implies that every object in X is either positive, negative, or null.

Completeness is, however, a rather strong assumption. A number of philosophers have argued that two objects a and b may both have value, although a is not at least as good as b, and b is not at least as good as a. Let us define a and b as *incomparable* iff $\neg(a \gtrsim b)$ & $\neg(b \gtrsim a)$. This definition of incomparability, although common, is not uncontroversial. For example, Ruth Chang has argued that two value bearers a and b may be "on a par", and hence comparable in an intuitive sense, even though $\neg(a \gtrsim b)$ & $\neg(b \gtrsim a)$ (Chang, 2002). Moreover, if Danielsson's objection to the definition of $\sim$ in terms of $\gtrsim$ is correct, and we therefore should adopt the suggested stipulative meaning of $\gtrsim$, neither $q \gtrsim (p \vee q)$ nor $(p \vee q) \gtrsim q$ holds

---

[7]Proofs are given in the Appendix.

in Danielsson's example, although q and $p \vee q$ are, intuitively, comparable. We may, however, regard our definition of incomparability as stipulative. It is immaterial whether it exactly corresponds to an intuitive or ordinary language notion of incomparability.

One popular argument for incomparability is the "small improvement argument". Let a and b be two rather different value bearers. For example, a may be an instance of pleasure, and b an instance of beauty. On reflection, we may judge that a is not better than b, and that b is not better than a. If completeness holds, this implies that a and b are equally good. Let a^+ be a marginally improved version of a, for instance a slightly more intense episode of pleasure. If a and b are equally good it follows, given transitivity, that a^+ is better than b. Very likely, however, we will find that we cannot judge a^+ better than b, either. If so, we have grounds for concluding that neither of a and b is at least as good as the other.

Moreover, I think there are positive reasons to doubt that every value bearer is good, bad, or neutral. Elsewhere, I have suggested that some value bearers may be evaluatively "indeterminate" (Carlson, 2005. See also Danielsson, 2005). If states of affairs can be bearers of value, the disjunction of a good and a bad state might be an example of an object with indeterminate value. In fact, it seems that if there are indeterminate value bearers, completeness cannot hold. Arguably, it is a conceptual truth that any object that is better (worse) than a neutral object is good (bad), and that any object equal in value to a neutral object is neutral. If so, an indeterminate object must be incomparable to any neutral object.

Conversely, if completeness fails to hold, there are probably evaluatively indeterminate value bearers. Suppose that beautiful objects and pleasant experiences are among the bearers of goodness, and that ugly objects and unpleasant experiences are among the bearers of badness. Suppose also that some beautiful objects are incomparable in value to some pleasant experiences. Then it appears highly likely that there are indeterminate objects. Consider the question whether there is, for a given beautiful object a, some ugly object b, whose ugliness exactly counterbalances the beauty of a, so that $a \circ b$ is neutral. If there is such a b, let c be a pleasant experience, whose value is incomparable to that of a. Barring holistic effects, it seems that $b \circ c$ must be indeterminate. (If $b \circ c$ were good, bad, or neutral, c would have to be, respectively, better, worse, or equally good as a, contrary to hypothesis.) If there is no ugly b that exactly counterbalances a, it is very plausible that there is an ugly object d, such that $a \circ d$ is indeterminate. The alternative possibility, that $a \circ b$ is either good or bad, for every ugly *b*, implies that there is a rather unlikely gap among the ugly objects in the betterness ordering, such that the concatenation of *a* and any object above this gap is good, while the concatenation of *a* and any object below the gap is bad.

Axioms 3.1-3.4 leave room for evaluative indeterminacy. Let $\parallel$ denote incomparability; i.e., $a \parallel b$ iff $\neg(a \succsim b)$ and $\neg(b \succsim a)$. We define $a \in X$ as *indefinite* iff $a \circ a \parallel a$. Indefiniteness obviously excludes positivity, negativity, and nullity. Definitions 3.1 can now be supplemented:

1*I*. An object in X is *indeterminate* iff it is indefinite.

The following proposition is easily seen to hold, given Axioms 3.1-3.4:

Proposition 3.5. Suppose that *a* is indefinite. Then if *b* is positive, $a \parallel b$ or $b \succ a$; if *b* is negative, $a \parallel b$ or $a \succ b$; and if *b* is null, $a \parallel b$.

Proposition 3.5 corresponds to the following plausible claim:

Fifth conceptual claim: Any good value bearer is incomparable to or better than any indeterminate one. Any bad value bearer is incomparable to or worse than any indeterminate one. Any neutral value bearer is incomparable to any indeterminate one.

Since it is not ruled out that indefinite objects are incomparable to positive or negative ones, Definitions 3.1 do not allow us to infer that any good object is better than any value bearer that is not good; or that any value bearer that is not bad is better than any bad object. I doubt, however, that these are conceptual truths. For example, let *a* be a moderately good state of affairs, and let *b* be the disjunction of a very good and a moderately bad state. I find it far from obvious that we must conclude either that *a* is better than *b*, that *b* is good, or that *b* is not a value bearer.

3.5. Giving Up Monotonicity

It is easy to come up with putative counterexamples to Axiom 3.4. Many such examples appear to trade on the existence of "organic unities", in G. E. Moore's sense. Moore defined an organic unity as a whole whose intrinsic value "is neither identical with nor proportional to the sum of the values of its parts" (Moore, 1903, p. 184). The statement that the value of a particular whole is, or is not, proportional to the sum of the values of its parts assumes that it is meaningful to add the values of the parts. Additive measurement, however, presupposes Axiom 3.4. If Moore's examples of organic unities are construed as counterexamples to this axiom, which is often a natural interpretation, it follows that his definition makes no sense. Arguably, therefore, the doctrine of organic unities is best understood simply as the rejection of Axiom 3.4 (See Carlson, 1997).

Suppose that great works of art are value bearers, and imagine that Leonardo painted two identical *Mona Lisas*. Letting a be *Mona Lisa 1*, letting b be Rembrandt's *Night Watch*, and letting c be *Mona Lisa 2*, the evaluations $a \succsim b$ and $b \circ c \succ a \circ c$ seem very plausible. Intuitively, the fact that $b \circ c$ offers more *variation* than $a \circ c$ is important. In the words of Chisholm, "other things being equal, it is better to combine two dissimilar goods than to combine two similar goods" (Chisholm, 2005, p. 306).

In response, a defender of Axiom 3.4 might suggest that, since there are in our example two qualitatively identical *Mona Lisas*, the value of each of them is less than the value of the *Night Watch*. Hence, the claim that $a \succsim b$ is false. If there were only one *Mona Lisa*, on the other hand, its value would be at least as great as that of the *Night Watch*. This response presupposes that an object's intrinsic or final value may depend on its externally relational properties. Traditionally, this view has been rejected by most philosophers discussing the issue, but it has recently gained popularity (see Korsgaard, 2005; Kagan, 2005; Rabinowicz & Rønnow-Rasmussen, 2005). If it is accepted, it appears that many alleged counterexamples to Axiom 3.4 can be rebutted.

On the other hand, the view that relational properties can be good-making opens up for other putative counterexamples. Thus, it has been maintained that a unique object may be good partly because of its very uniqueness.[8] If a is such a unique object, and b is some other value bearer, it might be that $a \succsim b$ and $b \circ a \succ a \circ a$, in violation of Axiom 3.4.

Since Axiom 3.4 is a controversial assumption, it is of interest to investigate the consequences of abandoning it. In the absence of this axiom, we need stronger notions of positivity and negativity than those defined above, in order to obtain adequate definitions of goodness and badness. Thus, let us define $a \in X$ as *universally positive* iff, for all $b \in X$, $a \circ b \succ b$. Similarly,

---

[8]Monroe Beardsley, 2005, pp. 61f., discusses an example involving a rare stamp. Beardsley doubts, however, that rarity confers value "for its own sake" to the stamp, although its (philatelic) value is not "for the sake of something else". Rabinowicz and Rønnow-Rasmussen, 2005, p. 120, n. 15, point out that Beardsley's reasons for these doubts are unclear.

a is *universally negative* iff, for all $b \in X$, $b \succ a \circ b$; and a is *universally null* iff, for all $b \in X$, $a \circ b \sim b$. Finally, a is *universally indefinite* iff, for all $b \in X$, $a \circ b \parallel b$.

We make the following observation:

Observation 3.1. Given Axioms 3.2 and 3.4, positivity (negativity, nullity, indefiniteness) implies universal positivity (universal negativity, universal nullity, universal indefiniteness).

Now, we may consider the following definitions:

Definitions 3.2:

2*G*. An object in X is *good* iff it is universally positive.

2*B*. An object in X is *bad* iff it is universally negative.

2*N*. An object in X is *neutral* iff it is universally null.

2*I*. An object in X is *indeterminate* iff it is universally indefinite.

The following three propositions hold, given Axioms 3.1-3.3:

Proposition 3.1*. Any object in X has at most one of the properties of universal positivity, universal negativity, and universal nullity.

Proposition 3.2*. If a is universally positive (universally null) and b is universally null or universally negative (universally negative), then $a \succ b$.

Proposition 3.4*. If a and b are universally null, then $a \sim b$.

Assuming Axioms 3.1-3.3, therefore, Definitions 3.2 imply the second and fourth conceptual claims. On the other hand, they do not imply the third conceptual claim. It may be the case, for example, that $a \succ b$, although b is universally positive and a is not. Nor does the fifth conceptual claim follow. Hence, Definitions 3.2 are acceptable only under the assumption that every object is either universally positive, universally null, universally negative, or universally indefinite:

Axiom 3.5. Every $a \in X$ is either universally positive, universally null, universally negative, or universally indefinite.

We note:

Observation 3.2. Axiom 3.5 follows from Axioms 3.2 and 3.4.

From Axioms 3.1-3.3 and 3.5 we can derive the following two propositions:

Proposition 3.3*. If a is universally positive and $b \succsim a$, then b is universally positive. Also, if a is universally negative and $a \succsim b$, then b is universally negative. Finally, if a is universally null and $a \sim b$, then b is universally null.

Proposition 3.5*. Suppose that a is indefinite. Then if b is positive, $a \parallel b$ or $b \succ a$; if b is negative, $a \parallel b$ or $a \succ b$; and if b is null, $a \parallel b$.

If Axioms 3.1-3.3 and 3.5 hold, therefore, all five conceptual claims follow.

Defining goodness as universal positivity, and badness as universal negativity, rhymes well with Chisholm and Sosa's idea about the intrinsically good or bad as that which rates any possible universe a plus or a minus, respectively. Assuming Definitions 3.2, Axiom 3.5 implies that if $a \circ b$ is good, at least one of a and b is good. Similarly, if $a \circ b$ is bad, at least one of a and b is bad. Further, if $a \circ b$ is neutral, then a and b are both neutral, or one of them is good and the other bad. Finally, if $a \circ b$ is indeterminate, then one of a and b is good and the other bad, or at least one of them is indeterminate.

I surmise that many putative counterexamples to these implications can be answered along the same lines as the reply to the first objection above to Axiom 3.4. However, the uniqueness example seems more recalcitrant. If a is good because of its uniqueness, it could be that $a \circ b \succ b$ and $a \succ a \circ a$. This would be a counterexample not only to Axiom 3.5 but, a being good and negative, to the very idea of defining goodness in terms of positivity or universal positivity. I shall not here try to assess the plausibility of this kind of example, but merely conclude that if there are indeed good but negative objects, the prospects for defining goodness and badness in terms of betterness look bleak.

It might be thought that goodness and badness could still be defined in terms of betterness and the monadic property of neutrality. Thus, we could define an object as good (bad) iff it is better (worse) than a neutral object. If there are good but negative (or bad but positive) objects, however, there is no reason to believe that neutrality can be defined in terms of betterness and concatenation. If not, a neutral object must, it seems, be defined as an object that is neither good nor bad (nor indeterminate, if indeterminate objects are acknowledged). The suggested definitions of goodness and badness will then be viciously circular.

3.6. Conclusion

We assumed that the relation "at least as good as" is a quasi-order, and that the value bearers can be concatenated by means of mereological summation. This operation was supposed to satisfy the properties of weak associativity and weak commutativity. Given this framework, it was shown that the properties of goodness, badness, neutrality, and indeterminacy can be defined in terms of the betterness relation, if every value bearer is either universally positive, universally negative, universally null, or universally indefinite. The definitions proposed do not assume that negation or other logical connectives are applicable to the value bearers.

To some extent, the definitions of the monadic value properties suggested in this chapter resemble the second definition format, mentioned in Section 3.1 and exemplified by Chisholm and Sosa's definitions. According to that format, a value bearer is good (bad) iff it is better (worse) than an evaluatively indifferent entity. Definitions 3.1 and 3.2 imply, similarly, that any good (bad) object is better (worse) than any neutral object. However, the concept of neutrality is not used in the *definientia* of Definitions 3.1 or 3.2. Unlike earlier proposals of the second format, therefore, our definitions do not presuppose the existence of a neutral object.

Acknowledgments

Earlier versions of this work were presented at seminars at Umeå University and Uppsala University. I wish to thank the participants, especially Jens Johansson, Sten Lindström and Frans Svensson, for helpful comments.

Appendix: Proofs.

Proposition 3.1. This proposition follows immediately from the definitions of $\sim$ and $\succ$ in terms of $\succsim$. □

Proposition 3.2. Suppose that a is positive and b is null or negative. This means that $a \circ a \succ a$ and $b \succsim b \circ b$. By Axiom 3.4, $a \circ a \succ a$ implies $(a \circ a) \circ b \succ a \circ b$, and by Axioms 3.1 and 3.2, $a \circ (a \circ b) \succ a \circ b$. By Axioms 3.1 and 3.3, this yields $(a \circ b) \circ a \succ b \circ a$. Hence, by Axiom 3.4, $a \circ b \succ b$. By a similar argument, $b \succsim b \circ b$ implies $a \succsim a \circ b$. Hence, by Axiom 3.1, $a \succ b$. The proof if a is positive or null and b is negative is similar. □

Proposition 3.3. Suppose that a is positive; that is, $a \circ a \succ a$. If $b \succsim a$, then, by Axiom 3.4, $b \circ a \succsim a \circ a$. Axiom 3.1 then implies $b \circ a \succ a$. Hence, by Axiom 3.4, $(b \circ a) \circ b \succ a \circ b$. By Axioms 3.1, 3.2, and 3.3, this yields $(b \circ b) \circ a \succ b \circ a$. Axiom 3.4 thus implies that $b \circ b \succ b$; that is, b is positive. This proves the first clause of the proposition. The proofs of the remaining two clauses are similar. □

Proposition 3.4. By Axiom 3.4, $a \circ a \sim a$ implies $(a \circ a) \circ b \sim a \circ b$. By Axioms 3.1 and 3.2, this yields $a \circ (a \circ b) \sim a \circ b$. Hence, by Axioms 3.1 and 3.3, $(a \circ b) \circ a \sim a \circ b$, and, by Axiom 3.4, $a \circ b \sim a$. A parallel argument shows that $b \circ b \sim b$ implies $a \circ b \sim b$. Hence, by Axiom 3.1, $a \sim b$. □

Proposition 3.5. Since $(a \parallel b$ or $b \succ a)$ is equivalent to $\neg(a \succsim b)$, and $(a \parallel b$ or $a \succ b)$ is equivalent to $\neg(b \succsim a)$, the first two clauses of the proposition follow from Proposition 3.3. The third clause follows from Propositions 3.2 and 3.3. □

Proposition 3.1*. This proposition follows immediately from the definitions of $\sim$ and $\succ$ in terms of $\succsim$. □

Proposition 3.2*. If a is universally positive and b is universally null or universally negative, then $a \circ b \succ b$ and $a \succsim b \circ a$. Hence, by Axioms 3.1 and 3.3, $a \succ b$. If a is universally null and b is universally negative, then $a \circ b \sim b$ and $a \succ b \circ a$. Again, Axioms 3.1 and 3.3 imply that $a \succ b$. □

Proposition 3.3*. If $a \circ b \succ b$ and $b \succsim a$, then, by Axiom 3.1, $a \circ b \succ a$, and, by Axiom 3.3, $b \circ a \succ a$. By Axiom 3.5, this means that b is universally positive. This proves the first clause of the proposition. The proofs of the other two clauses are similar. □

Proposition 3.4*. By Axioms 3.1 and 3.3, $a \circ b \sim b$ and $b \circ a \sim a$ imply $a \sim b$. □

Proposition 3.5*. Since $(a \parallel b$ or $b \succ a)$ is equivalent to $\neg(a \succsim b)$, and $(a \parallel b$ or $a \succ b)$ is equivalent to $\neg(b \succsim a)$, the first two clauses of the proposition follow from Proposition 3.3*. The third clause follows from Propositions 3.2* and 3.3*. □

Observation 3.1. If $a \circ a \succ a$, Axiom 3.4 implies $(a \circ a) \circ b \succ a \circ b$. Hence, by Axioms 3.2 and 3.4, $a \circ b \succ b$. Similarly, $a \circ a \sim a$ implies $a \circ b \sim b$, $a \succ a \circ a$ implies $b \succ a \circ b$, and $a \circ a \parallel a$ implies $a \circ b \parallel b$. □

Observation 3.2. By the definitions of $\succ$, $\sim$, and $\parallel$, every object in X is either positive, negative, null, or indefinite. By Observation 3.1, Axioms 3.2 and 3.4 imply that positivity (negativity, nullity, indefiniteness) is equivalent to universal positivity (universal negativity, universal nullity, universal indefiniteness). Hence, Axiom 3.5 follows from Axioms 3.2 and 3.4. □

References

Beardsley, M. (2005). Intrinsic value. In Rønnow-Rasmussen and Zimmerman (Eds.), *Recent Work on Intrinsic Value*. Dordrecht: Springer.

Brogan, A. P. (1919). The fundamental value universal. *Journal of Philosophy, Psychology and Scientific Method, 16*, 96–104.

Carlson, E. (1997). A note on Moore's organic unities. *The Journal of Value Inquiry, 31*, 55–59.

Carlson, E. (2001). Organic unities, non-trade-off, and the additivity of intrinsic value. *The Journal of Ethics, 5*, 335–360.

Carlson, E. (2005). The intrinsic value of non-basic states of affairs. In Rønnow-Rasmussen and Zimmerman (Eds.), *Recent Work on Intrinsic Value*. Dordrecht: Springer.

Chang, R. (2002). The possibility of parity. *Ethics, 112*, 659–688.

Chang, R. (2005). Parity, interval value, and choice. *Ethics, 115*, 331–350.

Chisholm, R. M. (2005). Organic unities. In Rønnow-Rasmussen and Zimmerman (Eds.), *Recent Work on Intrinsic Value*. Dordrecht: Springer.

Chisholm, R. M., & Sosa, E. (1966). On the logic of 'intrinsically better'. *American Philosophical Quarterly, 3*, 244–249.

Danielsson, S. (1997). Harman's equation and the additivity of intrinsic value. In L. Lindahl, P. Needham, and R. Sliwinski (Eds.), *For Good Measure: Philosophical Essays Dedicated to Jan Odelstad on the Occasion of His Fiftieth Birthday*. Uppsala: Department of Philosophy, Uppsala University.

Danielsson, S. (2005). Harman's equation and non-basic intrinsic value. In

Rønnow-Rasmussen and Zimmerman (Eds.), *Recent Work on Intrinsic Value*. Dordrecht: Springer.

Feldman, F. (2005). Basic intrinsic value. In Rønnow-Rasmussen and Zimmerman (Eds.), *Recent Work on Intrinsic Value*. Dordrecht: Springer.

Hansson, S. O. (1990). Defining 'good' and 'bad' in terms of 'better'. *Notre Dame Journal of Formal Logic, 31*, 136–149.

Kagan, S. (2005). Rethinking intrinsic value. In Rønnow-Rasmussen and Zimmerman (Eds.), *Recent Work on Intrinsic Value*. Dordrecht: Springer.

Korsgaard, C. M. (2005). Two distinctions in goodness. In Rønnow-Rasmussen and Zimmerman (Eds.), *Recent Work on Intrinsic Value*. Dordrecht: Springer.

Krantz, D. H., Luce, R. D., Suppes, P., & Tversky, A. (2007). *Foundations of Measurement, Vol. 1: Additive and Polynomial Representations*. Mineola, NY: Dover.

Lewis, D. (1991). *Parts of Classes*. Oxford: Basil Blackwell.

Moore, G. E. (1903). *Principia Ethica*. Cambridge, MA: Cambridge University Press.

O'Neill, J. (1992). The varieties of intrinsic value. *The Monist, 75*, 119–137.

Rabinowicz, W., & Rønnow-Rasmussen, T. (2005). A distinction in value: Intrinsic and for its own sake. In Rønnow-Rasmussen and Zimmerman (Eds.), *Recent Work on Intrinsic Value*. Dordrecht: Springer.

Rachels, S. (1998). Counterexamples to the transitivity of 'better than'. *Australasian Journal of Philosophy, 76*, 71–83.

Roberts, F. (2009). *Measurement Theory*. Cambridge, MA: Cambridge University Press.

Rønnow-Rasmussen, T., & Zimmerman, M. J. (2005). *Recent Work on Intrinsic Value*. Dordrecht: Springer.

Temkin, L. S. (1996). A continuum argument for intransitivity. *Philosophy & Public Affairs, 25*, 175–210.

Varzi, A. (2009). Mereology. In E. N. Zalta (Ed.), *Stanford Encyclopedia of Philosophy*. From http://plato.stanford.edu/entries/mereology/.

Zimmerman, M. J. (2001). *The Nature of Intrinsic Value*. Lanham: Rowman & Littlefield.

Chapter 4

Optimality in Multisensory Integration Dynamics: Normative and Descriptive Aspects

Hans Colonius and Adele Diederich

Carl von Ossietzky Universität Oldenburg and Jacobs University Bremen

"Temporal window of integration" has become a widely accepted concept in multisensory research: crossmodal information falling within this window will typically be integrated, whereas information falling outside will not. Specifically, an infinitely large time window will always allow integration, a zero-width time window would rule out integration entirely. From a decision-making point of view, neither case is likely to be optimal in the long run. In a noisy, complex, and potentially hostile environment exhibiting multiple sources of sensory stimulation, the issue of whether or not two given stimuli of different modality arise from a common source may become critical for a behaving organism. Making explicit assumptions about the arrival time difference between peripheral sensory processing times triggered by a crossmodal stimulus set, we derive a decision rule determining optimal window width as a function of the prior odds in favor of a common multisensory source, the likelihood of arrival time differences, and the payoff for making correct or wrong decisions. The empirical validity of the suggested approach is shown to be supported by data from a recent study measuring head saccades toward visual-auditory stimuli under varying prior probabilities for a common source.

4.1. Introduction

Evidence for multisensory integration is found in many different forms, notably as facilitation or inhibition of responses to a crossmodal stimulus set, compared to unimodal stimulation. Prime examples of such facilitation include, at the level of neurophysiology, an increased number of responses of a multisensory neuron, and, at the behavioral level, acceleration of manual or saccadic reaction time. Besides facilitation and inhibition, crossmodal stimulation may also result in no discernable effect at all, that is, beyond what would be expected from unimodal stimulation alone. This has sometimes been referred to as a situation where the crossmodal stimulus combination falls outside of a spatiotemporal "window of integration" (e.g., Meredith, 2002). This window concept has enjoyed increasing popularity over the past 5–10 years, as a major determinant of the dynamics of multisensory integration at both the neural and behavioral level of observation. In the following, it is demonstrated that *time window of integration* is not only a metaphor describing a host of crossmodal stimulation effects in various contexts but that it is a cornerstone of a quantitative approach in reaction time modeling of multisensory integration. We will not review the extensive experimental literature on the topic here (but see Diederich & Colonius, 2008a). Instead, we demonstrate, within the time window of integration (TWIN) model first proposed in Colonius and Diederich (2004), how the concept leads to empirically testable predictions across different response time paradigms. Second, it is shown that the time window of integration notion can be incorporated into current theories of multisensory integration based on principles of optimal statistical inference (see Ernst, 2005, for a review).

Although a window of integration has been defined for both spatial and temporal aspects of a crossmodal stimulus set (e.g., Wallace et al., 2004) and has even been suggested for higher-level aspects like semantic congruity (e.g., van Attefeldt et al., 2007), we will confine discussion to the temporal dimension within the reaction time context considered here. On a descriptive level, the *time window hypothesis* holds that information from different sensory modalities must not be too far apart in time so that integration into a multisensory (perceptual) unit may occur. Thus, an infinitely large time window affords the potential for unrestricted integration whereas a zero-width time window will rule out integration entirely. When a sensory event simultaneously produces both sound and light, we usually do not notice any temporal disparity between the two sensory inputs (within a distance of up to 20–26 m), even though the sound arrives with a small delay, a phenomenon referred to as 'temporal ventriloquism' (cf. Lewald and Guski, 2003; Morein-Zamir et al., 2003; Recanzone, 2003; Spence & Squire, 2003). In orienting responses towards a visual-auditory stimulus complex, acceleration of (saccadic) reaction times has been observed under a multitude of experimental settings within a time window of 150 to 250 ms (e.g., Frens et al., 1995; Corneil et al., 2002; Diederich & Colonius, 2004, 2008a,b; Diederich et al., 2009; van Opstal & Munoz, 2004; van Wanrooij et al., 2009; Romei et al., 2007). On the neural level, a temporal window of integration has been observed at multisensory convergence sites, such as the superior colliculus in cat and other animals: Enhanced spike response rates of multisensory neurons occur when periods of unimodal peak activity overlap within a certain time range (Meredith et al., 1987; Rowland & Stein, 2008). Note that the fact that estimates of the width of the time window differ widely, ranging from 40 to 600 ms, is not really surprising given that these estimates arise from rather different experimental contexts, i.e., single-cell recordings, simple manual or saccadic reaction time, and judgments of temporal order or simultaneity.

4.1.1. *Two basic experimental paradigms*

In the *redundant target* paradigm (RTP), stimuli from two (or more) different modalities are presented simultaneously or with certain stimulus onset asynchrony (SOA), and the participant is instructed to respond to the stimulus detected first. Typically, the time to respond in the crossmodal condition is faster than in either of the unimodal conditions. In the *focused attention* paradigm (FAP), crossmodal stimulus sets are presented in the same manner but now participants are instructed to respond only to the onset of a stimulus from a specifically defined target modality, such as the visual, and to ignore the remaining nontarget stimulus, the tactile or the auditory, say. The nontarget stimulus has been shown to modulate the saccadic response to the target: Depending on the exact spatiotemporal configuration of target and nontarget, the effect can be a speed-up or an inhibition of saccadic RT (see, e.g., Diederich & Colonius 2007b). Some striking similarities to human data have been found in cats trained to orient to visual or auditory stimuli utilizing either paradigm (Stein, Huneycutt, & Meredith, 1988; Stein, Jiang, & Stanford, 2004).

4.1.2. *The race model*

A classic explanation for a speed-up of responses to crossmodal stimuli is that subjects are merely responding to the first stimulus detected. Taking the detection times to be random variables and adding some technical assumptions, observed reaction time is representable by the minimum of the reaction times to the, say, visual or auditory signal leading to a purely statistical facilitation effect in response speed. Numerous studies have shown that the race model is not sufficient to explain the observed speedup in reaction time; see Diederich and Colonius (2004) for a review. Using the race model inequality (Miller, 1982) as a benchmark test, responses to bimodal stimuli have been found to be faster than predicted by statistical facilitation, in particular, when the stimuli were spatially aligned. Although the race model test has been applied to data from both types of paradigms, its validity for FAP data seems dubious because the effect of a stimulus from the nontarget modality winning the race is not specified in the model. Moreover, the race model gives no explanation for the decrease in facilitation observed with variations in many crossmodal stimulus properties, e.g, increasing spatial disparity between the stimuli, and it cannot predict inhibition.

4.2. Time Window of Integration (TWIN) Model: Assumptions and Predictions

Although the race model in its simple form cannot account for the empirical data, the basic concept of a race occurring at a very early stage of processing has considerable plausibility. Hence the time-window-of-integration (TWIN) model postulates that a crossmodal stimulus triggers a race mechanism in the very early, peripheral sensory pathways which is then followed by a compound stage of converging subprocesses that comprise neural integration of the input and preparation of a response (for details, see Diederich & Colonius, 2011; Colonius & Diederich, 2004). Note that this second stage is defined by default: it includes all subsequent, possibly temporally overlapping, processes that are not part of the peripheral processes in the first stage. The central assumption of the model concerns the temporal configuration needed for multisensory integration to occur: *Multisensory integration occurs only if the peripheral processes of the first stage all terminate within a given temporal interval, the "time window of integration"* (TWIN assumption). Thus, the window acts as a *filter* determining whether afferent information delivered from different sensory organs is registered close enough in time to trigger multisensory integration. Passing the filter is necessary, but not sufficient, for crossmodal interaction to occur since the amount of interaction may also depend on many other aspects of the stimulus set, like spatial configuration of the stimuli. The amount of crossmodal interaction manifests itself in an increase or decrease of second stage processing time, but it is assumed not to depend on the stimulus onset asynchrony (SOA) of the stimuli.

Although the TWIN model's assumptions oversimplify matters, they afford quite a number of experimentally testable predictions, many of which having found empirical support in recent studies (cf. Diederich & Colonius 2007a,b; 2008a,b). Here we focus on predictions related to the difference in the experimental paradigm utilized, FAP or RTP. Since in both paradigms physically identical stimuli can be presented under the same spatiotemporal configuration, any important differences observed in the corresponding reaction times have to be due to the instructions being different, thereby pointing to a possible separation of top-down from bottom-up processes in the underlying multisensory integration mechanism.

For FAP, the TWIN assumption is further constrained in one important aspect: *Crossmodal interaction occurs only if (i) a nontarget stimulus wins the race in the first stage, opening the time window of integration such that (ii) the termination of the target peripheral process falls in the window.* One interpretation is that the winning nontarget will keep the system in a state of heightened reactivity such that the upcoming target stimulus, if it falls into the time window, will trigger crossmodal interaction. For saccadic eye movements, in particular, this may correspond to a gradual inhibition of fixation neurons (in superior colliculus) and/or omnipause neurons (in midline pontine brain stem). If a stimulus from the target modality is the winner of the race in the peripheral channels, second stage processing is initiated without any multisensory integration mechanism being involved. For further discussion, we introduce these assumptions in a more formal way.

4.2.1. *Deriving TWIN predictions for RTP and FAP*

The race in the first stage is made explicit by assigning statistically independent, nonnegative random variables V and A to the peripheral processing times for a visual and an auditory stimulus, respectively. With τ as SOA value and ω as integration window width parameter, the TWIN assumption for FAP is that multisensory integration takes place whenever the event

$$I_{FAP} = \{A + \tau < V < A + \tau + \omega\},$$

occurs, assuming the visual is defined as target modality. Here, a positive τ value indicates that the visual is presented before the auditory stimulus, and a negative τ value indicates the reverse presentation order.

In both FAP and RTP, either the visual stimulus wins, $V < A + \tau$, or the auditory stimulus wins, $A + \tau < V$. Thus, in either case, $\min(V, A + \tau) < \max(V, A + \tau)$ (event $\{V = A + \tau\}$ having zero probability, due to assuming continuous distributions). The TWIN assumption for RTP is that multisensory integration takes place whenever the event

$$I_{RTP} = \{\max(V, A + \tau) < \min(V, A + \tau) + \omega\}.$$

occurs. Therefore, under identical stimulus conditions,

$$I_{FAP} = I_{RTP} \cap \{A + \tau \text{ is the winner of the race}\}.$$

It follows that any realization of the peripheral processing times V and A that leads to an opening of the time window under the focused attention instruction also leads to that event under the redundant target instruction, i.e., $I_{FAP} \subset I_{RTP}$. Thus, the probability of integration under redundant target instructions cannot be smaller than that under focused attention instruction: $P(I_{FAP}) \leq P(I_{RTP})$, given identical stimulus conditions hold (for a computation of both probabilities under an exponential distribution assumption, see Appendix 1). Although the events I_{FAP} and I_{RTP} are not observable, the ordering of their probabilities translates into an analogous prediction about mean crossmodal reaction times. To see this, consider the expression for mean crossmodal RT under the two different paradigms (e.g., Diederich & Colonius, 2008a):

$$\mathrm{E}[RT_{VA,\tau}] = \begin{cases} \mathrm{E}[\min(V, A + \tau)] + \mu - P(I_{RTP}) \cdot \Delta, & \text{for RTP,} \\ \mathrm{E}[V] + \mu - P(I_{FAP}) \cdot \Delta, & \text{for FAP,} \end{cases} \quad (4.1)$$

where E refers to 'expected value' and μ is the expected second-stage processing time when no integration occurs, assumed to be the same for both paradigms; Δ is the amount of facilitation ($\Delta > 0$) or inhibition ($\Delta < 0$) occurring in the second stage, also assumed to be the same. Given that $\mathrm{E}[\min(V, A + \tau)] \leq \mathrm{E}[V]$, it follows that, if facilitation occurs, mean crossmodal reaction time can never be longer in RTP than in FAP, under identical stimulus conditions:

$$\mathrm{E}[RT_{VA,\tau} \,|\, \mathrm{RTP}] \leq \mathrm{E}[RT_{VA,\tau} \,|\, \mathrm{FAP}].$$

Initial empirical support for this prediction was found in an unpublished experiment from our lab. Visual and auditory stimuli were presented with different SOAs both ipsi- and contralateral, under identical conditions in FAP and RTP tasks. Average manual reaction times were clearly shorter in RTP compared to FAP, except when the nontarget auditory stimulus was presented prior to the target (cf. Figure 4.1).

4.3. Towards an Optimal Time Window of Integration

In TWIN, the width of the time window of integration (ω) is a numerical parameter, which is typically estimated from the data to optimize model fit with respect to some deviance criterion. This may yield interpretable

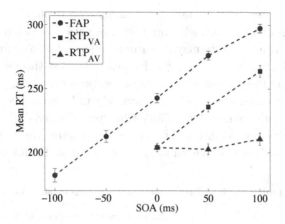

Fig. 4.1. Visual and auditory stimuli were presented both ipsi- and contralateral (±20° from fixation point), under identical conditions in separate FAP and RTP tasks (only ipsilateral data shown). Average manual reaction times were clearly shorter in RTP compared to FAP, except when the nontarget auditory stimulus was presented prior to the target (SOA = −50, −100). RTP$_{VA}$ refers to visual presented prior to auditory, RTP$_{AV}$ refers to auditory presented prior to visual. 200 data points were collected per condition.

results; for example, Diederich et al. (2008) found windows of different size for two different age groups which could be attributed to a slowing of peripheral processing in the old age group. Nevertheless, from a modeling point of view, it is desirable to derive window width from some basic principles that might yield an explanation for the various widths observed in different contexts.

4.3.1. *The basic decision situation*

It has been recognized that integrating crossmodal information always involves a possibly implicit decision about whether or not two (or more) sensory cues originate from the same event, i.e., have a common cause (e.g., Körding et al., 2007). For example, in a predator-prey situation, when the potential prey perceives a sudden movement in the dark, it may be vital to recognize whether this is caused by a predator or a harmless wind gust. If visual information is accompanied by some vocalization from a similar direction, it may be adequate to respond to the potential threat by assuming that the visual and auditory information are caused by the same source, i.e., to perform multisensory integration leading to a speeded escape reaction. On the other hand, in such a rich dynamic environment it may also be disadvantageous, e.g., leading to a depletion of resources, or even hazardous, to routinely combine information associated with sensory events which in reality may be entirely independent and unrelated. In Colonius and Diederich (2010), the following decision-theoretic approach towards finding the optimal time window has been proposed.

The basic decision situation just described can be presented in a simplified and schematic manner by the following "payoff matrix" (Table 4.1). It defines the gain (or cost) function U associated with the *states of nature* (C) and the *action* (I) of audiovisual integration: Variable C indicates whether

Table 4.1. Payoff matrix for the basic decision situation

	integration ($I = 1$)	no integration ($I = 0$)
common source ($C = 1$)	U_{11}	U_{10}
separate sources ($C = 2$)	U_{21}	U_{20}

visual and auditory stimulus information are generated by a common source ($C = 1$), i.e., an *audiovisual event*, or by two separate sources ($C = 2$), i.e., auditory and visual stimuli are unrelated to each other. Variable I indicates whether or not integration occurs ($I = 1$ or $I = 0$, respectively). The values U_{11} and U_{20} correspond to correct decisions and will in general be assumed to be positive numbers, while U_{21} and U_{10}, corresponding to incorrect decisions, will be negative. The organism's task is to balance these costs and benefits of multisensory integration by an appropriate optimizing strategy (cf. Körding et al., 2007).

We assume that *a priori* probabilities for the events $\{C = i\}_{i=1,2}$ exist, with $(C = 1) = 1 - (C = 2)$. In general, an optimal strategy may involve many different aspects of the empirical situation, like spatial and temporal contiguity, or more abstract aspects, like semantic relatedness of the information from different modalities (cf. van Attefeldt et al., 2007). For now, we limit the analysis to temporal information. In other words, the *only* perceptual evidence utilized by the organism is the temporal disparity between the 'arrival times' of the unimodal signals. Thus, computation of the optimal time window will be based on the prior probability of common cause and the likelihood of temporal disparities between the unimodal signals. Note that our approach does not presuppose existence of some high-level decision-making entity contemplating different action alternatives. We only assume that behavior can be assessed as being consistent or inconsistent with an optimal strategy with respect to the time window width.

4.3.2. *Deriving an optimal decision rule*

For concreteness, let A, V denote the auditory and visual arrival time, respectively, which we assume to be equivalent to the peripheral processing times of TWIN. For FAP with the visual modality being the target modality, the arrival time difference (ATD) is defined as $T \equiv V - A$ (for auditory targets, $T \equiv A - V$). Thus, the range of ATDs may include both positive and negative values. On the other hand, for RTP, the absolute ATD, $T \equiv |V - A|$, is assumed to represent the empirical evidence available to the decision mechanism.

For a realization t of T, we define the *likelihood function* $f(t|C)$, where f denotes the probability mass function or, if it exists, the density function of T given C. Using Bayes's rule, we immediately have the *posterior* probability of a common cause given the occurrence of an arrival time difference t,

$$(C = 1|t) = \frac{f(t|C = 1)(C = 1)}{f(t|C = 1)(C = 1) + f(t|C = 2)(C = 2)}.$$

On each trial, in order to maximize the expected value $E[U]$ of function U in the payoff matrix (Table 4.1), the decision-making mechanism is to choose that action alternative (i.e., to integrate or not) which contributes, on the average, more to $E[U]$ than the other action alternative. Introducing the *likelihood ratio* function

$$L(t) = f(t|C = 1)/f(t|C = 2),$$

results in the following decision rule (cf. Colonius & Diederich, 2010):

"If $\quad L(t) > \dfrac{(C = 2)}{(C = 1)} \times \dfrac{U_{20} - U_{21}}{U_{11} - U_{10}}, \quad$ *integrate, otherwise do not integrate."*

(4.2)

The *optimal time window* is then defined as the set of all arrival time differences t satisfying the inequality in this decision rule. Whenever $L(t)$ is a strictly decreasing function, the optimal time window is defined as

$$\left\{ t \middle| t < L^{-1} \left[\frac{(C = 2)}{(C = 1)} \times \frac{U_{20} - U_{21}}{U_{11} - U_{10}} \right] \right\}. \qquad (4.3)$$

Setting

$$t_0 = L^{-1} \left[\frac{(C = 2)}{(C = 1)} \times \frac{U_{20} - U_{21}}{U_{11} - U_{10}} \right],$$

the optimal window is composed of all ATDs shorter than t_0. Note that, since L^{-1} is strictly decreasing, increasing the prior probability $(C = 1)$ for a common cause will make the optimal window larger, as expected. In general, however, the likelihood ratio $L(t)$ need not be a monotone function of t, and the optimal time window does not necessarily have the intuitive form of a "window", i.e., of an interval of the reals.

The window also depends on the U-values in the payoff matrix as follows. Keeping the (negative) values U_{21} and U_{10} fixed, an increase in U_{11}, (the gain of integrating when there is a common cause) will decrease the ratio of U-differences occurring in the decision rule and leads to an increase of optimal window width; on the other hand, an increase in U_{20} (the gain of not integrating when there is no common cause) will increase the ratio of U-differences leading to a narrowing of the window. Both effects are to be expected, and a symmetric argument holds for the remaining values U_{21} and U_{10}.

One plausible scenario for a decreasing likelihood ratio is to assume that $f(t \mid C = 1)$ has a maximum at $t = 0$ and then decreases with t, i.e., larger arrival time differences become less likely under a common cause, whereas $f(t \mid C = 2)$ is constant across all t values that may occur in a trial. An exact value of t_0 can only be determined for explicit values of $(C = 1)$, the payoff matrix entries, and the likelihood ratio function. For the case of RTP, a numerical example with exponential and uniform likelihood functions was presented in Colonius and Diederich (2010) yielding optimal window width as a function of the a priori probability of a common event.

4.3.3. *Optimal time window in focused attention paradigm*

We now consider the computation of an optimal time window in the case of a focused attention paradigm (FAP) where both positive and negative ATDs may occur. The optimal time window is

$$\left\{ t \mid L(t) > \frac{(C = 2)}{(C = 1)} \times \frac{U_{20} - U_{21}}{U_{11} - U_{10}} \right\}. \tag{4.4}$$

As in the RTP case mentioned above, we assume the likelihood function under separate causes to be uniform. A plausible shape for the likelihood function under a common cause is to have a maximum at, or around, the zero point (i.e., ATD= 0) and to monotonically decrease both left and right from zero.

In order to derive explicit numerical predictions for an optimal time window, the likelihood functions must be specified. In TWIN model applications, the choice of exponential distributions for the first-stage processing times, although mainly based on reasons of ease of computation, has been quite successful in describing many data sets at the level of mean reaction times (cf. Diederich & Colonius, 2008a,b). In keeping with the exponential distribution assumption, the distribution of arrival time differences, $V - A$, in a focused attention paradigm with the visual as target modality under a common source, is assumed to be an *asymmetric Laplace distribution* (see Appendix 2):

$$f(t|\,C = 1) = \frac{\lambda_V \lambda_A}{\lambda_V + \lambda_A} \begin{cases} \exp(-\lambda_V t) & \text{if } t \geq 0, \\ \exp(\lambda_A t) & \text{if } t < 0. \end{cases} \tag{4.5}$$

Here λ_V and λ_A are the intensity parameters of the visual and auditory exponential distributions, respectively. The distribution of ATDs will generally also depend on the specific interstimulus presentation interval τ (ISI) between the visual and the auditory stimulus in the experiment, but there is no need to make that explicit for now.

For two separate sources, we assume a uniform law,

$$f(t|\,C = 2) = \begin{cases} 1/(t_1 - t_0) & \text{if } t_0 < t < t_1, \\ 0 & \text{otherwise.} \end{cases} \tag{4.6}$$

Here, $t_0, t_1 \in \Re$ constitute the *observation interval*, that is, the interval of time limiting all possible ATDs due to the construction of a trial by the experimenter. Thus, within the observation interval (t_0, t_1), any arrival time difference is assumed to occur with the same likelihood under two separate sources. For $t_0 \leq t \leq t_1$, the likelihood ratio becomes

$$\begin{aligned} L(t) &= f(t|\,C = 1)/f(t|\,C = 2) \\ &= \frac{\lambda_V \lambda_A}{\lambda_V + \lambda_A}(t_1 - t_0) \begin{cases} \exp(-\lambda_V t) & \text{if } t \in (t_0, t_1) \cap [0, t_1), \\ \exp(\lambda_A t) & \text{if } t \in (t_0, t_1) \cap (t_0, 0]. \end{cases} \end{aligned} \tag{4.7}$$

Figure 4.2 illustrates the likelihood ratio as a function of t under specified parameter values. Note that for t outside of the observation interval the likelihood ratio remains undefined. To simplify the exposition in the following, the ratio of utility differences in (4.4) will be set equal to one. Thus, according to the optimal decision rule, audiovisual integration should be performed if and only if

$$L(t) > \frac{1 - p}{p},$$

with $p \equiv (C = 1)$.

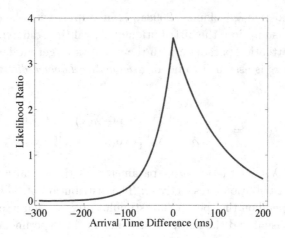

Fig. 4.2. Likelihood-ratio function for asymmetric-Laplace/uniform densities example. Parameter values are $\lambda_V = .01$, $\lambda_A = .025$, $t_1 = 200$, $t_0 = -300$; the function is defined for $t \in (t_0, t_1)$.

Inserting the expression for $L(t)$ from (4.7) and solving for t yields the following *optimal* time window for $t \in (t_0, t_1)$:

$$\left\{ t \;\middle|\; \frac{1}{\lambda_A} \log \left[\frac{(\lambda_V + \lambda_A)(1-p)}{\lambda_V \lambda_A (t_1 - t_0) p} \right] \leq t \leq \frac{1}{\lambda_V} \log \left[\frac{\lambda_V \lambda_A (t_1 - t_0) p}{(\lambda_V + \lambda_A)(1-p)} \right] \right\} \tag{4.8}$$

provided that

$$\frac{(\lambda_V + \lambda_A)(1-p)}{\lambda_V \lambda_A (t_1 - t_0) p} \leq 1. \tag{4.9}$$

This condition guarantees that the left side of the interval is non-positive and the right side is non-negative. For the width of the time window, we get immediately

$$W = \left(\frac{1}{\lambda_V} + \frac{1}{\lambda_A} \right) \log \left[\frac{\lambda_V \lambda_A (t_1 - t_0)}{\lambda_V + \lambda_A} \frac{p}{1-p} \right]. \tag{4.10}$$

Window width is seen to be an increasing function of the prior odds $p/(1-p)$ and of the observation interval (t_0, t_1). Figure 4.3 illustrates this dependency. It shows the corresponding upper and lower limits of the optimal time window as a function of the prior probability of a common source, $p = P(C = 1)$. Increasing $P(C = 1)$ leads to a widening of the time window, in this case approaching infinity in a nonlinear fashion. Moreover, the optimal time window disappears for values of the prior below a certain positive threshold value. Although the exact threshold value depends on

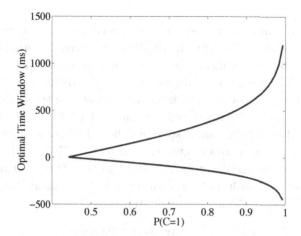

Fig. 4.3. Upper and lower limits of the optimal time window as a function of the prior probability for a common source $P(C = 1)$ under the asymmetric-Laplace/uniform likelihood functions of Figure 4.2. Optimal window width (upper minus lower limit) increases nonlinearly with prior probability.

the experimental context and may get closer to zero, this prediction may provide a strong model test: for a small enough value of $P(C = 1)$ there should be no multisensory integration effect at all.

4.4. Two Empirical Studies of Time Window Optimality

4.4.1. *The effect of prior probability: The van Wanrooij et al. 2010 study*

In a recent focused attention study, Van Wanrooij and colleagues (Van Wanrooij et al., 2010) measured human head saccades towards a visual target (50 ms flash) which, in the bimodal condition, was accompanied by a synchronous (white noise) sound (50 ms , 65 dB). Visual (V) and auditory (A) stimuli were presented either spatially aligned or vertically disparate in the midsagittal plane at 10 possible locations: $\pm 15, 20, 25, 30$ or $35°$ in elevation (with $0°$ at straight ahead). Consistent with the previous results on saccadic eye movements mentioned above, (i) responses to spatially aligned AV stimuli were, on average, faster than unimodal responses, and (ii) spatial disparity systematically delayed the AV responses, by up to about 38 ms, when the auditory nontarget was presented in the hemifield opposite to the V target. As reported in Van Wanrooij et al. (2010), the physical and perceived spatial disparities were indistinguishable under the experimental conditions.

Importantly, the bimodal responses were recorded under three different conditions of spatial disparity distribution, collected in separate experiments. In the first experiment, the visual stimuli were always accompanied by spatially (and temporally) aligned sounds (AV-$100/0$). In the second, the visual stimuli were aligned with the sound in 50% of the trials while the others had a spatial disparity of more than 45° (AV-$50/50$). In the third, only 10% of the trials contained spatially aligned stimulus pairs while the others had spatial disparities between 5° and 75° (AV-$10/90$). Although the actual prior probabilities for spatial alignment (thus, for a common audiovisual source) utilized by the subjects are not observable, the only plausible ordering of the prior probabilities is

$$p_{AV\text{-}100/0} > p_{AV\text{-}50/50} > p_{AV\text{-}10/90}.$$

According to the decision rule introduced in Section 4.3.2, this ordering implies a corresponding ordering of the time-window width such that under perfect spatial alignment facilitation of reaction time should reflect the ordering. This is exactly what Van Wanrooij and colleagues found (see Figure 4, redrawn from their data). Reaction times in conditions AV-$50/50$ and AV-$10/90$ were larger than in AV-$100/0$ for 9 out of 14 cases ($p < .001$) and the ordering was as predicted in all but one case.

4.4.2. The effect of age: The Diederich et al. 2008 study

A number of studies assessing the efficiency with which multisensory integration occurs as a function of aging have found the benefit of multisensory signals to perception to be greater in older than younger adults (e.g., Peiffer et al., 2007; Laurienti et al., 2006; but see Hugenschmidt et al., 2009). For an example in response speed effects, we consider the study by Diederich et al. (2008) measuring saccadic RT of aged (65-75 years) and young individuals (20-22 years) to the onset of a visual target stimulus with and without an accessory auditory stimulus occurring ipsi- or contralateral to the target (focused attention task). The response time pattern for both groups was similar: mean RT to bimodal stimuli was generally shorter than to unimodal stimuli, and mean bimodal RT was shorter when the auditory accessory was presented ipsilateral rather than contralateral to the target. The elderly participants were considerably slower than the younger participants under all conditions but showed a greater multisensory enhancement, that is, they seemed to benefit more from bimodal stimulus presentation relative to unimodal presentation.

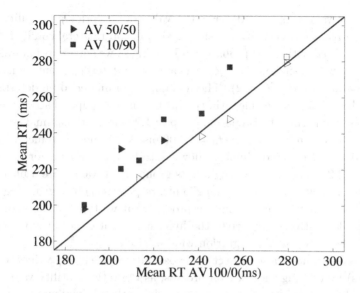

Fig. 4.4. Mean saccadic reaction time in the 50% alignment (triangles) and 10% alignment (squares) conditions against the mean in the 100% alignment condition for 7 subjects in van Wanrooij et al. (2010). Only one subject exhibited faster reactions in the 10% condition than in the 50% condition. Non-filled data points were not significantly different from the 100% condition.

Fitting the TWIN model to the data from both age groups separately, the resulting numerical parameter values reflected the slowing of the peripheral sensory processing in the elderly $(1/\lambda_V = 84$ ms for visual and $1/\lambda_A = 98$ ms for auditory stimuli compared to 48 ms and 18 ms for the younger group, respectively) but, importantly, the elderly seemed to have a larger time window of integration $(\omega = 450$ ms vs. 275 ms). Note that the slowing of the peripheral processes makes the arrival time for visual and auditory sensory information more variable and thereby diminishes the likelihood of them terminating within the time window. It seems, however, that a broader temporal window could only partially compensate for this since the older group was nevertheless slower in the bimodal conditions than the younger one[1].

This raises the question of whether or not the elderly in this study may have an enlarged time window that is close to the optimal value. Computing the optimal window is straightforward: we first insert the estimated λ

---

[1]What remains unclear is whether the slowing is only due to an age-related decrement in peripheral processing or also in a more centrally located processing stage where integration of visual and auditory input is taking place.

parameter values into (4.10). For the value of the prior probability of a single source, we choose $p = 0.5$ since ipsi- and contralateral configurations were presented with equal probability. In doing so we equate ipsilateral presentations with a single source ($C = 1$) and contralateral presentations with two separate sources ($C = 2$). The remaining parameter to be determined is the observation interval length, $t_1 - t_0$. Given the temporal structure of a trial in the experiment (foreperiod of up to 1,200 ms and maximal stimulus duration of 500 ms) we set $t_1 - t_0 = 1,700$ ms. With these parameter values inserted in (4.10), the optimal window width becomes 407 ms for the older group and 214 ms for the young adults group. Both values are a bit below the estimated widths (450 ms and 275 ms, respectively) but, given the variability in the parameter estimation process that we typically observe, this does not lend strong support to the hypothesis that either group deviates much from the optimal integration window size.

This computation also neglects the possibility of the utility values (costs/benefits) influencing the decision process, as the ratio of utility values was set equal to one in the decision rule underlying the optimal window width. The fact that a large integration window may come at a certain cost is nicely demonstrated in a recent study by Setti et al. (2011). When information coming from different sources is erroneously combined, this can result in distractibility and inefficient processing of the relevant stimulus in older adults possibly rendering older adults more accident prone (Poliakoff et al., 2006). In their study, Setti et al. investigated whether the causes of falling, a significant problem for older persons, are related to audiovisual integration ability. They measured susceptibility to the sound-induced flash illusion (Shams et al., 2000) for three groups, fall-prone, non-fall prone older adults, and young adults. The flash illusion is induced when two auditory stimuli (beeps) presented with a single brief visual stimulus (flash) results in the perception of two visual flashes. It occurs when the two sensory inputs are integrated due to their temporal proximity (Shams et al., 2002) but spatial proximity does not seem to play a role (Innes-Brown & Crewther, 2009). Specifically, Setti et al. found that older adults with a history of falling integrated audiovisual stimuli over a longer delay between the onset of cross-sensory stimulations than either older adults without a history of falling or younger adults. In particular, fall-prone older adults experienced higher overall rates of the sound induced flash illusion at longer SOAs (from 110 to 270 ms) than non-fallers. Setti et al. (ibid., p. 382) speculate that there may be either an indirect or a direct effect of a wider temporal window of integration on balance maintenance and posture control in fall-prone older adults.

4.5. Summary and Outlook

Generalizing our previous approach (Colonius & Diederich, 2010), we have demonstrated that an optimal time window, based on a decision rule maximizing expected gain, can also be computed for a focused attention task, utilizing the prior probability of a common source and the likelihood function of arrival time differences between peripheral sensory processing times triggered by a crossmodal stimulus set. This is illustrated introducing an asymmetric Laplace distribution for the arrival time differences, which is defined for both positive and negative arrival time differences.

The time-window-of-integration (TWIN) model framework permits estimating window width as a numerical model parameter. This suggests the possibility of examining whether humans are able to adapt their time window width to the optimal value. The results on the aging study from our lab (Diederich et al., 2008) suggest that this is a feasible task. The results of the van Wanrooij et al. study are consistent with the hypothesis that the time window changes with varying prior probability. The next step would be to modify systematically the likelihood functions of arrival time differences. Although arrival time differences are not observable, one can indirectly manipulate their likelihood distribution by changing the stimulus onset asynchrony (SOA): large SOA values should generate, on average, large arrival time differences.

Acknowledgments

We are most grateful to Marc van Wanrooij for providing us with data on the effect of prior probability. This research is supported by grant SFB/TR31 (Project B4) from Deutsche Forschungsgemeinschaft (DFG) to H.C. and by a grant from Nowetas Foundation to both authors.

Appendix 1

Deriving the probability of interaction in TWIN

The peripheral processing times V for the visual and A for the visual stimulus have an exponential distribution with parameters λ_V and λ_A, respectively. That is,

$$f_V(t) = \lambda_V\, e^{-\lambda_V\, t},$$
$$f_A(t) = \lambda_A\, e^{-\lambda_A\, t}$$

for $t \geq 0$, and $f_V(t) = f_A(t) \equiv 0$ for $t < 0$. The corresponding distribution functions are referred to as $F_V(t)$ and $F_A(t)$.

Focused Attention Paradigm

The visual stimulus is the target and the auditory stimulus is the nontarget. By definition,

$$P(I_{FAP}) = Pr(A + \tau < V < A + \tau + \omega)$$
$$= \int_0^\infty f_A(x)\{F_V(x + \tau + \omega) - F_V(x + \tau)\}\, dx,$$

where τ denotes the SOA value and ω is the width of the integration window. Computing the integral expression requires that we distinguish between three cases for the sign of $\tau + \omega$:

(i) $\tau < \tau + \omega < 0$

$$P(I_{FAP}) = \int_{-\tau-\omega}^{-\tau} \lambda_A\, e^{-\lambda_A\, x}\{1 - e^{-\lambda_V\,(x+\tau+\omega)}\}\, dx$$

$$+ \int_{-\tau}^\infty \lambda_A\, e^{-\lambda_A\, x}\{e^{-\lambda_V\,(x+\tau)} - e^{-\lambda_V\,(x+\tau+\omega)}\}\, dx$$

$$= \frac{\lambda_V}{\lambda_V + \lambda_A}\, e^{\lambda_A\tau}(-1 + e^{\lambda_A\,\omega});$$

(ii) $\tau < 0 < \tau + \omega$

$$P(I_{FAP}) = \int_0^{-\tau} \lambda_A\, e^{-\lambda_A\, x}\{1 - e^{-\lambda_V\,(x+\tau+\omega)}\}\, dx$$

$$+ \int_{-\tau}^\infty \lambda_A\, e^{-\lambda_A\, x}\{e^{-\lambda_V\,(x+\tau)} - e^{-\lambda_V\,(x+\tau+\omega)}\}\, dx$$

$$= \frac{1}{\lambda_V + \lambda_A}\, \{\lambda_A(1 - e^{-\lambda_V\,(\omega+\tau)}) + \lambda_V(1 - e^{\lambda_A\,\tau})\};$$

(iii) $0 < \tau < \tau + \omega$

$$P(I_{FAP}) = \int_0^\infty \lambda_A\, e^{-\lambda_A\, x}\{e^{-\lambda_V\,(x+\tau)} - e^{-\lambda_V\,(x+\tau+\omega)}\}\, dx$$

$$= \frac{\lambda_A}{\lambda_V + \lambda_A}\{e^{-\lambda_V\,\tau} - e^{-\lambda_V\,(\omega+\tau)}\}$$

Redundant Target Paradigm

The visual stimulus is presented first and the auditory stimulus second. By definition,

$$P(I_{RTP}) = Pr\{\max(V, A + \tau) < \min(V, A + \tau) + \omega\}$$

If the visual stimulus wins:
(i) $0 \le \tau \le \omega$

$$P(I_{RTP_V}) = \int_0^\tau \lambda_V e^{-\lambda_V x}(1 - e^{-\lambda_A(x+\omega-\tau)})\, dx$$

$$+ \int_\tau^\infty \lambda_V e^{-\lambda_V x}(1 - e^{-\lambda_A(x+\omega-\tau)} - (1 - e^{-\lambda_A(x-\tau)}))\, dx$$

$$= \frac{1}{\lambda_V + \lambda_A}\{\lambda_V(1 - e^{\lambda_A(-\omega+\tau)}) + \lambda_A(1 - e^{(-\lambda_V \tau)})\};$$

(ii) $0 < \omega \le \tau$

$$P(I_{RTP_V}) = \int_{\tau-\omega}^\tau \lambda_V x(1 - e^{-\lambda_A(x+\omega-\tau)})\, dx$$

$$+ \int_\tau^\infty \lambda_V e^{-\lambda_V x}(1 - e^{-\lambda_A(x+\omega-\tau)} - (1 - e^{-\lambda_A(x-\tau)}))\, dx$$

$$= \frac{\lambda_A}{\lambda_V + \lambda_A}\{e^{-\lambda_V \tau} \cdot (-1 + e^{\lambda_V \omega})\}.$$

If the auditory stimulus wins: $0 < \tau \le \tau + \omega$ and

$$P(I_{RTP_A}) = \int_0^\infty \lambda_A\, e^{-\lambda_A x}\{e^{-\lambda_V(x+\tau)} - e^{-\lambda_V(x+\tau+\omega)}\}\, dx$$

$$= \frac{\lambda_A}{\lambda_V + \lambda_A}\{e^{-\lambda_V \tau} - e^{-\lambda_V(\omega+\tau)}\}.$$

The probability that the visual or the auditory stimulus wins is therefore

$$P(I_{RTP}) = P(I_{RTP_V}) + P(I_{RTP_A}).$$

Appendix 2

The asymmetric Laplace density $\mathbf{AL}(\theta, \kappa, \sigma)$ is defined as (cf. Kotz et al., 2001, p. 137)

$$f(t \mid \theta, \sigma, \kappa) = \frac{\sqrt{2}}{\sigma} \frac{\kappa}{1 + \kappa^2} \begin{cases} \exp\left(-\frac{\sqrt{2}\kappa}{\sigma} |x - \theta|\right) & \text{if } t \geq \theta, \\ \exp\left(-\frac{\sqrt{2}}{\sigma\kappa} |x - \theta|\right) & \text{if } t < \theta, \end{cases}$$

where θ is location parameter, $\kappa = 1$ implies symmetry, and σ is a scale parameter. In deriving the distributional form used in the text, the following parameter mappings have been made:

$$\sigma^2 = \frac{2}{\lambda_V \lambda_A}; \quad \kappa^2 = \lambda_V / \lambda_A; \quad \theta = 0.$$

References

Colonius, H., & Diederich, A. (2010). The optimal time window of visual-auditory integration: A reaction time analysis. *Frontiers in Integrative Neuroscience, 4.* doi:10.338/fnint.2010.00011.

Colonius, H., & Diederich, A. (2004). Multisensory interaction in saccadic reaction time: A time-window-of-integration model. *Journal of Cognitive Neuroscience, 16*, 1000–1009.

Corneil, B. D., Van Wanrooij, M., Munoz, D. P., & van Opstal, A. J. (2002). Auditory-visual interactions subserving goal-directed saccades in a complex scene. *Journal of Neurophysiology, 88*, 438–454.

Diederich, A., & Colonius, H. (2011). Modeling multisensory processes in saccadic responses: Time-window-of-integration model. In M. T. Wallace and M. M. Murray (Eds.), *Frontiers in the Neural Bases of Multisensory Processes.* Boca Raton, FL: CRC Press.

Diederich, A., & Colonius, H. (2004). Modeling the time course of multisensory interaction in manual and saccadic responses. In G. Calvert, C. Spence, and B. E. Stein (Eds.), *Handbook of Multisensory Processes.* Cambridge, MA: MIT Press.

Diederich, A., & Colonius, H. (2007a). Why two "distractors" are better than one: Modeling the effect of non-target auditory and tactile stimuli on visual saccadic reaction time. *Experimental Brain Research, 179*, 43–54.

Diederich, A., & Colonius, H. (2007b). Modeling spatial effects in visual-tactile saccadic reaction time. *Perception & Psychophysics, 69*, 56–67.

Diederich, A., & Colonius, H. (2008a). Crossmodal interaction in saccadic reaction time: Separating multisensory from warning effects in the time window of integration model. *Experimental Brain Research, 186*, 1–22.

Diederich, A., & Colonius, H. (2008b). When a high-intensity "distractor" is better then a low-intensity one: Modeling the effect of an auditory or tactile nontarget stimulus on visual saccadic reaction time. *Brain Research, 1242*, 219–230.

Diederich, A., Colonius, H., & Schomburg, A. (2008). Assessing age-related multisensory enhancement with the time-window-of-integration model. *Neuropsychologia, 46*, 2556–2562.

Ernst, M. O. (2005). A Bayesian view on multimodal cue integration. In G. Knoblich, I. Thornton, M. Grosjean, and M. Shiffrar (Eds.), *Human Body Perception From the Inside Out*. New York, NY: Oxford University Press.

Frens, M. A., van Opstal, A. J., & van der Willigen, R. F. (1995). Spatial and temporal factors determine auditory-visual interactions in human saccadic eye movements. *Perception & Psychophysics, 57*, 802–816.

Hugenschmidt, C. E., Mozolic, J. L., Tan, H., Kraft, R. A., & Laurienti, P. J. (2009). Age-related increase in cross-sensory noise in resting and steady-state cerebral perfusion. *Brain Topography, 21*, 241–251.

Innes-Brown, H., & Crewther, D. (2009). The impact of spatial incongruence on an auditory-visual illusion. PLoS ONE 4:e6450. doi:10.1371/journal.pone.0006450.

Körding, K. P., Beierholm, U., Ma, W. J., Quartz, S., Tenenbaum, J. B., & Shams, L. (2007). Causal inference in multisensory perception. PLoS ONE 2(9): e943. doi:10.1371/journal.pone.0000943.

Kotz, S., Kozubowski, T. J., & Podgórski, K. (2001). *The Laplace Distribution and Generalizations. A Revisit with Applications to Communications, Economics, Engineering, and Finance*. Boston, MA: Birkhäuser.

Laurienti, P. J., Burdette, J. H., Maldjian, J. A., & Wallace, M. T. (2006). Enhanced multisensory integration in older adults. *Neurobiology of Aging, 27*, 1155–1163.

Lewald, J., & Guski, R. (2003). Cross-modal perceptual integration of spatially and temporally disparate auditory and visual stimuli. *Cognitive Brain Research, 16*, 468–478.

Meredith, M. A., Nemitz, J. W., & Stein, B. E. (1987). Determinants of multisensory integration in superior colliculus neurons. I. Temporal factors. *Journal of Neuroscience, 10*, 3215–3229.

Meredith, M. A. (2002). On the neural basis for multisensory convergence: A brief overview. *Cognitive Brain Research, 14*, 31–40.

Miller, J. O. (1982). Divided attention: Evidence for coactivation with redundant signals. *Cognitive Psychology, 14*, 247–279.

Morein-Zamir, S., Soto-Faraco, S., & Kingstone, A. (2003). Auditory capture of vision: Examining temporal ventriloquism. *Cognitive Brain Research, 17*, 154–163.

Peiffer, A. M., Mozolica, J. L., Hugenschmidt, C. E., & Laurienti, P. J. (2007). Age-related multisensory enhancement in a simple audiovisual detection task. *NeuroReport, 18*, 1077–1081.

Poliakoff, E., Shore, D. I., Lowe, C., & Spence, C. (2006). Visuotactile temporal order judgments in ageing. *Neuroscience Letters, 396*, 207–211.

Romei, V., Murray, M. M., Merabet, L. B., & Thut, G. (2007). Occipital transcranial magnetic stimulation has opposing effects on visual and auditory stimulus detection: Implications for multisensory interactions. *Journal of Neuroscience, 7*, 11465–11472.

Rowland, B. A., & Stein, B. E. (2008). Temporal profiles of response enhancement in multisensory integration. *Frontiers in Neuroscience, 2*, 218–224.

Shams, L., Kamitani, Y., & Shimojo, S. (2002). Visual illusion induced by sound. *Cognitive Brain Research, 14*, 147–152.

Shams, L., Kamitani, Y., & Shimojo, S. (2000). Illusions. What you see is what you hear. *Nature, 408*, 788.

Setti, A., Burke, K. E., Kenny, R. A., & Newell, F. N. (2011). Is inefficient multisensory processing associated with falls in older people? *Experimental Brain Research, 209*, 375–384.

Spence, C., & Squire, S. (2003). Multisensory integration: Maintaining the perception of synchrony. *Current Biology, 13*, R519–R521.

Stein, B. E., Huneycutt, W. S., & Meredith, M. A. (1988). Neurons and behavior: The same rules of multisensory integration apply. *Brain Research, 448*, 355–358.

Stein, B. E., Jiang, W., & Stanford, T. R. (2004). Multisensory integration in single neurons in the midbrain. In G. Calvert, C. Spence, and B. E. Stein (Eds.), *Handbook of Multisensory Processes*. Cambridge, MA: MIT Press.

van Attefeldt, N. M., Formisano, E., Blomert, L., & Goebel, R. (2007). The effect of temporal asynchrony on the multisensory integration of letters and speech sounds. *Cerebral Cortex, 17*, 962–974.

van Wanrooij, M. M., Bell, A. H., Munoz, D. P., & van Opstal, A. J. (2009). The effect of spatial-temporal audiovisual disparities on saccades in a complex scene. *Experimental Brain Research, 198*, 425–437.

van Wanrooij, M. M., Bremen, P., & van Opstal, A. J. (2010). Acquired prior knowledge modulates audiovisual integration. *European Journal of Neuroscience, 31*, 1763–1771.

Wallace, M. T., Roberson, G. E., Hairston, W. D., Stein, B. E., Vaughan, J. W., & Schirillo, J. A. (2004). Unifying multisensory signals across time and space. *Experimental Brain Research, 158*, 252–258.

Chapter 5

On the Reverse Problem of Fechnerian Scaling

Ehtibar N. Dzhafarov

Purdue University

Fechnerian Scaling imposes metrics on two sets of stimuli related to each other by a discrimination function subject to Regular Minimality. The two sets of stimuli usually represent the same set of stimulus values presented to an observer in two distinct observation areas. A discrimination function associates with every pair of stimuli a nonnegative number interpretable as the degree or probability with which, "from the observer's point of view," the two stimuli differ from each other, overall or in a specified respect. Regular Minimality is a principle according to which the relation "to be the best match for" across the two stimulus areas has certain uniqueness and symmetry properties. Fechnerian distances are computed by means of a dissimilarity cumulation procedure: discrimination values are first converted into an appropriately defined dissimilarity function, the sums of the latter's values are computed for all finite chains of stimuli connecting a given pair of stimuli, and the infimum of such sums (cumulative dissimilarities) is taken to be the distance from one element of this pair to the other. The Fechnerian distance between two stimuli in one observation area is the same as the Fechnerian distance between the corresponding (best matching) stimuli in the other observation area. This chapter deals with the reverse problem of Fechnerian Scaling: under which conditions one can compute the discrimination function values given the Fechnerian distances and the discrimination values between the best matching stimuli.

5.1. Background

Some familiarity with the modern theory of (generalized) Fechnerian Scaling is desirable but not necessary for reading this chapter: the background information needed will be recapitulated. The reader interested in the latest published version of the theory is referred to Dzhafarov and Colonius (2007) and Dzhafarov (2008a, 2008b, 2010). For historical details and the origins of the adjective "Fechnerian," the reader can consult Dzhafarov

(2001) and Dzhafarov and Colonius (2011).

A prototypical example of an experiment to which Fechnerian Scaling pertains is this: an observer is presented various pairs of stimuli (sounds, color patches, drawings, photographs of faces) and asked to indicate for every pair whether the two stimuli are the same or different (possibly, in a specified respect, as in "do these photographs depict the same person?"). The assumption is that each pair $(\mathbf{a}, \mathbf{b})$ is associated with the probability

$$\psi(\mathbf{a}, \mathbf{b}) = \Pr[\mathbf{a} \text{ and } \mathbf{b} \text{ are judged to be different}].$$

Every stimulus is characterized by its *value* (e.g., the shape of a line drawing) and its *observation area* (or *stimulus area*), usually a spatiotemporal location, serving to distinguish the two stimuli to be compared (e.g., two line drawings can be presented in distinct locations, one to the left of the other, or successively, one before the other). The pairs $(\mathbf{a}, \mathbf{b})$ therefore are defined as

$$(\mathbf{a}, \mathbf{b}) = \left\{ \begin{array}{l} \text{value of } \mathbf{a} \text{ in observation area 1,} \\ \text{value of } \mathbf{b} \text{ in observation area 2} \end{array} \right\},$$

so that $(\mathbf{a}, \mathbf{b}) \neq (\mathbf{b}, \mathbf{a})$.

Assuming, as we do throughout this chapter, that the two observation areas are fixed, $\mathbf{a}$ and $\mathbf{b}$ in $(\mathbf{a}, \mathbf{b})$ belong to different sets even if their values are identical. We denote these sets $\mathcal{A}$ and $\mathcal{B}$, so that the *discrimination probability function* ψ is

$$\psi : \mathcal{A} \times \mathcal{B} \to [0, 1].$$

This definition can be immediately generalized. First, we can replace $[0, 1]$ with the set of all nonnegative reals,

$$\psi : \mathcal{A} \times \mathcal{B} \to \mathbb{R}^+.$$

This allows us to include discrimination functions which are not probabilities, such as expected values of numerical estimates of dissimilarity given in response to pairs of stimuli. Although the present treatment is confined to probabilistic ψ, its extension to arbitrary (bounded or unbounded) functions is straightforward. Second, we can interpret $\mathcal{A}$ and $\mathcal{B}$ as being arbitrary sets, not necessarily

$$\mathcal{A} = \mathcal{V} \times \{1\},$$
$$\mathcal{B} = \mathcal{V} \times \{2\},$$

with $\mathcal{V}$ being a common set of stimulus values. Thus, one can consider

$$\mathcal{A} = \mathcal{V}_1 \times \{1\},$$
$$\mathcal{B} = \mathcal{V}_2 \times \{2\},$$

where $\mathcal{V}_1$ and $\mathcal{V}_2$ are different subsets of a set $\mathcal{V}$ of possible stimulus values. This may be convenient if the matching pairs, as defined below, involve "constant error," that is, if $\mathcal{V}_2$ is the set of matches for the elements of $\mathcal{V}_1$ and $\mathcal{V}_2 \neq \mathcal{V}_1$. We can even consider the possibility that $\mathcal{A}$ and $\mathcal{B}$ are sets of different nature, with the relation "are the same" being defined in special ways. For instance, $\mathcal{A}$ may be a set of examinees and $\mathcal{B}$ the set of tests, with the relation "a and b are the same" meaning that the problem **b** is neither too difficult nor too easy for the examinee **a** (in the opinion of a judge, or as computed from performance data). For other non-traditional examples of pairwise comparisons, see Dzhafarov and Colonius (2006).

Without loss of generality, let us assume that no two distinct stimuli in $\mathcal{A}$ or in $\mathcal{B}$ are *equivalent*, in the following sense: if $\mathbf{a}_1 \neq \mathbf{a}_2$ in $\mathcal{A}$, then

$$\psi(\mathbf{a}_1, \mathbf{b}) \neq \psi(\mathbf{a}_2, \mathbf{b})$$

for some $\mathbf{b} \in \mathcal{B}$; and if $\mathbf{b}_1 \neq \mathbf{b}_2$ in $\mathcal{B}$, then

$$\psi(\mathbf{a}, \mathbf{b}_1) \neq \psi(\mathbf{a}, \mathbf{b}_2)$$

for some $\mathbf{a} \in \mathcal{A}$. (If this is not the case, then $\mathcal{A}$ and $\mathcal{B}$ can always be "reduced" to the requisite form.) We say that $\mathbf{a} \in \mathcal{A}$ *is matched by* $\mathbf{b} \in \mathcal{B}$ and write **aMb** if

$$\psi(\mathbf{a}, \mathbf{b}) = \min_{\mathbf{y} \in \mathcal{B}} \psi(\mathbf{a}, \mathbf{y}).$$

Analogously, if

$$\psi(\mathbf{a}, \mathbf{b}) = \min_{\mathbf{x} \in \mathcal{A}} \psi(\mathbf{x}, \mathbf{b}),$$

we say that $\mathbf{b} \in \mathcal{B}$ *is matched by* $\mathbf{a} \in \mathcal{A}$ and write **bMa**. The space $(\mathcal{A}, \mathcal{B}, \psi)$, or the discrimination function ψ itself, is said to satisfy *Regular Minimality* (Dzhafarov, 2002b; Dzhafarov & Colonius, 2006; Kujala & Dzhafarov, 2008) if

($\mathcal{RM}1$) for every $\mathbf{a} \in \mathcal{A}$ there is one and only one $\mathbf{b} \in \mathcal{B}$ such that **aMb**;
($\mathcal{RM}2$) for every $\mathbf{b} \in \mathcal{B}$ there is one and only one $\mathbf{a} \in \mathcal{A}$ such that **bMa**;
($\mathcal{RM}3$) **aMb** if and only if **bMa**, for all $(\mathbf{a}, \mathbf{b}) \in \mathcal{A} \times \mathcal{B}$.

If this is the case, we can relabel the stimuli in $\mathcal{A}$ and $\mathcal{B}$ so that any two matching stimuli receive the same label. Formally, if Regular Minimality holds, then one can find (non-uniquely) bijective functions

$$\mathbf{h} : \mathcal{A} \to \mathfrak{S}$$

and

$$\mathbf{g} : \mathcal{B} \to \mathfrak{S}$$

such that

$$a\mathbf{M}b \iff \mathbf{h}(\mathbf{a}) = \mathbf{g}(\mathbf{b}).$$

Any such mapping $(\mathbf{h}, \mathbf{g})$ is called a *canonical transformation* of the space $(\mathcal{A}, \mathcal{B}, \psi)$, and it creates a *canonical discrimination space* $(\mathfrak{S}, \psi^*)$, with the function

$$\psi^* : \mathfrak{S} \times \mathfrak{S} \to \mathbb{R}^+$$

defined by

$$\psi^*(\mathbf{a}, \mathbf{b}) = \psi\left(\mathbf{h}^{-1}(\mathbf{a}), \mathbf{g}^{-1}(\mathbf{b})\right).$$

Clearly, the function ψ^* satisfies Regular Minimality in the simplest (canonical) form: for any distinct $\mathbf{a}, \mathbf{b} \in \mathfrak{S}$,

$$\psi^*(\mathbf{a}, \mathbf{a}) < \min\{\psi^*(\mathbf{a}, \mathbf{b}), \psi^*(\mathbf{b}, \mathbf{a})\}.$$

Although the canonical discrimination space $(\mathfrak{S}, \psi^*)$ is not uniquely determined by $(\mathcal{A}, \mathcal{B}, \psi)$, it can be viewed as being essentially unique, in the following sense. Any two canonical spaces, $(\mathfrak{S}_1, \psi_1^*)$ and $(\mathfrak{S}_2, \psi_2^*)$, are related to each other by a bijective transformation $\mathbf{t} : \mathfrak{S}_1 \to \mathfrak{S}_2$ such that

$$\psi_1^*(\mathbf{a}, \mathbf{b}) = \psi_2^*(\mathbf{t}(\mathbf{a}), \mathbf{t}(\mathbf{b})).$$

In other words, the canonical transformation is unique up to trivial renaming of the stimuli.

Henceforth we will deal with canonical discrimination spaces $(\mathfrak{S}, \psi^*)$ only. We drop the asterisk for the canonical discrimination function and write ψ in place of ψ^*. We also use the notational conventions adopted in most of the author's previous publications on generalized Fechnerian Scaling:

(1) for any binary function $f : \mathfrak{S} \times \mathfrak{S} \to \mathbb{R}$, we write $f\mathbf{ab}$ instead of $f(\mathbf{a}, \mathbf{b})$ (in particular, the discrimination function is written as $\psi\mathbf{ab}$);

(2) if a binary function f is followed by a string (or *chain*) $\mathbf{X} = \mathbf{x}_1 \ldots, \mathbf{x}_n$ of more than one point, then

$$f\mathbf{X} = f\mathbf{x}_1 \ldots \mathbf{x}_n = \sum_{i=1}^{n-1} f\mathbf{x}_i\mathbf{x}_{i+1};$$

(3) for a chain $\mathbf{X} = \mathbf{x}_1, \ldots, \mathbf{x}_n$ with $n = 1$ or $n = 0$ the expression $f\mathbf{X}$ is set equal to zero;

(4) any two chains of points $\mathbf{X}$ and $\mathbf{Y}$ can be concatenated into a chain $\mathbf{XY}$ (e.g., if $\mathbf{X} = \mathbf{x}_1 \ldots, \mathbf{x}_n$, then $\mathbf{aXb}$ is the chain $\mathbf{ax}_1 \ldots, \mathbf{x}_n\mathbf{b}$).

A *dissimilarity function* $D : \mathfrak{S} \times \mathfrak{S} \to \mathbb{R}$ is defined by the following properties: for any $\mathbf{a}, \mathbf{b} \in \mathfrak{S}$, any sequences $\mathbf{a}_n, \mathbf{a'}_n, \mathbf{b}_n, \mathbf{b'}_n$ in $\mathfrak{S}$, and any sequence of chains $\mathbf{X}_n$ with elements in $\mathfrak{S}$ ($n = 1, 2, \ldots$),

(𝒟1) $D\mathbf{ab} \geq 0$;
(𝒟2) $D\mathbf{ab} = 0$ if and only if $\mathbf{a} = \mathbf{b}$;
(𝒟3)   if $\max\{D\mathbf{a}_n\mathbf{a'}_n, D\mathbf{b}_n\mathbf{b'}_n\} \to 0$ then $D\mathbf{a}_n\mathbf{b}_n - D\mathbf{a'}_n\mathbf{b'}_n \to 0$;
(𝒟4)   if $D\mathbf{a}_n\mathbf{X}_n\mathbf{b}_n \to 0$ then $D\mathbf{a}_n\mathbf{b}_n \to 0$.

Given any chain $\mathbf{X} = \mathbf{x}_1, \ldots, \mathbf{x}_n$, the quantity

$$DX = \sum_{i=1}^{n-1} D\mathbf{x}_i\mathbf{x}_{i+1}$$

is called the *cumulative dissimilarity* for this chain.

In Fechnerian Scaling the role of dissimilarity functions is played by the *psychometric increments* of the first and second kind, defined as, respectively,

$$\Psi^{(1)}\mathbf{ab} = \psi\mathbf{ab} - \psi\mathbf{aa},$$
$$\Psi^{(2)}\mathbf{ab} = \psi\mathbf{ba} - \psi\mathbf{aa}.$$

(It is clearly unnecessary to consider separately the versions $\psi\mathbf{ba} - \psi\mathbf{bb} = \Psi^{(1)}\mathbf{ba}$ and $\psi\mathbf{ab} - \psi\mathbf{bb} = \Psi^{(2)}\mathbf{ba}$.) In other words, a canonical discrimination space $(\mathfrak{S}, \psi)$ induces a *double-dissimilarity space* $\left(\mathfrak{S}, \Psi^{(1)}, \Psi^{(2)}\right)$.

We use the notation

$$\mathbf{a}_n \leftrightarrow \mathbf{b}_n$$

(as $n \to \infty$) to designate any of the pairwise equivalent convergences

$$\Psi^{(1)}\mathbf{a}_n\mathbf{b}_n \to 0,$$
$$\Psi^{(1)}\mathbf{b}_n\mathbf{a}_n \to 0,$$
$$\Psi^{(2)}\mathbf{a}_n\mathbf{b}_n \to 0,$$
$$\Psi^{(2)}\mathbf{b}_n\mathbf{a}_n \to 0.$$

The discrimination function ψ is uniformly continuous with respect to the uniformity induced by this convergence:

$$\left.\begin{array}{c} \mathbf{a'}_n \leftrightarrow \mathbf{a}_n \\ \mathbf{b'}_n \leftrightarrow \mathbf{b}_n \end{array}\right\} \implies \psi\mathbf{a'}_n\mathbf{b'}_n - \psi\mathbf{a}_n\mathbf{b}_n \to 0.$$

In particular, $\mathbf{a}_n \leftrightarrow \mathbf{b}_n$ implies

$$\psi \mathbf{b}_n \mathbf{b}_n - \psi \mathbf{a}_n \mathbf{a}_n \to 0.$$

A *metric* (or *distance function*) $G : \mathfrak{S} \times \mathfrak{S} \to \mathbb{R}$ can be defined as a dissimilarity function satisfying the triangle inequality: for all $\mathbf{a}, \mathbf{b}, \mathbf{c} \in \mathfrak{S}$,

$$G\mathbf{ab} + G\mathbf{bc} \geq G\mathbf{ac}.$$

This definition differs from the classical Frechét's definition in that it does not require global symmetry,

$$G\mathbf{ab} = G\mathbf{ba}.$$

However, it is more specific than the notion of a *quasimetric* (defined by dropping from the classical definition of a metric the global symmetry requirement). Namely, since G is a dissimilarity it has the following *symmetry-in-the-small* property: for all sequences $\mathbf{a}_n, \mathbf{b}_n$ in $\mathfrak{S}$,

$$G\mathbf{a}_n \mathbf{b}_n \to 0 \Longleftrightarrow G\mathbf{b}_n \mathbf{a}_n \to 0.$$

In Fechnerian Scaling, the *Fechnerian metric* G_1 induced by the dissimilarity function $\Psi^{(1)}$ is defined as

$$G_1 \mathbf{ab} = \inf_{\mathbf{X}} \Psi^{(1)} \mathbf{aXb}.$$

Analogously,

$$G_2 \mathbf{ab} = \inf_{\mathbf{X}} \Psi^{(2)} \mathbf{aXb}$$

is the Fechnerian metric induced by the dissimilarity function $\Psi^{(2)}$. Both G_1 and G_2 are well-defined metrics, and

$$\left. \begin{array}{c} G_1 \mathbf{ab} + G_1 \mathbf{ba} \\ \shortparallel \\ G_2 \mathbf{ab} + G_2 \mathbf{ba} \end{array} \right\} = G\mathbf{ab}.$$

This sum, $G\mathbf{ab}$, is a conventional (symmetric) metric. It is referred to as the *overall Fechnerian metric*.

The asymmetric (*"oriented"*) metrics G_1 and G_2 are also related to each other by the identities

$$\left. \begin{array}{c} G_1 \mathbf{ab} - G_2 \mathbf{ba} \\ \shortparallel \\ G_2 \mathbf{ab} - G_1 \mathbf{ba} \end{array} \right\} = \psi \mathbf{bb} - \psi \mathbf{aa}.$$

This follows from the procedure of computing G_1 and G_2 from $\Psi^{(1)}$ and $\Psi^{(2)}$ and from the immediately verifiable identities

$$\left.\begin{array}{c} \Psi^{(1)}\mathbf{ab} - \Psi^{(2)}\mathbf{ba} \\ \shortparallel \\ \Psi^{(2)}\mathbf{ab} - \Psi^{(1)}\mathbf{ba} \end{array}\right\} = \psi\mathbf{bb} - \psi\mathbf{aa}.$$

For any sequences $\mathbf{a}_n$ and $\mathbf{b}_n$ in $\mathfrak{S}$, we have, as $n \to \infty$,

$$\mathbf{a}_n \leftrightarrow \mathbf{b}_n \iff G_1\mathbf{a}_n\mathbf{b}_n \to 0 \iff G_2\mathbf{a}_n\mathbf{b}_n \to 0.$$

It follows that G_1 and G_2 are uniformly continuous with respect to the uniformity induced by $\leftrightarrow$.

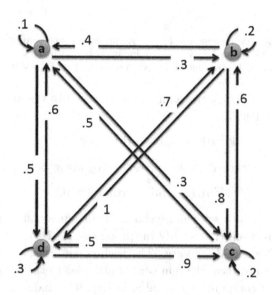

Fig. 5.1.   Stimulus space of Example 5.1. For each stimulus pair $(\mathbf{x}, \mathbf{y})$, the number attached to the arrow from $\mathbf{x}$ to $\mathbf{y}$ is the value of $\psi\mathbf{xy}$.

Example 5.1. Let $\mathfrak{S}$ be the set of four stimuli, $\{\mathbf{a}, \mathbf{b}, \mathbf{c}, \mathbf{d}\}$,[1] with the values of ψ shown in Figure 5.1. Fechnerian computations are illustrated

---

[1] As a rule we use symbols $\mathbf{a}$ and $\mathbf{b}$ (interchangeably with $\mathbf{x}$ and $\mathbf{y}$, respectively) to generically refer to stimuli in, respectively, the first and second observation areas. Thus, in an expression like $\Psi^{(1)}\mathbf{ab} > 0$, $\mathbf{a}$ and $\mathbf{b}$ are variables, arbitrary members of $\mathfrak{S}$. However, in some of our examples $\mathbf{a}$ and $\mathbf{b}$ are used to designate specific stimuli, together with other specific stimuli (here, $\mathbf{c}$ and $\mathbf{d}$). The two uses of $\mathbf{a}$ and $\mathbf{b}$ should be easily distinguishable by the context.

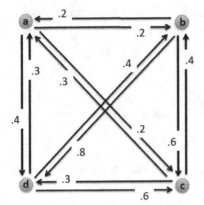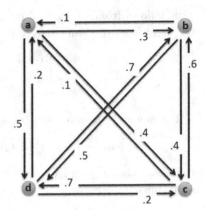

Fig. 5.2. Psychometric increments of the first and second kind computed for the stimulus space shown in Figure 5.1. For each stimulus pair $(\mathbf{x}, \mathbf{y})$, the number attached to the arrow from $\mathbf{x}$ to $\mathbf{y}$ is the value of $\Psi^{(1)}\mathbf{xy}$ (on the left) or $\Psi^{(2)}\mathbf{xy}$ (on the right).

in Figures 5.2-5.5. Thus, the number 0.4 attached to the arrow from $\mathbf{d}$ to $\mathbf{b}$ in Figure 5.2, left, is

$$\Psi^{(1)}\mathbf{db} = \psi\mathbf{db} - \psi\mathbf{dd} = 0.7 - 0.3.$$

The number 0.7 attached to the corresponding arrow on the right is

$$\Psi^{(2)}\mathbf{db} = \psi\mathbf{bd} - \psi\mathbf{dd} = 1 - 0.3.$$

For any finite stimulus set, the psychometric increments $\Psi^{(1)}$ and $\Psi^{(2)}$ are dissimilarity functions (i.e., satisfy the properties $\mathcal{D}1$-$\mathcal{D}4$ above) if and only if ψ satisfies Regular Minimality. The Fechnerian distances are shown in Figure 5.3. For instance, the number 0.2 attached to the arrow from $\mathbf{a}$ to $\mathbf{b}$ on the left is computed by forming all possible chains leading from $\mathbf{a}$ to $\mathbf{b}$, calculating their cumulative dissimilarities and choosing the smallest. Omitting chains containing loops, as we obviously do not need them in searching for the minimum, we get the list

chain	cumulative $\Psi^{(1)}$
ab	0.2
acb	0.2+0.4
adb	0.4+0.4
acdb	0.2+0.3+0.4
abcb	0.4+0.6+0.4

in which the direct link **ab** is clearly a *geodesic* (a shortest path). We conclude therefore that $G_1\mathbf{ab} = 0.2$. Note that geodesics generally are not unique if they exist (they have to exist in finite stimulus sets but not generally). The analogous calculations for, say, the stimuli **c** and **d** on the right yield

chain	cumulative $\Psi^{(2)}$
cd	0.7
cad	0.1+0.5
cbd	0.6+0.5
cabd	0.1+0.3+0.5
cbad	0.6+0.1+0.5

Here, the geodesic is **cad** and we conclude that $G_2\mathbf{cd} = 0.6$. This geodesic is shown in Figure 5.4, right, together with the two other "indirect" geodesic paths, those consisting of more than two stimuli. Figure 5.5 presents the values of the overall Fechnerian distance, obtained as

$$G_1\mathbf{xy} + G_1\mathbf{yx} = G_2\mathbf{xy} + G_2\mathbf{yx}$$

for every stimulus pair $(\mathbf{x}, \mathbf{y})$. Thus, for $(\mathbf{x}, \mathbf{y}) = (\mathbf{c}, \mathbf{d})$, we have

$$G\mathbf{cd} = G\mathbf{dc} = \begin{cases} G_1\mathbf{cd} + G_1\mathbf{dc} = 0.3 + 0.5 \\ \quad\quad \| \\ G_2\mathbf{cd} + G_2\mathbf{dc} = 0.6 + 0.2. \end{cases}$$

Note that $G\mathbf{cd}$ can be viewed as the cumulative dissimilarity

$$\Psi^{(1)}\mathbf{cdac} = \Psi^{(2)}\mathbf{cadc}$$

for the *geodesic loop* **cdac** obtained by concatenating the geodesic paths from **c** to **d** and back; the loop should be read in the opposite directions for $\Psi^{(1)}$ and $\Psi^{(2)}$.

5.2. Problem

We have seen that a canonical discrimination space $(\mathfrak{S}, \psi)$ induces a *double-metric* space $(\mathfrak{S}, G_1, G_2)$ from which one can form the space $(\mathfrak{S}, G)$ with a conventional, symmetric metric G. It is easy to see that these computations cannot generally be reversed. The following example shows that $(\mathfrak{S}, G)$ does not allow one to reconstruct $(\mathfrak{S}, \psi)$ uniquely.

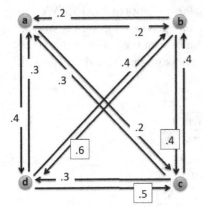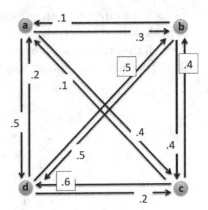

Fig. 5.3. Fechnerian distances of the first and second kind computed from the psychometric increments in Figure 5.2. The number attached to the arrow from **x** to **y** is the value of G_1**xy** (on the left) or G_2**xy** (on the right). The framed numbers indicate Fechnerian distances that are smaller than the corresponding psychometric increments: the geodesics for them are shown in Figure 5.4.

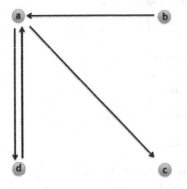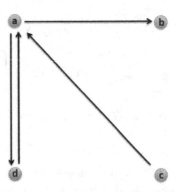

Fig. 5.4. Geodesic paths of the first and second kind corresponding to the framed values of the Fechnerian distances in Figure 5.3. Each geodesic consists of three stimuli connected by two consecutive arrows.

Example 5.2. Let $(\mathfrak{S}, \psi)$ induce $(\mathfrak{S}, G_1, G_2)$ and $(\mathfrak{S}, G)$, and let ψ**aa** be some nonconstant function of **a**. Denote by $\omega_\mathbf{a}$ an arbitrary function such that

$$\omega_\mathbf{a} \not\equiv \psi\mathbf{aa}$$

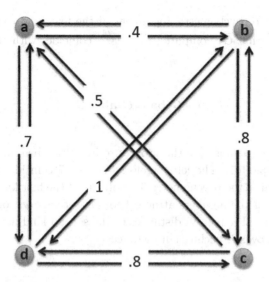

Fig. 5.5. The overall (symmetric) Fechnerian distances computed from the Fechnerian distances shown in Figure 5.3.

(that is, $\omega_{\mathbf{a}}$ and $\psi\mathbf{aa}$ are not identical),

$$0 \leq \omega_{\mathbf{a}} \leq \min\left\{1 - \sup_{\mathbf{b}} \Psi^{(1)}\mathbf{ab}\right\},$$

and

$$\omega_{\mathbf{b}} - \omega_{\mathbf{a}} \leq \Psi^{(1)}\mathbf{ab}.$$

All three inequalities can always be achieved, for instance, by putting $\omega_{\mathbf{a}} \equiv 0$ (a nonconstant $\psi\mathbf{aa}$ has to be positive at some $\mathbf{a}$, and then $1 - \sup_{\mathbf{b}} \Psi^{(1)}\mathbf{ab} > 0$). The function

$$\overline{\psi}\mathbf{ab} = \Psi^{(1)}\mathbf{ab} + \omega_{\mathbf{a}}$$

is clearly bounded by 1 from above. It satisfies Regular Minimality because

$$\overline{\psi}\mathbf{aa} = \omega_{\mathbf{a}} \leq \overline{\psi}\mathbf{ab}$$

and

$$\omega_{\mathbf{a}} \leq \overline{\psi}\mathbf{ba} = \Psi^{(1)}\mathbf{ba} + \omega_{\mathbf{b}}.$$

The function $\overline{\psi}$ induces precisely the same metric space $(\mathfrak{S}, G)$ as the original function ψ. Indeed, from the definition of $\overline{\psi}$ it follows that

$$\overline{\Psi}^{(1)}\mathbf{ab} = \Psi^{(1)}\mathbf{ab},$$

where $\overline{\Psi}^{(1)}$ is the psychometric increment of the first kind computed from $\overline{\psi}$. This means that $\overline{G}_1$ computed from $\overline{\Psi}^{(1)}$ coincides with G_1. But then also

$$\overline{G}_1\mathbf{ab} + \overline{G}_1\mathbf{ba} = G_1\mathbf{ab} + G_1\mathbf{ba}.$$

Figures 5.6 and 5.7 illustrate the procedure just described on the stimulus space of Example 5.1. The left panel of Figure 5.7 coincides with that of Figure 5.3 because the psychometric increments of the first kind remain the same. By adding the numbers attached to opposite arrows, one can verify that although the Fechnerian distances of the second kind do change, they yield the same overall Fechnerian distances.

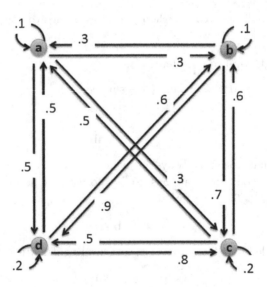

Fig. 5.6.   Stimulus space of Example 5.1 modified in accordance with Example 5.2.

Is it possible then that ψ (or at least the psychometric increments $\Psi^{(1)}$ and $\Psi^{(2)}$) can be reconstructed if one knows both G_1 and G_2? The next example shows that this is not the case.

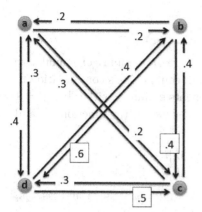 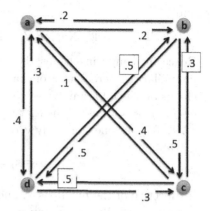

Fig. 5.7. Fechnerian distances of the first and second kind computed from the psychometric increments in Figure 5.6.

Example 5.3. Let $\mathfrak{S}$ be a countable set of stimuli enumerated $\mathbf{s}_1, \mathbf{s}_2, \ldots,$ and let

$$\psi\mathbf{s}_i\mathbf{s}_j = \begin{cases} 0 & \text{if } i = j \\ 1/3 & \text{if } |i - j| = 1 \\ 2/3 + \gamma_i & \text{if } i - j = 2 \\ 2/3 + \gamma_i' & \text{if } j - i = 2 \\ 1 & \text{if } |i - j| \geq 3, \end{cases}$$

where $0 \leq \gamma_i, \gamma_i' < 1/3$. This is an example of a uniformly discrete stimulus space, considered in Section 5.4.3. The psychometric increments

$$\Psi^{(1)} \equiv \Psi^{(2)} \equiv \psi$$

in this case are dissimilarity functions because ψ satisfies Regular Minimality. For every chain containing $\mathbf{s}_i$ and $\mathbf{s}_{i+2}$ as successive elements,

$$\mathbf{X}\mathbf{s}_i\mathbf{s}_{i+2}\mathbf{Y} \text{ or } \mathbf{X}\mathbf{s}_{i+2}\mathbf{s}_i\mathbf{Y},$$

the cumulative dissimilarity

$$\Psi^{(\iota)}\mathbf{X}\mathbf{s}_i\mathbf{s}_{i+2}\mathbf{Y} \text{ or } \Psi^{(\iota)}\mathbf{X}\mathbf{s}_{i+2}\mathbf{s}_i\mathbf{Y}$$

where ι stands for 1 or 2, cannot increase if one replaces this chain with

$$\mathbf{X}\mathbf{s}_i\mathbf{s}_{i+1}\mathbf{s}_{i+2}\mathbf{Y} \text{ or } \mathbf{X}\mathbf{s}_{i+2}\mathbf{s}_{i+1}\mathbf{s}_i\mathbf{Y},$$

respectively:

$$\Psi^{(\iota)}\mathbf{X}\mathbf{s}_i\mathbf{s}_{i+2}\mathbf{Y} - \Psi^{(\iota)}\mathbf{X}\mathbf{s}_i\mathbf{s}_{i+1}\mathbf{s}_{i+2}\mathbf{Y} = \gamma_i$$

and

$$\Psi^{(\iota)}\mathbf{X}\mathbf{s}_{i+2}\mathbf{s}_i\mathbf{Y} - \Psi^{(\iota)}\mathbf{X}\mathbf{s}_{i+2}\mathbf{s}_{i+1}\mathbf{s}_i\mathbf{Y} = \gamma_i'.$$

Hence in computing G_ι as the infimum (here, minimum) of cumulative dissimilarities across a set of chains, one can confine one's consideration to chains in which for any two successive elements $\mathbf{s}_i$ and $\mathbf{s}_j$, either $|i - j| = 1$ or $|i - j| \geq 3$. This means that G_ι cannot depend on the functions γ_i and γ_i'. In fact, as one can easily verify,

$$G_1\mathbf{s}_i\mathbf{s}_j = G_2\mathbf{s}_i\mathbf{s}_j = \begin{cases} |j-i|/3 & \text{if } |i - j| < 3 \\ 1 & \text{if } |i - j| \geq 3, \end{cases}$$

irrespective of the functions γ_i and γ_i'. The values of $\psi\mathbf{s}_i\mathbf{s}_{i+2}$ and $\psi\mathbf{s}_{i+2}\mathbf{s}_i$ can be arbitrarily chosen between $2/3$ and 1.

The next example demonstrates the same point, that $(\mathfrak{S}, G_1, G_2)$ determines neither ψ nor the psychometric increments uniquely, for a continuous stimulus space.

Example 5.4. Let $\mathfrak{S}$ be the interval $[0, 1]$, and let

$$\psi\mathbf{ab} = \begin{cases} \frac{2b-a}{4} + \frac{|a-b|^p}{2} & \text{if } a \leq b, \\[2mm] \frac{3a-2b}{4} + \frac{|a-b|^p}{4} & \text{if } a > b, \end{cases}$$

where a and b are the numerical values of stimuli $\mathbf{a}$ and $\mathbf{b}$, respectively, and $p > 1$. We have here

$$\Psi^{(1)}\mathbf{ab} = \begin{cases} \frac{b-a}{2} + \frac{|a-b|^p}{2} & \text{if } a \leq b, \\[2mm] \frac{a-b}{2} + \frac{|a-b|^p}{4} & \text{if } a > b, \end{cases}$$

and

$$\Psi^{(2)}\mathbf{ab} = \begin{cases} \frac{3(b-a)}{4} + \frac{|a-b|^p}{4} & \text{if } a < b, \\[2mm] \frac{a-b}{4} + \frac{|a-b|^p}{2} & \text{if } a \geq b. \end{cases}$$

We omit a demonstration that these psychometric increments are dissimilarity functions: it can be done using the methods presented in Dzhafarov (2010). For any $a < m < b$ in $[0, 1]$,

$$\Psi^{(1)}\mathbf{ab} > \Psi^{(1)}\mathbf{amb}$$

and

$$\Psi^{(2)}\mathbf{ab} > \Psi^{(2)}\mathbf{amb}.$$

This is an example of a space with intermediate points, considered in Dzhafarov (2008a). The "inverse triangle inequalities" imply that, for any **a** and **b**, the cumulative dissimilarities $\Psi^{(1)}\mathbf{aXb}$ and $\Psi^{(2)}\mathbf{aXb}$ decrease as one progressively refines the chain $\mathbf{X} = \mathbf{x}_1 \ldots \mathbf{x}_k$ whose elements' values partition the interval between a and b (i.e., for $a < b$, $a < x_1 < \ldots < x_k < b$, and analogously for $a > b$). As $n \to \infty$ and the maximal gap in $\left(a, x_1^{(n)}, \ldots, x_{k_n}^{(n)}, b \right)$ for chains $\mathbf{X}_n = \mathbf{x}_1^n \ldots \mathbf{x}_{k_n}^n$ tends to zero,

$$\Psi^{(1)}\mathbf{aX}_n\mathbf{b} \to G_1\mathbf{ab} = \frac{|a - b|}{2}$$

and

$$\Psi^{(2)}\mathbf{aX}_n\mathbf{b} \to G_2\mathbf{ab} = \begin{cases} \frac{3(b-a)}{4} & \text{if } a < b, \\[2mm] \frac{a-b}{4} & \text{if } a \geq b. \end{cases}$$

As these values do not depend on p, the function ψ cannot be reconstructed even if one knows all values of $G_1\mathbf{ab}$ and $G_2\mathbf{ab}$.

The question arises: can one impose on the function ψ certain constraints under which ψ can be uniquely restored from $(\mathfrak{S}, G_1, G_2)$ and the set of "self-discrimination" probabilities[2]

$$\{\omega_\mathbf{a} = \psi\mathbf{aa} : \mathbf{a} \in \mathfrak{S}\}?$$

On this level of generality the problem is too difficult, however. Its formulation does not exclude the possibility that $\psi\mathbf{ab}$ for a given pair of stimuli $(\mathbf{a}, \mathbf{b})$ is determined by the values of G_1, G_2 and ω on some subset of pairs in $\mathfrak{S} \times \mathfrak{S}$, if not the entire Cartesian product. The problem we pose in this chapter is more restricted: under what conditions can one compute $\psi\mathbf{ab}$ from the quantities

$$G_1\mathbf{ab}, G_2\mathbf{ab}, G_1\mathbf{ba}, G_2\mathbf{ba}, \omega_\mathbf{a}, \omega_\mathbf{b}?$$

We refer to this as the *reverse problem of Fechnerian Scaling (in the restricted sense)*. The so-called "Fechner's problem" (as formulated in Luce & Galanter, 1963) is closely related to the reverse problem in the restricted sense but is left outside the scope of this chapter. The reader interested in the issue is referred to Falmagne (1985) and Dzhafarov (2002a).

---

[2]One should keep in mind that due to a canonical transformation of the space, the first and the second **a** in $\psi\mathbf{aa}$ may be stimuli physically different in value, and even if not, they always have different observation areas. Therefore the "self" in "self-discrimination" is a convenient but potentially misleading prefix.

5.3. General Considerations

The formulation of the reverse problem immediately suggests the following representation for ψab.

Theorem 5.1. *The reverse problem has a solution if and only if ψ can be presented in either of the two equivalent forms,*

$$\psi\mathbf{ab} = \omega_{\mathbf{a}} + G_1\mathbf{ab} + R\left(G_1\mathbf{ab}, G_2\mathbf{ab}, G_1\mathbf{ba}, G_2\mathbf{ba}, \omega_{\mathbf{a}}, \omega_{\mathbf{b}}\right)$$

or

$$\psi\mathbf{ab} = \omega_{\mathbf{b}} + G_2\mathbf{ba} + R\left(G_1\mathbf{ab}, G_2\mathbf{ab}, G_1\mathbf{ba}, G_2\mathbf{ba}, \omega_{\mathbf{a}}, \omega_{\mathbf{b}}\right),$$

where R is uniformly continuous, nonnegative, and vanishing at $\mathbf{a} = \mathbf{b}$.

Proof. It is obvious that any function of

$$G_1\mathbf{ab}, G_2\mathbf{ab}, G_1\mathbf{ba}, G_2\mathbf{ba}, \omega_{\mathbf{a}}, \omega_{\mathbf{b}}$$

can be presented in either of the two forms. That they are equivalent follows from

$$\omega_{\mathbf{a}} + G_1\mathbf{ab} = \omega_{\mathbf{b}} + G_2\mathbf{ba},$$

which is a consequence of

$$\left.\begin{array}{c} G_1\mathbf{ab} - G_2\mathbf{ba} \\ \shortparallel \\ G_2\mathbf{ab} - G_1\mathbf{ba} \end{array}\right\} = \omega_{\mathbf{b}} - \omega_{\mathbf{a}}.$$

Since

$$(\psi\mathbf{ab} - \omega_{\mathbf{a}}) - G_1\mathbf{ab} = \Psi^{(1)}\mathbf{ab} - G_1\mathbf{ab}$$

is the difference of uniformly continuous functions, R is uniformly continuous. R is nonnegative because

$$\psi\mathbf{ab} - \omega_{\mathbf{a}} = \Psi^{(1)}\mathbf{ab} \geq G_1\mathbf{ab}$$

and

$$\psi\mathbf{ab} - \omega_{\mathbf{b}} = \Psi^{(2)}\mathbf{ba} \geq G_2\mathbf{ba}.$$

R vanished at $\mathbf{a} = \mathbf{b}$ because

$$\psi\mathbf{ab} = \omega_{\mathbf{a}} = \omega_{\mathbf{b}}$$

and

$$G_1\mathbf{ab} = G_2\mathbf{ba} = 0.$$

This proves the "only if" part of the theorem. The "if" part is obvious. $\square$

The examples in the previous section show that the representation given in this theorem does not have to exist. Moreover, despite its formulation in the form of a necessary and sufficient condition, it is not obvious whether a function ψ satisfying this condition can in fact be constructed: the condition in question relates ψ to G_1 and G_2, which are themselves computed from ψ. The situation is remedied by the following examples.

Example 5.5. Let $\mathfrak{S}$ be the interval $[0, 1]$, and

$$\psi\mathbf{ab} = \frac{a}{6} + \frac{|a - b|}{3} + \frac{95}{216}(a - b)^2 (a + b).$$

The function is easily checked to be between 0 and 1 and satisfy Regular Minimality. The psychometric increments of the first kind are

$$\Psi^{(1)}\mathbf{ab} = \frac{1}{3}|a - b| + \frac{1}{2}(a^2 - b^2)(a - b),$$

and they can be shown to form a dissimilarity function (we skip this demonstration). Since, for any $a, b \in [0, 1]$ and m between a and b,

$$\Psi^{(1)}\mathbf{amb} < \Psi^{(1)}\mathbf{ab},$$

we use the same argument as in Example 5.4 to arrive at

$$G_1\mathbf{ab} = \frac{|a - b|}{3}.$$

One can check now that

$$\frac{a}{6} + \frac{|a-b|}{3} + \frac{95}{216}(a - b)^2 (a + b)$$

$$= \frac{a}{6} + \frac{|a-b|}{3} + \left(\frac{2}{3}|a - b| + \left(\frac{a}{6} - \frac{b}{6}\right)\right)\left(\left(\frac{b}{6} - \frac{a}{6}\right) + \frac{2}{3}|a - b|\right)\left(\frac{a}{6} + \frac{b}{6}\right)$$

$$= \omega_\mathbf{a} + G_1\mathbf{ab} + (2G_1\mathbf{ab} + (\omega_\mathbf{a} - \omega_\mathbf{b}))((\omega_\mathbf{b} - \omega_\mathbf{a}) + 2G_1\mathbf{ba})(\omega_\mathbf{a} + \omega_\mathbf{b}).$$

That is, ψ can be presented in a form required in Theorem 5.1, with

$$R = \omega_\mathbf{a} + G_1\mathbf{ab} + (2G_1\mathbf{ab} + (\omega_\mathbf{a} - \omega_\mathbf{b}))((\omega_\mathbf{b} - \omega_\mathbf{a}) + 2G_1\mathbf{ba})(\omega_\mathbf{a} + \omega_\mathbf{b}).$$

This can be rewritten to involve G_1 and G_2 symmetrically: using the identities

$$2G_1\mathbf{ab} + (\omega_\mathbf{a} - \omega_\mathbf{b}) = G_1\mathbf{ab} + G_2\mathbf{ba}$$

and

$$(\omega_\mathbf{b} - \omega_\mathbf{a}) + 2G_1\mathbf{ba} = G_1\mathbf{ba} + G_2\mathbf{ab},$$

we get

$$\psi ab = \begin{cases} \omega_{\mathbf{a}} + G_1\mathbf{ab} + (G_1\mathbf{ab} + G_2\mathbf{ba})\,(G_2\mathbf{ab} + G_1\mathbf{ba})\,(\omega_{\mathbf{a}} + \omega_{\mathbf{b}}) \\ \qquad\qquad\qquad\qquad || \\ \omega_{\mathbf{b}} + G_2\mathbf{ba} + (G_2\mathbf{ba} + G_1\mathbf{ab})\,(G_1\mathbf{ba} + G_2\mathbf{ab})\,(\omega_{\mathbf{b}} + \omega_{\mathbf{a}})\,. \end{cases}$$

This representation is of interest in view of the symmetry considerations to be invoked below.

The next example provides another demonstration of the same nature, arguably the simplest possible because it corresponds to $R \equiv 0$.

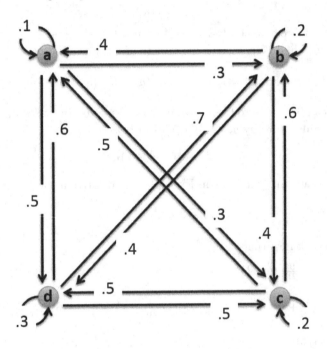

Fig. 5.8.   Stimulus space of Example 5.6.

Example 5.6. Consider the space shown in Figure 5.8. One can check that, for any pair of stimuli $\mathbf{x}, \mathbf{y} \in \{\mathbf{a}, \mathbf{b}, \mathbf{c}, \mathbf{d}\}$ and any chain $\mathbf{X}$ with elements in $\{\mathbf{a}, \mathbf{b}, \mathbf{c}, \mathbf{d}\}$,

$$\Psi^{(1)}\mathbf{xy} \le \Psi^{(1)}\mathbf{xXy}$$

and

$$\Psi^{(2)}\mathbf{xy} \le \Psi^{(2)}\mathbf{xXy}.$$

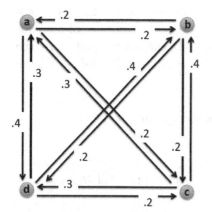 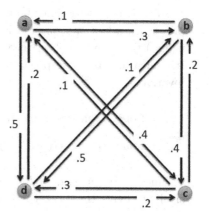

Fig. 5.9. Psychometric increments (in this case coinciding with Fechnerian distances) of the first and second kind computed for the stimulus space shown in Figure 5.8.

This means that the psychometric increments shown in Figure 5.9 coincide with the corresponding Fechnerian distances, and we have

$$\psi ab = \begin{cases} \omega_a + G_1 ab \\ \| \\ \omega_b + G_2 ba, \end{cases}$$

which are special cases of the representations in Theorem 5.1.

Theorem 5.1 encompasses a wealth of possible special cases. One can restrict this class by considering stimulus spaces with special properties (as we do in the next section) or by imposing certain symmetry constraints on the function R directly. The latter can be done as follows.

First, we can eliminate from the expression

$$R\left(G_1 ab, G_2 ab, G_1 ba, G_2 ba, \omega_a, \omega_b\right)$$

quantities determinable from other quantities. Knowing $G_1 ab$, $G_1 ba$, ω_b ω_a one can compute $G_2 ab$, $G_2 ba$ as

$$G_2 ba = G_1 ab - (\omega_b - \omega_a)$$

and

$$G_2 ab = G_1 ba + (\omega_b - \omega_a)$$

(these identities were used in Example 5.5). This leads to

$$R\left(G_1 ab, G_2 ab, G_1 ba, G_2 ba, \omega_a, \omega_b\right) = \begin{cases} R_1\left(G_1 ab, G_1 ba, \omega_a, \omega_b\right) \\ \| \\ R_2\left(G_2 ba, G_2 ab, \omega_b, \omega_a\right). \end{cases}$$

Now, it may sometimes be natural to posit (e.g., in psychophysical applications, when two observation areas contain the same set of stimulus values) that the two observation areas are interchangeable, and so are the stimuli **a** and **b**. More precisely, one can assume that R_1 and R_2 remain invariant

(1) if one exchanges **a** and **b** in all their arguments; and

(2) if one replaces G_1 with G_2 or vice versa.

Since the arguments of R_2 can be obtained from those of R_1 by successively applying these two rules, we have

$$R_1 \equiv R_2 \equiv R^*.$$

The assumption in question can now be formulated by saying that any asymmetry between the two observation areas is only in the first two summands of the four equivalent representations for ψ:

$$\psi\mathbf{ab} = \begin{cases} \omega_{\mathbf{a}} + G_1\mathbf{ab} + R^*\left(G_1\mathbf{ab}, G_1\mathbf{ba}, \omega_{\mathbf{a}}, \omega_{\mathbf{b}}\right) \\ \omega_{\mathbf{a}} + G_1\mathbf{ab} + R^*\left(G_1\mathbf{ba}, G_1\mathbf{ab}, \omega_{\mathbf{b}}, \omega_{\mathbf{a}}\right) \\ \omega_{\mathbf{b}} + G_2\mathbf{ba} + R^*\left(G_2\mathbf{ba}, G_2\mathbf{ab}, \omega_{\mathbf{b}}, \omega_{\mathbf{a}}\right) \\ \omega_{\mathbf{b}} + G_2\mathbf{ba} + R^*\left(G_2\mathbf{ab}, G_2\mathbf{ba}, \omega_{\mathbf{a}}, \omega_{\mathbf{b}}\right). \end{cases}$$

Let us call such a function R^* *symmetric*.

Theorem 5.2. *If a reverse problem has a solution with a symmetric function R^*, then*

$$G_1\mathbf{ab} - G_2\mathbf{ab} = \psi\mathbf{ab} - \psi\mathbf{ba},$$

$$G_1\mathbf{ab} - G_1\mathbf{ba} = \Psi^{(1)}\mathbf{ab} - \Psi^{(1)}\mathbf{ba},$$

$$G_2\mathbf{ab} - G_2\mathbf{ba} = \Psi^{(2)}\mathbf{ab} - \Psi^{(2)}\mathbf{ba}.$$

Proof. The first of these identities is obtained by subtracting

$$\psi\mathbf{ba} = \omega_{\mathbf{a}} + G_2\mathbf{ab} + R\left(G_2\mathbf{ab}, G_2\mathbf{ba}, \omega_{\mathbf{a}}, \omega_{\mathbf{b}}\right)$$

from

$$\psi\mathbf{ab} = \omega_{\mathbf{a}} + G_1\mathbf{ab} + R\left(G_1\mathbf{ab}, G_1\mathbf{ba}, \omega_{\mathbf{a}}, \omega_{\mathbf{b}}\right).$$

The representations for the differences of the psychometric increments follow then from the identities

$$G_2\mathbf{ba} = G_1\mathbf{ab} - (\omega_{\mathbf{b}} - \omega_{\mathbf{a}})$$

and

$$G_2 \mathbf{ab} = G_1 \mathbf{ba} + (\omega_\mathbf{b} - \omega_\mathbf{a}).$$

$\square$

Example 5.7. The following functions satisfy the symmetry requirement for R^*:

$$\psi \mathbf{ab} = \begin{cases} \omega_\mathbf{a} + G_1 \mathbf{ab} + (G_1 \mathbf{ab} + G_2 \mathbf{ba})(G_2 \mathbf{ab} + G_1 \mathbf{ba})(\omega_\mathbf{a} + \omega_\mathbf{b}) \\ \quad\quad || \\ \omega_\mathbf{b} + G_2 \mathbf{ba} + (G_2 \mathbf{ba} + G_1 \mathbf{ab})(G_1 \mathbf{ba} + G_2 \mathbf{ab})(\omega_\mathbf{b} + \omega_\mathbf{a}) \end{cases}$$

and

$$\psi \mathbf{ab} = \begin{cases} \omega_\mathbf{a} + G_1 \mathbf{ab} + f(G \mathbf{ab}, S(\omega_\mathbf{a}, \omega_\mathbf{b})) \\ \quad\quad || \\ \omega_\mathbf{b} + G_2 \mathbf{ba} + f(G \mathbf{ab}, S(\omega_\mathbf{a}, \omega_\mathbf{b})), \end{cases}$$

where S is some commutative function (which may, as a special case, be identically constant, say, zero).

5.4. Special Stimulus Spaces

5.4.1. *Directly linked spaces*

Let us say that a point $\mathbf{a}$ in a stimulus space $(\mathfrak{S}, \psi)$ is *directly 1-linked* to point $\mathbf{b}$ (or directly linked to it in the first observation area) if

$$\Psi^{(1)} \mathbf{ab} = G_1 \mathbf{ab}.$$

Analogously, a point $\mathbf{a}$ is *directly 2-linked* to point $\mathbf{b}$ (or directly linked to it in the second observation area) if

$$\Psi^{(2)} \mathbf{ab} = G_2 \mathbf{ab}.$$

Theorem 5.3. *A point $\mathbf{a}$ is directly 1-linked to a point $\mathbf{b}$ if and only if $\mathbf{b}$ is directly 2-linked to $\mathbf{a}$.*

Proof. The equality

$$\Psi^{(1)} \mathbf{ab} = G_1 \mathbf{ab}$$

means that, for every chain $\mathbf{X} = \mathbf{x}_1 \ldots \mathbf{x}_k$,

$$\Psi^{(1)} \mathbf{aXb} = \Psi^{(1)} \mathbf{ax}_1 + \sum_{i=1}^{k-1} \Psi^{(1)} \mathbf{x}_i \mathbf{x}_{i+1} + \Psi^{(1)} \mathbf{x}_k \mathbf{b} \geq \Psi^{(1)} \mathbf{ab}.$$

This can be written as

$$(\psi\mathbf{a}\mathbf{x}_1 - \omega_{\mathbf{a}}) + \sum_{i=1}^{k-1}(\psi\mathbf{x}_i\mathbf{x}_{i+1} - \omega_{\mathbf{x}_i}) + (\psi\mathbf{x}_k\mathbf{b} - \omega_{\mathbf{x}_k}) \geq (\psi\mathbf{a}\mathbf{b} - \omega_{\mathbf{a}}).$$

Replacing $\omega_{\mathbf{a}}$ with $\omega_{\mathbf{b}}$ on both sides and rearranging the ω-terms as

$$\begin{array}{ccccccc}
\omega_{\mathbf{b}} & \omega_{\mathbf{x}_1} & \cdots & \omega_{\mathbf{x}_i} & \cdots & \omega_{\mathbf{x}_{k-1}} & \omega_{\mathbf{x}_k} \\
\downarrow & \downarrow & \cdots & \downarrow & \cdots & \downarrow & \downarrow \\
\omega_{\mathbf{x}_1} & \omega_{\mathbf{x}_2} & \cdots & \omega_{\mathbf{x}_{i+1}} & \cdots & \omega_{\mathbf{x}_k} & \omega_{\mathbf{b}},
\end{array}$$

we get

$$(\psi\mathbf{a}\mathbf{x}_1 - \omega_{\mathbf{x}_1}) + \sum_{i=1}^{k-1}\left(\psi\mathbf{x}_i\mathbf{x}_{i+1} - \omega_{\mathbf{x}_{i+1}}\right) + (\psi\mathbf{x}_k\mathbf{b} - \omega_{\mathbf{b}}) \geq (\psi\mathbf{a}\mathbf{b} - \omega_{\mathbf{b}}).$$

In other words, for every chain $\mathbf{Y} = \mathbf{x}_k \ldots \mathbf{x}_1$,

$$\Psi^{(2)}\mathbf{b}\mathbf{Y}\mathbf{a} = \Psi^{(2)}\mathbf{b}\mathbf{x}_k + \sum_{i=1}^{k-1}\Psi^{(2)}\mathbf{x}_{i+1}\mathbf{x}_i + \Psi^{(2)}\mathbf{x}_1\mathbf{a} \geq \Psi^{(1)}\mathbf{b}\mathbf{a}.$$

But this means

$$\Psi^{(2)}\mathbf{b}\mathbf{a} = G_2\mathbf{b}\mathbf{a},$$

proving the theorem.   □

Corollary 5.1. *Points* $\mathbf{a}$ *and* $\mathbf{b}$ *are directly 1-linked to each other if and only if they are directly 2-linked to each other. In a space with any two points directly 1-linked any two points are directly 2-linked (and vice versa).*

The space referred to in this corollary is called a *directly linked* space. The reverse problem for such a space has its simplest possible solution:

$$\psi\mathbf{a}\mathbf{b} = \begin{cases} G_1\mathbf{a}\mathbf{b} + \omega_{\mathbf{a}} \\ \quad \| \\ G_2\mathbf{b}\mathbf{a} + \omega_{\mathbf{b}}. \end{cases}$$

Example 5.8. Consider a space $(\mathfrak{S}, \psi)$ such that

$$\psi\mathbf{a}\mathbf{b} = M\mathbf{a}\mathbf{b} + r_1(\mathbf{a}) + r_2(\mathbf{b}),$$

where M is a symmetric metric and r_1, r_2 some nonnegative functions. Then

$$\Psi^{(1)}\mathbf{a}\mathbf{b} = M\mathbf{a}\mathbf{b} + r_2(\mathbf{b}) - r_2(\mathbf{a})$$

and

$$\Psi^{(2)}\mathbf{a}\mathbf{b} = M\mathbf{a}\mathbf{b} + r_1(\mathbf{a}) - r_1(\mathbf{b}).$$

To ensure Regular Minimality, we posit

$$|r_1(\mathbf{a}) - r_1(\mathbf{b})| < M\mathbf{ab}$$

and

$$|r_2(\mathbf{a}) - r_2(\mathbf{b})| < M\mathbf{ab}.$$

We verify that for any $\mathbf{a}, \mathbf{b}, \mathbf{m}$ in $\mathfrak{S}$,

$$\begin{aligned}
\Psi^{(1)}\mathbf{am} &= M\mathbf{am} + r_2(\mathbf{m}) - r_2(\mathbf{a}) \\
&+ \qquad\qquad\qquad\qquad\qquad \geq M\mathbf{ab} + r_2(\mathbf{b}) - r_2(\mathbf{a}) = \Psi^{(1)}\mathbf{ab}, \\
\Psi^{(1)}\mathbf{mb} &= M\mathbf{mb} + r_2(\mathbf{b}) - r_2(\mathbf{m})
\end{aligned}$$

proving thereby

$$\Psi^{(1)}\mathbf{ab} = G_1\mathbf{ab}.$$

Analogously we prove

$$\Psi^{(2)}\mathbf{ab} = G_2\mathbf{ab}.$$

Since

$$\left.\begin{array}{c} G_1\mathbf{ab} + G_1\mathbf{ba} \\ \shortparallel \\ G_2\mathbf{ab} + G_2\mathbf{ba} \end{array}\right\} = \left.\begin{array}{c} \Psi^{(1)}\mathbf{ab} + \Psi^{(1)}\mathbf{ba} \\ \shortparallel \\ \Psi^{(1)}\mathbf{ab} + \Psi^{(1)}\mathbf{ba} \end{array}\right\} = 2M\mathbf{ab},$$

we conclude that

$$M\mathbf{ab} = \frac{1}{2}G\mathbf{ab},$$

so that the definition of ψ can be given as

$$\psi\mathbf{ab} = \frac{1}{2}G\mathbf{ab} + r_1(\mathbf{a}) + r_2(\mathbf{b}).$$

This is one possible form of presenting the "quadrilateral dissimilarity model" (Dzhafarov & Colonius, 2006).

The statement of the following theorem is obvious and given without proof. We denote (referring to Theorem 5.1)

$$R(G_1\mathbf{ab}, G_2\mathbf{ab}, G_1\mathbf{ba}, G_2\mathbf{ba}, \omega_\mathbf{a}, \omega_\mathbf{b}) = R_\mathbf{ab}.$$

Theorem 5.4. *If the reverse problem has a solution, then point* $\mathbf{a}$ *is directly 1-linked to point* $\mathbf{b}$ *(and* $\mathbf{b}$ *is directly 2-linked to point* $\mathbf{a}$*) if and only if* $R_\mathbf{ab} = 0$.

This observation agrees with Theorem 5.3: both direct linkages are equivalent to $R_\mathbf{ab} = 0$. Example 5.6 shows that a directly linked space can be easily constructed.

5.4.2. *Spaces with metric-in-the-small dissimilarities*

A dissimilarity function D is said to be *metric-in-the-small* if, whenever $a_n \leftrightarrow b_n$ with $a_n \neq b_n$,

$$\frac{Da_n b_n}{Ga_n b_n} \to 1.$$

The convergence is from the right because

$$Da_n b_n \geq Ga_n b_n.$$

Applying this definition to psychometric increments,

$$\left.\begin{array}{c} a_n \leftrightarrow b_n \\ \& \\ a_n \neq b_n \end{array}\right\} \implies \frac{\Psi^{(1)}a_n b_n}{G_1 a_n b_n} \to 1$$

and

$$\left.\begin{array}{c} a_n \leftrightarrow b_n \\ \& \\ a_n \neq b_n \end{array}\right\} \implies \frac{\Psi^{(2)}a_n b_n}{G_2 a_n b_n} \to 1.$$

Clearly, these convergences imply

$$\frac{\Psi^{(\iota)}a_n b_n + \Psi^{(\iota)}b_n a_n}{Ga_n b_n} \to 1.$$

Recall that $a_n \leftrightarrow b_n$ means any of the equivalent convergences

$$\Psi^{(\iota)}a_n b_n \to 0,$$
$$\Psi^{(\iota)}b_n a_n \to 0,$$

where ι stands for 1 or 2.

Theorem 5.5. *If the reverse problem has a solution, then the dissimilarities $\Psi^{(1)}$ and $\Psi^{(2)}$ are metric-in-the-small if and only if $R(x, y, u, v, a, b)$ is of a higher degree of infinitesimality than either of the arguments x and v.*

Proof. Rewrite the expressions

$$\psi a_n b_n = \omega_{a_n} + G_1 a_n b_n + R_{a_n b_n}$$
$$= \omega_{b_n} + G_2 b_n a_n + R_{a_n b_n}$$

as

$$\frac{\Psi^{(1)}a_n b_n}{G_1 a_n b_n} = 1 + \frac{R\left(G_1 a_n b_n, G_2 a_n b_n, G_1 b_n a_n, G_2 b_n a_n, \omega_{a_n}, \omega_{b_n}\right)}{G_1 a_n b_n}$$

and

$$\frac{\Psi^{(2)}\mathbf{b}_n\mathbf{a}_n}{G_2\mathbf{b}_n\mathbf{a}_n} = 1 + \frac{R\left(G_1\mathbf{a}_n\mathbf{b}_n, G_2\mathbf{a}_n\mathbf{b}_n, G_1\mathbf{b}_n\mathbf{a}_n, G_2\mathbf{b}_n\mathbf{a}_n, \omega_{\mathbf{a}_n}, \omega_{\mathbf{b}_n}\right)}{G_2\mathbf{b}_n\mathbf{a}_n}.$$

We know that, as $\mathbf{a}_n \leftrightarrow \mathbf{b}_n$, both

$$G_1\mathbf{a}_n\mathbf{b}_n \to 0$$

and

$$G_2\mathbf{b}_n\mathbf{a}_n \to 0.$$

The left-hand sides tend to 1 if and only if the ratios on the right tend to zero, proving the theorem. □

If we make use of the symmetry constraint of Section 5.3 and present the function R in the above theorem as

$$R^*\left(G_1\mathbf{ab}, G_1\mathbf{ba}, \omega_{\mathbf{a}}, \omega_{\mathbf{b}}\right) = R^*\left(G_2\mathbf{ba}, G_2\mathbf{ab}, \omega_{\mathbf{b}}, \omega_{\mathbf{a}}\right),$$

then the condition of the higher-order infinitesimality acquires a simpler form.

Corollary 5.2. *If the reverse problem has a solution with a symmetric R^*, then $\Psi^{(1)}$ is metric-in-the-small if and only if so is $\Psi^{(2)}$ and if and only if $R^*(x, y, a, b)$ is of a higher degree of infinitesimality than both x and y.*

Example 5.9. In particular, if R^* can be presented as in the second function of Example 5.7,

$$R^* = f\left(G\mathbf{ab}, S\left(\omega_{\mathbf{a}}, \omega_{\mathbf{b}}\right)\right),$$

the condition of the higher-order infinitesimality reduces to

$$\frac{R^*(x, a)}{x} \to 0$$

as $x \to 0$ $(x \neq 0)$. Such functions as

$$\psi\mathbf{ab} = \begin{cases} \omega_{\mathbf{a}} + G_1\mathbf{ab} + \frac{|\omega_{\mathbf{a}} - \omega_{\mathbf{b}}|}{\omega_{\mathbf{a}} + \omega_{\mathbf{b}}}\left(G\mathbf{ab}\right)^2 \\ \| \\ \omega_{\mathbf{b}} + G_2\mathbf{ba} + \frac{|\omega_{\mathbf{a}} - \omega_{\mathbf{b}}|}{\omega_{\mathbf{a}} + \omega_{\mathbf{b}}}\left(G\mathbf{ab}\right)^2 \end{cases}$$

and

$$\psi\mathbf{ab} = \begin{cases} \omega_{\mathbf{a}} + G_1\mathbf{ab} + k_2\left(G\mathbf{ab}\right)^2 + k_3\left(G\mathbf{ab}\right)^3 + ... + k_r\left(G\mathbf{ab}\right)^r \\ \| \\ \omega_{\mathbf{b}} + G_2\mathbf{ba} + k_2\left(G\mathbf{ab}\right)^2 + k_3\left(G\mathbf{ab}\right)^3 + ... + k_r\left(G\mathbf{ab}\right)^r \end{cases}$$

provide examples.

5.4.3. *Uniformly discrete spaces*

A space $(\mathfrak{S}, \psi)$ is *uniformly discrete* if

$$\inf_{\mathbf{x} \neq \mathbf{y}} \Psi^{(1)} \mathbf{x} \mathbf{y} > 0.$$

This condition is equivalent to

$$\inf_{\mathbf{x} \neq \mathbf{y}} \Psi^{(2)} \mathbf{x} \mathbf{y} > 0$$

because, as we know,

$$\Psi^{(1)} \mathbf{x}_n \mathbf{y}_n \to 0 \iff \Psi^{(2)} \mathbf{x}_n \mathbf{y}_n \to 0.$$

Any finite space is uniformly discrete.

In the following we will tacitly assume, with no loss of generality, that all chains $\mathbf{X} = \mathbf{x}_1 ... \mathbf{x}_k$ considered are *non-wasteful*, in the following sense: for no $i = 1, ..., k - 1$, $\mathbf{x}_i = \mathbf{x}_{i+1}$.

A chain $\mathbf{X} = \mathbf{x}_1 ... \mathbf{x}_k$ is called 1-*basic* if, for any $1 \leq i < k$, $\mathbf{x}_i$ is directly 1-linked to $\mathbf{x}_{i+1}$. A 2-basic chain is defined analogously.

The class of all 1-basic (2-basic) chains connecting $\mathbf{a}$ to $\mathbf{b}$ and containing k elements, not counting $\mathbf{a}$ and $\mathbf{b}$, is denoted by $\mathcal{C}_k^1 \mathbf{a} \mathbf{b}$ (respectively, $\mathcal{C}_k^2 \mathbf{a} \mathbf{b}$). Clearly, $k = 0, 1, ...$.

Theorem 5.6. *For any* $\mathbf{a}, \mathbf{b}$ *in a uniformly discrete space one can find* k_1 *and* k_2 *such that*

$$G_1 \mathbf{a} \mathbf{b} = \inf_{\mathbf{a} \mathbf{X} \mathbf{b} \in \mathcal{C}_{k_1}^1 \mathbf{a} \mathbf{b}} \Psi^{(1)} \mathbf{a} \mathbf{X} \mathbf{b}$$

and

$$G_2 \mathbf{a} \mathbf{b} = \inf_{\mathbf{a} \mathbf{X} \mathbf{b} \in \mathcal{C}_{k_2}^2 \mathbf{a} \mathbf{b}} \Psi^{(2)} \mathbf{a} \mathbf{X} \mathbf{b}.$$

Proof. Let ι stand for 1 or 2. Consider all chains $\mathbf{X}^{(k)}$ containing $k = 0, 1, ...$, elements, and define

$$G_\iota^{(k)} \mathbf{a} \mathbf{b} = \inf_{\mathbf{X}^{(k)}} \Psi^{(\iota)} \mathbf{a} \mathbf{X}^{(k)} \mathbf{b}.$$

Denoting

$$s_\iota = \inf_{\mathbf{x} \neq \mathbf{y}} \Psi^{(\iota)} \mathbf{x} \mathbf{y} > 0,$$

we have

$$\Psi^{(\iota)} \mathbf{a} \mathbf{X}^{(k)} \mathbf{b} \geq (k + 1) s_\iota,$$

hence also

$$G_\iota^{(k)} \mathbf{ab} \geq (k+1)\, s_\iota.$$

Therefore, for some $K > 0$, we have to have

$$G_\iota \mathbf{ab} < G_\iota^{(K)} \mathbf{ab}.$$

Consider a number k with the following property: in some sequence of chains $\mathbf{aX}_n\mathbf{b}$ such that

$$\Psi^{(\iota)} \mathbf{aX}_n\mathbf{b} \to G_\iota \mathbf{ab}$$

the chains with k elements occur infinitely often. Denote by k_0 the largest number with this property (which exists because $k < K$). By construction, there is a sequence of chains $\mathbf{aX}_n^{(k_0)}\mathbf{b}$ such that

$$\Psi^{(\iota)} \mathbf{aX}_n^{(k_0)}\mathbf{b} \to G_\iota \mathbf{ab},$$

but there is no sequence of chains $\mathbf{aX}_n^{(>k_0)}\mathbf{b}$ in which each $\mathbf{X}_n^{(>k_0)}$ has more than k_0 elements such that

$$\Psi^{(\iota)} \mathbf{aX}_n^{(>k_0)}\mathbf{b} \to G_\iota \mathbf{ab}.$$

We will show that all but a finite number of these chains $\mathbf{aX}_n^{(k_0)}\mathbf{b}$ are ι-basic. Suppose this is not the case. Then one can choose a sequence of chains $\mathbf{aX}_n^{(k_0)}\mathbf{b}$ all of which are not ι-basic, with

$$\Psi^{(\iota)} \mathbf{aX}_n^{(k_0)}\mathbf{b} \to G_\iota \mathbf{ab}.$$

Let $\mathbf{x}_{i_n,n}\mathbf{x}_{i_n+1,n}$ be a link in each of these chains with $\mathbf{x}_{i_n,n}$ not directly ι-linked to $\mathbf{x}_{i_n+1,n}$ (where i_n may be 0 or k_0, with $\mathbf{x}_{0,n} = \mathbf{a}$ and $\mathbf{x}_{k_0+1,n} = \mathbf{b}$). Then one can find nonempty chains $\mathbf{Y}_n$ such that

$$\Psi^{(\iota)} \mathbf{ax}_{1,n}...\mathbf{x}_{i_n,n}\mathbf{Y}_n\mathbf{x}_{i_n+1,n}...\mathbf{x}_{k_0,n}\mathbf{b} < \Psi^{(\iota)}\mathbf{aX}_n^{(k_0)}\mathbf{b},$$

and since the convergence of $\Psi^{(\iota)}\mathbf{aX}_n^{(k_0)}\mathbf{b}$ to $G_\iota \mathbf{ab}$ is from the right,

$$\Psi^{(\iota)} \mathbf{ax}_{1,n}...\mathbf{x}_{i_n,n}\mathbf{Y}_n\mathbf{x}_{i_n+1,n}...\mathbf{x}_{k_0,n}\mathbf{b} \to G_\iota \mathbf{ab}.$$

Clearly, $\mathbf{Y}_n$ must contain an element

$$\mathbf{m}_n \notin \left\{\mathbf{x}_{i_n,n}, \mathbf{x}_{i_n+1,n}\right\},$$

whence the chains

$$\mathbf{x}_{1,n}...\mathbf{x}_{i_n}\mathbf{Y}_n\mathbf{x}_{i_n+1}...\mathbf{x}_{k_0,n}$$

contain more than k_0 elements. But this contradicts the definition of k_0.$\square$

Corollary 5.3. *With ι standing for 1 or 2, in a uniformly discrete space any point can be directly ι-linked to some other point, and any two points can be connected by an ι-basic chain.*

To formulate another immediate consequence of Theorem 5.6, we need a new concept. A *base* for G_1 is a subset

$$\mathfrak{D} \subseteq \mathfrak{S} \times \mathfrak{S}$$

such that if

$$G_1\mathbf{ab} = G_1^*\mathbf{ab}$$

for all $(\mathbf{a}, \mathbf{b}) \in \mathfrak{D}$, then

$$G_1\mathbf{ab} = G_1^*\mathbf{ab}$$

for all $(\mathbf{a}, \mathbf{b}) \in \mathfrak{S} \times \mathfrak{S}$. The definition of a base for G_2 is analogous.

Corollary 5.4. *With ι standing for 1 or 2, in a uniformly discrete space the set of all directly ι-linked ordered pairs of points forms a base for G_ι.*

We conclude this chapter by considering the reverse problem for a special case of uniformly discrete spaces, those *with geodesics*. All finite spaces fall within this category.

A uniformly discrete space is said to be with geodesics if, for any points $\mathbf{a}$ and $\mathbf{b}$ in it,

$$G_1\mathbf{ab} = \min_{\mathbf{X}} \Psi^{(1)}\mathbf{aXb}$$

and

$$G_2\mathbf{ab} = \min_{\mathbf{X}} \Psi^{(2)}\mathbf{aXb}.$$

In other words, the requirement is that for any $\mathbf{a}$ and $\mathbf{b}$ one be able to find chains $\mathbf{X}_1$ and $\mathbf{X}_2$ such that

$$G_1\mathbf{ab} = \Psi^{(1)}\mathbf{aX}_1\mathbf{b}$$

and

$$G_2\mathbf{ab} = \Psi^{(2)}\mathbf{aX}_2\mathbf{b}.$$

The chains $\mathbf{aX}_1\mathbf{b}$ and $\mathbf{aX}_2\mathbf{b}$ in this definition are referred to as *geodesics of the first and second kind*, respectively. That geodesics need not exist in all uniformly discrete spaces is shown by the following example.

Example 5.10. Let $\mathfrak{S}$ consist of **a**, **b**, and $\mathbf{c}_1, \mathbf{c}_2, \dots$. Let $\Psi^{(1)}$ be a symmetric function with the following values: for $i, j \in \{1, 2, \dots\}$,

$$\Psi^{(1)}\mathbf{ab} = 2,$$
$$\Psi^{(1)}\mathbf{ac}_i = 1 + 1/i,$$
$$\Psi^{(1)}\mathbf{bc}_i = 1/2,$$
$$\Psi^{(1)}\mathbf{c}_i\mathbf{c}_j = |i - j|.$$

The space is uniformly discrete because

$$\inf_{\mathbf{x} \neq \mathbf{y}} \Psi^{(1)}\mathbf{xy} = \frac{1}{2}.$$

For the sequence of chains $\mathbf{ac}_n\mathbf{b}$, as $n \to \infty$,

$$\Psi^{(1)}\mathbf{ac}_n\mathbf{b} = \frac{3}{2} + \frac{1}{n} \to \frac{3}{2},$$

and it is easy to see that

$$G_1\mathbf{ab} = \frac{3}{2}.$$

There is, however, no chain **X** such that

$$\Psi^{(1)}\mathbf{aXb} = \frac{3}{2}.$$

Therefore, this uniformly discrete space is not with geodesics.

In a uniformly discrete space with geodesics, we say that **a** is *strongly 1-linked* (*strongly 2-linked*) to **b** if **ab** is the only geodesic of the first (respectively, second) kind connecting **a** to **b**. Clearly, strong ι-linkage implies direct ι-linkage ($\iota = 1, 2$).

A chain $\mathbf{X} = \mathbf{x}_1 \dots \mathbf{x}_k$ is called *strongly ι-basic* if, for any $1 \leq i < k$, $\mathbf{x}_i$ is strongly ι-linked to $\mathbf{x}_{i+1}$ ($\iota = 1, 2$).

Theorem 5.7. *With ι standing for 1 or 2, in a uniformly discrete space with geodesics,*

(i) any two points can be connected by a strongly ι-basic geodesic chain;

(ii) any point can be strongly ι-linked to some other point; and

(iii) the set of all strongly ι-linked ordered pairs of points forms a base for G_ι.

Proof. Assume (i) is not true. Then, for some **a** and **b**, all geodesics **aXb** are not strongly ι-basic. Choose a geodesic

$$\mathbf{ax}_1 \dots \mathbf{x}_k\mathbf{b}$$

with the largest number of elements k. It should exist by the same argument as in Theorem 5.6:

$$\Psi^{(\iota)} \mathbf{a} \mathbf{x}_1 \dots \mathbf{x}_k \mathbf{b} \geq (k+1) \inf_{\mathbf{x} \neq \mathbf{y}} \Psi^{(\iota)} \mathbf{x} \mathbf{y} > 0.$$

Let $\mathbf{x}_i \mathbf{x}_{i+1}$ be a link in $\mathbf{a}\mathbf{X}\mathbf{b}$ with $\mathbf{x}_i$ not strongly ι-linked to $\mathbf{x}_{i+1}$. Then there exists a geodesic chain $\mathbf{x}_i \mathbf{Y} \mathbf{x}_{i+1}$ with a nonempty $\mathbf{Y}$, whence

$$\mathbf{a} \mathbf{x}_1 \dots \mathbf{x}_i \mathbf{Y} \mathbf{x}_{i+1} \dots \mathbf{x}_k \mathbf{b}$$

is a geodesic chain from $\mathbf{a}$ to $\mathbf{b}$ with more than k elements. This contradiction proves (i). The statements (ii) and (iii) are immediate corollaries. $\square$

Theorem 5.8. *With ι standing for 1 or 2, in a uniformly discrete space with geodesics, a point $\mathbf{a}$ is strongly ι-linked to a point $\mathbf{b}$ if and only if, for any point $\mathbf{m}$ distinct from $\mathbf{a}$ and $\mathbf{b}$,*

$$G_\iota \mathbf{a} \mathbf{m} \mathbf{b} > G_i \mathbf{a} \mathbf{b}.$$

Proof. We prove the equivalence of the negations of the two statements:

(1) "$\mathbf{a}$ is not strongly ι-linked to $\mathbf{b}$" means $\mathbf{a}$ is connected to $\mathbf{b}$ by a geodesic other than $\mathbf{a}\mathbf{b}$;

(2) the negation of the inequality in the formulation is (since G_ι is a metric) the equality

$$G_\iota \mathbf{a} \mathbf{m} \mathbf{b} = G_i \mathbf{a} \mathbf{b}.$$

If $\mathbf{a}$ can be connected to $\mathbf{b}$ by a geodesic $\mathbf{a}\mathbf{X}\mathbf{b}$ with a nonempty $\mathbf{X}$, then choosing an element $\mathbf{m}$ of this chain, we get the equality above. Conversely, if this equality is satisfied for some point $\mathbf{m}$, then the concatenation of the geodesics from $\mathbf{a}$ to $\mathbf{m}$ and from $\mathbf{m}$ to $\mathbf{b}$ is a geodesic other than $\mathbf{a}\mathbf{b}$. $\square$

We demonstrate the use of these results by the following example.

Example 5.11. Consider Figure 5.3 again. Inspecting, say, the link from $\mathbf{a}$ to $\mathbf{d}$ in the left panel we see that

$$G_1 \mathbf{a} \mathbf{d} = 0.4 < \begin{cases} G_1 \mathbf{a} \mathbf{b} \mathbf{d} = 0.2 + 0.6 \\ G_1 \mathbf{a} \mathbf{c} \mathbf{d} = 0.2 + 0.3. \end{cases}$$

Since this exhausts all triads of the form $\mathbf{a}\mathbf{m}\mathbf{d}$ in this space, we conclude that $\mathbf{a}$ is strongly 1-linked to $\mathbf{d}$, and

$$\Psi^{(1)} \mathbf{a} \mathbf{d} = G_1 \mathbf{a} \mathbf{d} = 0.4.$$

We come to the same conclusion and restore the values of $\Psi^{(1)}$ and $\Psi^{(2)}$ for all links in Figure 5.3 with unframed values of distance. For the links with framed values, say, from **c** to **d** in the right panel, we have

$$G_2\mathbf{cd} = 0.6 = G_1\mathbf{cad} = 0.1 + 0.5,$$

whence we conclude that **a** is not strongly 2-linked to **d**. The value of $\Psi^{(2)}\mathbf{cd}$ therefore cannot be reconstructed uniquely: it can be any value ≥ 0.6.

Acknowledgments

This research has been supported by AFOSR grant FA9550-09-1-0252 to Purdue University.

References

Dzhafarov, E. N. (2001). Fechnerian psychophysics. In N. J. Smelser and P. B. Baltes (Eds.), *International Encyclopedia of the Social and Behavioral Sciences* (vol. 8, pp. 5437–5440). NY: Pergamon Press.

Dzhafarov, E. N. (2002a). Multidimensional Fechnerian scaling: Probability-distance hypothesis. *Journal of Mathematical Psychology, 46*, 352–374.

Dzhafarov, E. N. (2002b). Multidimensional Fechnerian scaling: Pairwise comparisons, regular minimality, and nonconstant self-similarity. *Journal of Mathematical Psychology, 46*, 583–608.

Dzhafarov, E. N. (2008a). Dissimilarity cumulation theory in arc-connected spaces. *Journal of Mathematical Psychology, 52*, 73–92. [see Dzhafarov, E. N. (2009). Corrigendum to "Dissimilarity cumulation theory in arc-connected spaces". *Journal of Mathematical Psychology, 53*, 300.]

Dzhafarov, E. N. (2008b). Dissimilarity cumulation theory in smoothly connected spaces. *Journal of Mathematical Psychology, 52*, 93–115.

Dzhafarov, E. N. (2010). Dissimilarity, quasidistance, distance. *Journal of Mathematical Psychology, 54*, 334–337.

Dzhafarov, E. N., & Colonius, H. (2006). Regular Minimality: A fundamental law of discrimination. In H. Colonius and E. N. Dzhafarov (Eds.), *Measurement and Representation of Sensations* (pp. 1–46). Mahwah, NJ: Erlbaum.

Dzhafarov, E. N., & Colonius, H. (2007). Dissimilarity Cumulation theory and subjective metrics. *Journal of Mathematical Psychology, 51*, 290–304.

Dzhafarov, E. N., & Colonius, H. (2011). The Fechnerian idea. *American Journal of Psychology, 124*, 127–140.

Falmagne, J. C. (1985). *Elements of Psychophysical Theory*. Oxford: Oxford University Press.

Kujala, J. V., & Dzhafarov, E. N. (2008). On minima of discrimination functions. *Journal of Mathematical Psychology, 52*, 116–127.

Luce, R. D., & Galanter, E. (1963). Discrimination. In R. D. Luce, R. R. Bush, and E. Galanter (Eds.), *Handbook of Mathematical Psychology*, vol. 1 (pp. 191–244). NY: John Wiley & Sons.

Chapter 6

Bayesian Adaptive Estimation: A Theoretical Review

Janne V. Kujala

University of Jyväskylä

In this chapter, we review Bayesian adaptive estimation methodology under a formal decision theoretic view, extending on the work reported in Kujala (2010). We present simple concrete examples that illustrate the efficiency of the ubiquitous greedy information maximization algorithm, compare it to the optimal non-adaptive and batch strategies, and consider certain pitfalls of the greedy strategy. In particular, we show that the greedy strategy can be *arbitrarily much* worse than the optimal strategy for a certain number of trials. Then, we generalize the discussion to the situation where each observable variable is associated with a certain random cost of observation. For example, the cost could be defined as the random time taken by each trial in an experiment, and one might wish to maximize the expected total information gain over as many trials as can be completed in 15 minutes.

6.1. Introduction

In this chapter, we consider Bayesian estimation of an unobservable random variable Θ based on a sequence $y_{x_1}, \ldots, y_{x_T}$ of independent (given θ) realizations from some conditional densities $p(y_{x_t} \mid \theta)$ indexed by trial *placements* x_t, each of which can be adaptively chosen from some set $X_t \subset X$ based on the outcomes $\mathbf{y} := (y_{x_1}, \ldots, y_{x_{t-1}})$ of the earlier observations.[1]

We assume that the goal of choosing the placements is to maximize the expected value $E(U_T)$ of some *utility function* (DeGroot, 1970) $U_t := u(Y_{X_1}, \ldots, Y_{X_t})$ of the knowledge about Θ under some constraints such as

---

[1]It is often assumed that one can observe multiple independent (given θ) copies of the same random variable Y_x. However, instead of complicating the general notation with something like $Y_{x_t}^{(t)}$, we rely on the fact that the set X can explicitly include separate indices for any identically distributed copies, for example, one might have $[Y_{(x,t)} \mid \theta] \overset{\text{i.i.d.}}{\sim} [Y_{(x,t')} \mid \theta]$ for all $t, t' \in \mathbb{N}$, $t \neq t'$. Hence, we can use the simple notation with no loss of generality.

a given total number T of trials. The same decision theoretic framework applies to other problems as well, such as model discrimination (Myung & Pitt, 2009; Cavagnaro, Myung, Pitt, & Kujala, 2010), where the utility function is based on the discriminability of two or more models based on the observed data.

In the rest of this section, we review relevant literature under the decision theoretic view. In Section 6.2, we further illustrate the concepts with simple concrete examples and consider different placement strategies. In Section 6.3, we generalize the discussion to situations where the observation of each random variable is associated with a certain random cost and consider the problem of maximizing the expected utility of an experiment that terminates when the costs overrun a certain predetermined budget. This section closely follows Kujala (2010). In Appendix A, we prove some auxiliary results that are used in the main text, and finally, in Appendix B, we go through some measure theoretic technicalities that are mostly avoided in the main text.

6.1.1. *Psychophysics*

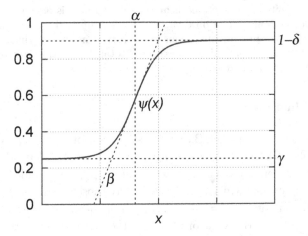

Fig. 6.1.   An illustration of a typical psychometric function relating stimulus intensity x to the observer's probability of detecting it.

In psychophysics, Bayesian adaptive estimation was first considered by Watson and Pelli (1983) for estimation of an observer's "threshold" α of detecting a stimulus of given intensity x. The dichotomous result of detecting (1) or not detecting (0) the stimulus is assumed to be distributed as

$$p(y_x \mid \theta) = \begin{cases} \psi_\theta(x), & y_x = 1, \\ 1 - \psi_\theta(x), & y_x = 0, \end{cases}$$

where ψ_θ is some sigmoidal function described by the four parameters $\theta = (\alpha, \beta, \gamma, \delta)$, where α is the *threshold*, β determines the *slope* at the threshold, γ is the *guessing rate*, and δ is the *lapsing rate*, see Figure 6.1.

Watson and Pelli (1983) assume that only the α component is unknown and define the *loss function* (negative of a utility function) as the variance of the posterior distribution of α. Their adaptive method places each test intensity x_t at the mode of the posterior distribution $p(\alpha \mid \mathbf{y})$. However, King-Smith (1984; King-Smith, Grigsby, Vingrys, Benes, & Suposit, 1994) discovered that placement at the mean is more efficient and that it is even more efficient to use the implicit rule of choosing x_t so as to minimize the expected posterior variance of α after the observation of Y_{x_t}.

Kontsevich and Tyler (1999) consider simultaneous estimation of both the threshold α and the slope β. They argue that the variance does not easily generalize into a loss function of multivariate uncertainty as it would require arbitrary weighting of the uncertainties along each dimension[2]. Therefore, they instead use the (differential[3]) Shannon (1948) entropy

$$H(\Theta) = - \int p(\theta) \log p(\theta) d\theta, \qquad (6.1)$$

which is well-defined for multivariate θ, too. Their adaptive method chooses each placement so as to minimize the expected posterior entropy of (α, β) after the observation of Y_{x_t}. This is a more elegant solution than the asymptotically justified explicit rules (e.g., King-Smith & Rose, 1997; Snoeren & Puts, 1997) that have been used before.

In more recent works (Kujala, 2004; Kujala & Lukka, 2006; Lesmes, Jeon, Lu, & Dosher, 2006) the same algorithm is generalized to multivariate

---

[2]There is in fact a natural way to generalize the variance loss function to multivariate Θ: Instead of a linear combination of the posterior variances of $\Theta_1, \ldots, \Theta_n$, one can use the *determinant* of the posterior covariance matrix as the loss function (a criterion generally referred to as *D-optimality* in the optimal design literature). This yields the relative utilities of different placements invariant w.r.t. linear reparametrizations of Θ. Still, the entropy is more general as it yields insensitivity w.r.t. *all* one-to-one reparametrizations (cf. Section 6.1.3).

[3]We use the same notation $\int f(\theta)p(\theta)d\theta$ for both the continuous case and the discrete case, in which it corresponds to a sum. This is measure-theoretically justified as "$d\theta$" can be considered as the counting measure in the discrete case. Thus, following Lindley (1956), even though we use the familiar notation, we are in fact working in full measure-theoretic generality, allowing the density $p(\theta)$ to be w.r.t. any measure "$d\theta$". See Appendix B for the technical details.

placements $x \in \mathsf{X} = \mathbb{R}^2$ as well. We do not go into any details here, as in the present theoretical framework X is considered just as an unstructured set and so there is no conceptual difference to the univariate placement case.

6.1.2. The greedy strategy

The implicit placement rules mentioned above are examples of a *greedy strategy*, which chooses each placement so as to optimize the expected immediate gain in utility after the next observation:

$$
\begin{aligned}
x_t &:= \arg\max_{x \in \mathsf{X}_t} \mathrm{E}(U_t \mid \mathbf{y}, \, X_t = x) \\
&= \arg\max_{x \in \mathsf{X}_t} \mathrm{E}(U_t - U_{t-1} \mid \mathbf{y}, \, X_t = x).
\end{aligned}
$$

As the value of U_{t-1} will be known at the time of choosing x_t, maximizing the expected change of utility is equivalent to maximizing its expected value. This strategy can be applied in any model for any utility function, but it is generally not the globally optimal strategy except when the experiment consists of exactly one trial.

Remark 6.1. Strictly speaking it is possible that no maximum of the expected utility exists, in which case one should generally choose such a placement that yields an expected gain sufficiently close to the supremum (DeGroot, 1970).

6.1.3. The entropy loss function and mutual information

The entropy loss function (6.1) has some remarkable properties. Most importantly, the expected difference between the prior entropy and the posterior entropy given Y_x is parametrization invariant (Lindley, 1956). Hence, even though the value U_t of the utility function defined as the (negative of the) entropy as well as its change $U_t - U_{t-1}$ do depend on the parametrization chosen for θ, the expected change $\mathrm{E}(U_t - U_{t-1} \mid \mathbf{y}, \, X_t = x)$ is parametrization invariant; it corresponds to the information-theoretic *mutual information*[4]

$$I(\Theta; Y_x \mid \mathbf{y}) = H(\Theta \mid \mathbf{y}) - E[H(\Theta \mid \mathbf{y}, Y_x)] \qquad (6.2)$$

of Θ and Y_x (given $\mathbf{y}$). Indeed, the mutual information generally defined as

$$I(A; B) := \iint p(a, b) \log \frac{p(a, b)}{p(a)p(b)} \, da \, db$$

is insensitive to any one-to-one transformations of A and B, and as defined, it is obviously symmetric, which yields the identity

$$H(A) - E[H(A \mid B)] = I(A; B) = I(B; A) = H(B) - E[H(B \mid A)],$$

which holds whenever the differences are defined, i.e., not $\infty - \infty$ or $-\infty - (-\infty)$ (see Cover & Thomas, 2006). Thus, the mutual information represents the expected amount of information that the observation of one random variable gives about the other. This expected amount is always nonnegative even though the entropy might actually increase in case an "unexpected" outcome is observed.

6.1.4. *Calculating expected information gain*

Making use of the symmetry of the mutual information, Kujala and Lukka (2006) write the objective function (6.2) as

$$I(\Theta; Y_x \mid \mathbf{y}) = H(Y_x \mid \mathbf{y}) - E[H(Y_x \mid \Theta, \mathbf{y})]. \qquad (6.3)$$

This formulation is usually more convenient than (6.2) as the distribution of Y_x given θ is typically much simpler than that of Θ given y_x. For example, in the case of dichotomous results, this yields the convenient expression

$$I(\Theta; Y_x \mid \mathbf{y}) = h\left(\int \Pr\{Y_x = 1 \mid \theta\} p(\theta \mid \mathbf{y}) d\theta\right)$$

$$- \int h(\Pr\{Y_x = 1 \mid \theta\}) p(\theta \mid \mathbf{y}) d\theta,$$

where $h(p) := -p \log p - (1 - p) \log(1 - p)$ is the entropy of a Bernoulli distribution with probabilities p and $1-p$.[5] Thus, only expectations over the

---

[4]In our notation $H(A \mid \ldots)$ always denotes the conditional entropy of A given the (possibly random) conditioning values $[\ldots]$. Thus, $H(A \mid \ldots)$ will be a random variable if any of its conditioning values are random variables unlike in the unfortunate standard notation where an expectation is implicitly taken over the conditioning values. Also, $I(\Theta; Y_x \mid \mathbf{y})$ denotes the mutual information of the random variables $\Theta \mid \mathbf{y}$ and $Y_x \mid \mathbf{y}$, that is, both Θ and Y_x are conditioned on $\mathbf{y}$. This is standard notation.

[5]We assume base e logarithm in expressions, but numerical results are given in bits, i.e., we define bit $:= \log 2$.

(sequential) prior $p(\theta \mid \mathbf{y})$ are needed which allows for efficient computation. For example, given an (approximately) i.i.d. sample $\{\theta_i\}_{i=1}^N$ drawn from $p(\theta \mid \mathbf{y})$, the objective function can be approximated as

$$I(\Theta; Y_x \mid \mathbf{y}) \approx h\left(\frac{1}{N}\sum_{i=1}^N \Pr\{Y_x = 1 \mid \theta_i\}\right) - \frac{1}{N}\sum_{i=1}^N h(\Pr\{Y_x = 1 \mid \theta_i\}).$$

This was used by Kujala and Lukka (2006) in a sequential Monte Carlo implementation of the greedy algorithm.

The same idea generalizes to any finite number of outcomes:

$$I(\Theta; Y_x \mid \mathbf{y}) = \sum_{y_x} g\left(\int p(y_x \mid \theta)p(\theta \mid \mathbf{y})d\theta\right) - \int \sum_{y_x} g(p(y_x \mid \theta))p(\theta \mid \mathbf{y})d\theta$$

$$\approx \sum_{y_x} g\left(\frac{1}{N}\sum_{i=1}^N p(y_x \mid \theta_i)\right) - \frac{1}{N}\sum_{i=1}^N \sum_{y_x} g(p(y_x \mid \theta_i)),$$

where $g(p) = -p\log(p)$. This formulation could be applied in, for example, the choice model and MCMC algorithm used in (Kujala, Richardson, & Lyytinen, 2010b).

Not only is (6.3) usually computationally more convenient than (6.2), but there is also a theoretical advantage. If Y_x is dichotomous or has a finite number of possible values, then the entropies in the right side of (6.3) will always be finite, and the expression is well-defined unlike (6.2) which may come out as $\infty - \infty$ or $-\infty - (-\infty)$ for some parametrizations of θ. Of course, if both Y_x and Θ have an infinite number of possible values, then either formulation may fail for some parametrization. Nonetheless, the mutual information itself is always well-defined (see Section B.2 for the measure-theoretic details). Hence, Kolmogorov (1956) argues that it is in fact the mutual information that is the fundamental concept of the theory of information.

Remark 6.2. Although the information gain may theoretically be infinite (e.g., $Y = \Theta \sim \text{Uniform}[0,1]$ yields $I(\Theta; Y) = \infty$), that will never happen in a realistic model as the observation of any real quantity Y is always subject to some measurement error.

6.1.5. *Nuisance variables and utility weights*

In some cases, one might be interested only in the value of some component Θ_1 of $\Theta = (\Theta_1, \Theta_2)$ even if its other components Θ_2 are unknown, too.

In that case, one can define the utility function as the (negative of the) marginal entropy

$$U_t = -\operatorname{H}(\Theta_1 \mid Y_{X_1}, \ldots, Y_{X_t})$$

of the interesting variables. The expected change of entropy still corresponds to the mutual information

$$\operatorname{E}(U_t - U_{t-1} \mid \mathbf{y}) = \operatorname{I}(Y_x; \Theta_1 \mid \mathbf{y})$$

and hence enjoys the same parametrization invariance properties.

More generally, a parametrization invariant objective function can always be defined as an arbitrary function

$$f(\operatorname{I}(Y_x; T_1(\Theta) \mid \mathbf{y}), \ldots, \operatorname{I}(Y_x; T_n(\Theta) \mid \mathbf{y}))$$

of mutual informations, where $T_1, \ldots, T_n$ can be any functions (e.g., T_k can be the component mappings $T_k(\Theta) = \Theta_k$). In particular, for any linear combination of marginal entropies

$$U_t = -\sum_k w_k \operatorname{H}(\Theta_k \mid Y_{X_1}, \ldots, Y_{X_t}),$$

a class of utility functions used by Tanner (2008) and also mentioned by Paninski (2005), the maximizer x of the expected utility $\operatorname{E}(U_t \mid \mathbf{y}, X_t = x)$ is insensitive to the parametrization of Θ because the expected gain in utility

$$\operatorname{E}(U_t - U_{t-1} \mid \mathbf{y}) = \sum_k w_k \operatorname{I}(Y_x; \Theta_k \mid \mathbf{y})$$

is a (linear) function of parametrization invariant mutual informations.

However, the expected change of any nonlinear function of entropies no longer corresponds to a function of mutual informations and hence does not inherit the parametrization invariance. Conversely, maximization of a nonlinear combination of mutual informations generally does not correspond to maximization of the expected value of any utility function.

While nuisance variables do not pose any conceptual problems, they can complicate the practical computations. It is usually computationally easier to apply (6.3) to Θ than any subset of its components, as integration out of the nuisance variables from the model at each trial interacts badly with the computational conveniences that (6.3) provides over (6.2). Furthermore, as we shall see in the following, the greedy algorithm typically works best when information about the nuisance variables is included in the utility function (although the greedy strategy can fail in that case, too, as we shall see).

6.1.6. *Global strategies*

Any adaptive (or non-adaptive) placement strategy defines a *decision function*

$$d : \mathsf{Y}_d \to \mathsf{X} \cup \{\lambda\} : (y_{x_1}, \ldots, y_{x_{t-1}}) \mapsto x_t,$$

whose domain Y_d is the set of possible sequences of trial results (including the empty sequence) and whose value is the next placement or the special value λ which flags the end of the experiment. Now we can define the random variable

$$Y_d := (Y_{X_1}, \ldots, Y_{X_T}) \in \mathsf{Y}_d$$

denoting the outcome of the whole adaptive experiment governed by the decision function d, where T is the possibly random time index of the trial that ends the experiment.

The fact that the whole adaptive experiment can be seen as just one observation Y_d implies that all the parametrization invariance results of the entropy loss function apply to the whole-experiment strategies as well, regardless of whether the termination rule is adaptive or not, as long as the experiment eventually terminates (and even that is not strictly necessary if the value of the utility function converges with probability one). One could even allow randomized decision functions mapping to distributions over $\mathsf{X} \cup \{\lambda\}$ instead of deterministic values. However, as randomized decisions generally gain nothing over deterministic ones (DeGroot, 1970), we shall only consider deterministic decision functions except for one reference to a random termination rule in Section 6.3.1 (where the randomness is outside the experimenter's control).

6.1.7. *The globally optimal strategy*

The optimal strategy is obviously to maximize the expected utility after the observation of the whole experiment result Y_d w.r.t. the decision function d. For example, in the case of dichotomous results and constant experiment length T, the decision function defines a binary decision tree with $2^T - 1$ nodes, each denoting the next placement after a certain sequence of observations. Thus, the optimal strategy can in theory be found by optimizing over these $2^T - 1$ parameters. In practice, the exponentially growing number of parameters makes the optimal strategy generally intractable very soon as T grows (although in some special cases the globally optimal strategy can be found analytically).

To improve the efficiency of the greedy algorithm in practice, one could try to apply it to the observable variables Y_d indexed by d varying in the set of decision trees of a certain small depth. For example, King-Smith et al. (1994) have implemented this strategy for a two-step "look-ahead" in a psychometric estimation procedure. However, their simulations indicated that the improvement over the one-step greedy strategy was generally small (compared to the one-step method after 16 trials, the two-step method yielded the same accuracy at 15.97 trials for $\gamma = 0.03$, $\delta = 0.01$ and at 15.6 trials for $\gamma = 0.5$, $\delta = 0.01$). Still, much larger improvements may be possible in other models.

There are some special cases where progress has been made in overcoming the exponential scaling of the optimal solution w.r.t. T. For example, if the prior data $\mathbf{y}$ can always be sufficiently well summarized by some low-dimensional statistic, then the globally optimal strategy can be approximated by a dynamic programming algorithm whose running time grows only linearly with the depth T (Brockwell & Kadane, 2003).

6.1.8. *Non-adaptive and batch strategies*

Lindley (1956) considers the case where i.i.d. copies of the same random variable Y_x are observed sequentially and shows that the expected information gain $I(Y_x^{(1)}, \ldots, Y_x^{(t)}; \Theta)$ over t observations is a concave function of t, i.e., the expected information gains from each additional observation are non-increasing. Note that it is possible that the first observation of some Y_x is expected to yield more information than the first observation of $Y_{x'}$, but that over repeated i.i.d. observations, $Y_{x'}$ would be more informative than the same number of observations of Y_x (this is the case in the example of Section 6.2.1).

In more recent works, Müller, Sansó, and Iorio (2004) and Amzal, Bois, Parent, and Robert (2006) consider Monte Carlo simulation methods for optimal experiment design under a given set of adjustable parameters, such as the placements $x_1, \ldots, x_T$ of all trials in a non-adaptive experiment. These methods may also be useful for such adaptive designs where one has to choose the placements for a batch of n trials simultaneously. In that case, one could apply the optimal design algorithm for each batch within a greedy strategy that considers each batch as a single trial. In theory, the same simulation methods could be applied to the complete design d of an adaptive experiment, too, although the exponential number of parameters would become a problem soon as discussed above.

6.2. Examples

The greedy algorithm of optimizing the expected immediate gain is ubiquitous in practical applications of Bayesian adaptive estimation. However, little is known about its relative efficiency compared to other strategies. The only definite result so far appears to be that under certain regularity conditions, the greedy strategy can be shown to be asymptotically more efficient than any non-adaptive strategy (Paninski, 2005).

In this section, through simple concrete examples, we demonstrate in particular that

(1) the per-trial efficiency of batch strategies generally deteriorates rapidly as the batch size increases (Section 6.2.1),
(2) although the greedy algorithm usually works, it can be *arbitrarily much* worse than the globally optimal strategy (Section 6.2.2), and
(3) even the globally optimal strategy can apparently fail due to the fact that the entropy loss function can be inappropriate in certain situations (Section 6.2.3).

6.2.1. *A simplified psychometric model*

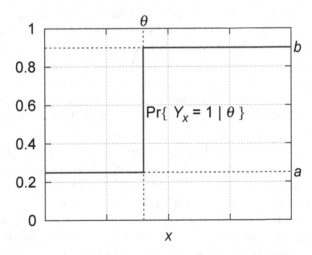

Fig. 6.2. A simplified version of the psychometric model shown in Figure 6.1 with an infinite slope at the threshold θ.

Example 6.1. Suppose that for all $x \in \mathbb{R}$, $Y_x \in \{0, 1\}$ is a dichotomous random variable defined by

$$\Pr\{Y_x = 1 \mid \theta\} = \begin{cases} a, & x < \theta, \\ b, & x \geq \theta \end{cases}$$

for some $0 < a < b < 1$, see Figure 6.2. While this model is simple, it does have the important feature that there is uncertainty of the results due to both the fact that Θ is unknown and the fact that for a given θ, the result $[Y_x \mid \theta]$ is random. This model differs from the typical psychometric model only in that, due to the infinite slope at the threshold, the obtainable information gains do not decrease over time as the scale of the uncertainty reduces. However, this detail is not very important during the first few trials, and therefore this example serves to illustrate the relative efficiencies of different placement strategies that can be expected in a typical psychophysical experiment and explain the success of the greedy strategy in psychophysical measurement.

6.2.1.1. *Success of the greedy strategy*

Using (6.3), the expected information gain of observing Y_x given any prior data $\mathbf{y} = (y_{x_1}, \dots, y_{x_{t-1}})$ can be calculated as

$$\begin{aligned} I(Y_x; \Theta \mid \mathbf{y}) &= H(Y_x \mid \mathbf{y}) - E[H(Y_x \mid \Theta, \mathbf{y})] \\ &= h(a + (b - a)\Pr\{\Theta \leq x \mid \mathbf{y}\}) \\ &\quad - (h(a) + (h(b) - h(a))\Pr\{\Theta \leq x \mid \mathbf{y}\}), \end{aligned} \tag{6.4}$$

which depends on the placement x and the prior data $\mathbf{y}$ only through $z := \Pr\{\Theta \leq x \mid \mathbf{y}\}$. Assuming that the prior on Θ is absolutely continuous w.r.t. the Lebesgue measure, the same will be true for the posterior given $\mathbf{y}$, and one can always choose x so as to attain any value of $z \in [0, 1]$. As (6.4) is continuous on the compact interval $z \in [0, 1]$, it attains a maximum value (which is independent of any prior data). It follows that the greedy strategy yields the maximum expected total information gain over any given constant number of trials.

Let us then find the value(s) of z that maximize (6.4). Obviously the expression is smooth and positive (assuming $a \neq b$) for $z \in (0, 1)$ and zero for $z = 0$ or $z = 1$. Hence, the expression can attain a maximum value only at critical points $z \in (0, 1)$. The derivative w.r.t. z is

$$h'(a + (b - a)z)(b - a) - (h(b) - h(a)),$$

where $h'(p) = \log(1/p - 1)$. The derivative is zero iff

$$\log\left(\frac{1}{a + (b-a)z} - 1\right) = \frac{h(b) - h(a)}{b - a},$$

and so the maximum value is attained at the unique point

$$z^* = \frac{\dfrac{1}{1 + \exp\left(\frac{h(b)-h(a)}{b-a}\right)} - a}{b - a}. \tag{6.5}$$

The value at this point is (after some algebra)

$$I(Y_{x^*}; \Theta \mid \mathbf{y}) = \log\left(1 + \exp\left(\frac{h(b) - h(a)}{b - a}\right)\right) - \frac{(1-a)h(b) - (1-b)h(a)}{b - a}. \tag{6.6}$$

6.2.1.2. *Non-adaptive and batch strategies*

Let us then consider non-adaptive strategies with a set of n placements $x_1 \leq \cdots \leq x_n$ to be chosen before the experiment. Denoting $g(p) = -p\log p$, $x_0 = -\infty$, $x_{n+1} = \infty$, and $z_k := \Pr\{\Theta \leq x_k\}$, the expected information gain can be written as

$$I(Y_{x_1}, \ldots, Y_{x_n}; \Theta) = H(Y_{x_1}, \ldots, Y_{x_n}) - E[H(Y_{x_1}, \ldots, Y_{x_n} \mid \Theta)]$$

$$= \sum_{y_{x_1}} \cdots \sum_{y_{x_n}} g\left(\int p(\theta)d\theta \prod_{j=1}^{n} p(y_{x_j} \mid \theta)\right) - \sum_{j=1}^{n} E[H(Y_{x_j} \mid \Theta)]$$

$$= \sum_{y_{x_1}} \cdots \sum_{y_{x_n}} g\left(\sum_{k=0}^{n} \underbrace{\Pr\{x_k < \Theta \leq x_{k+1}\}}_{= z_{k+1} - z_k} \prod_{j=1}^{n} \begin{cases} a, & j \leq k, \ y_{x_j} = 1, \\ 1 - a, & j \leq k, \ y_{x_j} = 0, \\ b, & j > k, \ y_{x_j} = 1, \\ 1 - b, & j > k, \ y_{x_j} = 0, \end{cases}\right)$$

$$\quad - \sum_{j=1}^{n}[z_j h(b) + (1 - z_j)h(a)]$$

$$= -(Mz + v) \cdot \log(Mz + v) - u \cdot z - c =: f(z)$$

for some matrix $M \in \mathbb{R}^{2^n \times n}$, vectors $v \in \mathbb{R}^{2^n}$ and $u \in \mathbb{R}^n$, and constant $c \in \mathbb{R}$, where $z = (z_1, \ldots, z_n) \in \mathbb{R}^n$ and the log function is applied elementwise to the vector $Mz + v \in (0, 1]^{2^n}$, which gives the probabilities of the 2^n

different outcomes. The gradient of the function is

$$\nabla f(z) = -M^T \left(\log(Mz + v) + \begin{bmatrix} 1 \\ \vdots \\ 1 \end{bmatrix} \right) - u$$

and the Hessian matrix is

$$Hf(z) = -M^T \mathrm{diag}(\underbrace{\mathrm{inv}(Mz + v)}_{\in [1,\infty)^{2^n}})M,$$

where inv denotes elementwise inverse. As the Hessian is negative definite (M has full rank given $a \neq b$), the function is strictly concave and therefore has a unique maximizer in the compact convex set $0 \leq z_1 \leq \cdots \leq z_n \leq 1$, which can be found using, for example, Newton's iteration $z^{(t+1)} = z^{(t)} - \lambda[Hf(z^{(t)})]^{-1}\nabla f(z^{(t)})$, (where at each step t we try $\lambda = 1$, 0.1, 0.01, ... until $z^{(t+1)}$ is within the convex feasible set).

While this linear algebraic formulation appears simple, it should be noted that the number of rows of the matrix M grows exponentially with n. Thus, this deterministic approach will only work up to around $n = 20$. Beyond that, one may have to resort to a Monte Carlo approach to simulate the expected results of each possible set of placements as discussed in Section 6.1.8.

Finally, let us consider a non-adaptive design where n i.i.d. trials are conducted with the *same* placement x. Conceptually, this is just a restricted special case of the general non-adaptive design considered above, but instead of the 2^n different outcomes, one can now consider a binomial distribution of the number of 1-results. However, as this simplification still does not lead to a closed form expression of the optimal placement, we do not go into any further details.

6.2.1.3. *Comparison*

The optimal placements of the three strategies for $n = 1, \ldots, 9$ are shown in Figure 6.3. Apparently the efficiency of the non-adaptive strategies decreases rapidly relative to the optimal strategy as n increases. The reasons for this are intuitively simple: when a is close to zero and b close to one, the adaptive strategy can sequentially bisect the range down to one of 2^n distinct sections, yielding n bits of information, while the non-adaptive strategy can only divide the range to $n + 1$ distinct sections yielding at most $\log_2(n+1)$ bits of information and the single-placement strategy with only 2 sections tops off at 1 bit. However, as the guessing and lapsing rates

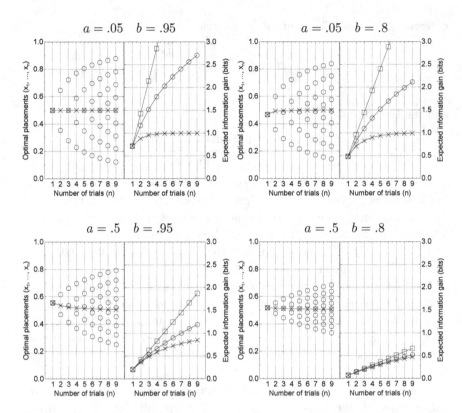

Fig. 6.3. Illustrations of the placements and efficiency of the optimal adaptive design (squares), optimal non-adaptive design (circles), and optimal single-placement repeated observations design (crosses) in the model shown in Figure 6.2. These are the optimal placements assuming a uniform prior for Θ on $[0, 1]$. For any other (absolutely continuous) prior, the placements are obtained by interpreting the values on the y-axis as the fractiles $\Pr\{\Theta \le x_k \mid \mathbf{y}\}$, which also yields the not shown optimal adaptive placements for $n > 1$. The circles are almost but not exactly evenly spaced for each n. The crosses approach a gain of 1 bit and placement at 0.5 as $n \to \infty$.

increase, the differences between the strategies become smaller. For example, with $a = .5$ and $b = .8$, one could present batches of 3 identical trials with only a small loss in efficiency compared to the fully adaptive strategy.

The optimal adaptive placement (6.5) appears to be close to the median of the distribution of $[\Theta \mid \mathbf{y}]$ under several values of a and b. With two-alternative forced choice design (i.e., guessing rate $a = 1/2$), the optimal placement tends to $z = 0.6$ as $b \to 1$, and for any a and b, the placement z is within $[1/e, 1 - 1/e] \approx [.3679, .6321]$.

6.2.2. *Exploration versus exploitation*

In this section, we give examples of sequential estimation problems where the greedy strategy is clearly suboptimal.

Example 6.2. Suppose Y_n is a dichotomous random variable depending on the random variables Θ and M through the conditional distribution

$$\Pr\{Y_n = 1 \mid \theta, m\} = \begin{cases} \theta + (1 - \theta)\frac{1}{n}, & n \leq m, \\ \frac{1}{n}, & n > m. \end{cases}$$

Here n is the number of alternatives in a multiple choice task and Θ represents the probability that the observer knows the correct answer to each question. If the correct answer is not known, the observer is assumed to guess. Increasing n decreases the probability of guessing correctly, but we also assume that there is an unknown maximum number $M \geq 2$ of choices that the observer can handle before being overwhelmed in which case the answer will be random again.

We assume that Θ and M are independent in the prior and consider for M the prior

$$p(m) \propto \begin{cases} 1/m, & m = 2, \ldots, 7, \\ 0, & \text{otherwise.} \end{cases}$$

Under these assumptions, it turns out that maximization of $I(Y_n; \Theta)$ yields $n = 2$ regardless of the distribution of Θ (see Section A.1 for details). But as Y_2 does not depend on M, the posterior distribution of M given the result y_2 remains unchanged and so $I(Y_n; \Theta \mid y_2^{(1)}, \ldots, y_2^{(t)})$ is always maximized by $n = 2$. Thus, the greedy algorithm will only present trials with $n = 2$. This is clearly suboptimal in the long run as there is a positive probability that $M > 2$. Presenting some trials with $n > 2$ first would allow estimating M. Once the value is known with high enough confidence, the rest of the trials can be presented with the optimal number of choices $n = M$.

It is often suggested that the objective function of the greedy algorithm should combine both the expected gain of utility as well as the expected gain of information about all unknowns as it might lead to better gains of utility in the following trials (e.g., Verdinelli & Kadane, 1992). Thus, even though M is a nuisance variable we are not interested in estimating, it turns out that it would still be more efficient in the long run to apply the greedy algorithm to minimization of $H(\Theta, M \mid \mathbf{y})$ instead of $H(\Theta \mid \mathbf{y})$. Indeed, this strategy appears to be asymptotically optimal in the sense that the distribution of the placements eventually converges to M provided that $\Theta > 0$. (We do not have a rigorous proof of this result, but it is what happens in simulations.)

However, even if the utility function is defined as the information gained about all unobservable variables, the greedy strategy can still be arbitrarily much worse than the optimal strategy:

Example 6.3. Suppose that each of $\Theta \sim$ Uniform$[0,1]$ and $\Phi_n \sim$ Uniform$\{1,\ldots,n\}$ for $n \in \{1,2,\ldots\}$ are independent, unobservable variables. The observable variables are given by $Y_{n,x,a,b} \in \{-1,0,\ldots,2^n-1\}$, $n, x \in \{1,2,\ldots\}$, $a, b \in [0,1]$, where

$$Y_{n,x,a,b} = \begin{cases} 0, & x = \Phi_n,\ \Theta < a, \\ \left\lfloor 2^n \frac{\Theta-a}{b-a} \right\rfloor, & x = \Phi_n,\ a \le \Theta < b, \\ 2^n - 1, & x = \Phi_n,\ \Theta \ge b, \\ -1, & \text{otherwise.} \end{cases}$$

In this model, the posterior of Θ given any observations $\mathbf{y}$ will always be a uniform distribution on some interval $[a,b]$. For any given n, if one knows or guesses correctly the value of Φ_n, then observation of $Y_{n,x,a,b}$ with $x = \Phi_n$ reduces the posterior interval $[a,b]$ to some of its 2^n subdivisions and thus yields n new bits of information about Θ (as well as confirms the value of Φ_n if it was uncertain). An incorrect guess yields no information about Θ, but decreases the set of possible values of Φ_n by one. However, as there is no randomness in the observed variables given the hidden state, the expected information gain of both Θ and $\Phi := (\Phi_1, \Phi_2, \ldots)$ for a given n and any untried value of x is most conveniently calculated as

$$I(Y_{n,x,a,b}; \Theta, \Phi \mid \mathbf{y}) = H(Y_{n,x,a,b} \mid \mathbf{y}) - \underbrace{E[H(Y_{n,x,a,b} \mid \Theta, \Phi, \mathbf{y})]}_{=0}$$

$$= \frac{1}{n - m_n} \log 2^n + h\left(\frac{1}{n - m_n}\right), \qquad (6.7)$$

where m_n denotes the number of incorrect values of Φ_n already tried, and where we have split the computation of the entropy into two cases according as the outcome is -1 or ≥ 0 (the entropy of the first case is zero, the entropy of the latter case is $\log 2^n$, and the uncertainty between these two cases corresponds to the latter term of the expression). Initially $m_n = 0$ for all n, in which case (6.7) is maximized by $n = 2$. This maximum value is 2 bits (.5 chance of guessing the correct value of Φ_2, in which case two bits of Θ are obtained, and in either case, the value of $\Phi_2 \in \{1,2\}$ is learned which yields another bit). Observing this variable does not change the optimal informativity for $n \neq 2$ nor does it change it for $n = 2$ as the optimal gain is still 2 bits after the value of Φ_2 has been learned. Thus, if $I(Y_{n,x,a,b}; \Theta, \Phi \mid \mathbf{y})$ is maximized at every step, only trials with $n = 2$ will ever be presented, each yielding exactly 2 bits of new information.

However, the optimal measurement strategy in this model strongly depends on the total number of trials that the experiment will include. If there are T trials, then a good strategy would be to spend (at most) $T/2$ trials to find the value of $\Phi_{T/2}$ after which the remaining trials will each yield $T/2$ bits of information. The total information gain will thus be at least $(T/2)(T/2) = T^2/4$ bits while the greedy one-step strategy will only yield $2T$ bits (see Figure 6.4 for more details).

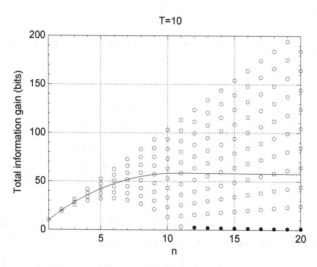

Fig. 6.4. An illustration of the distribution of the information gain obtained by observing $T = 10$ trials of type $Y_{n,\dots}$ in the model of Example 6.3. For each n, each empty circle has probability $1/n$ and the filled circle has the remaining probability $1 - T/n$; the curve shows the expectation of this distribution. The picture is qualitatively the same for any value of T: the global maximum of the expected gain is at $n = T + 1$ after which the expected gain starts to slowly decrease. However, the variance is very high at this maximum and therefore it may be a better strategy to instead maximize the minimum gain with $n \approx T/2$ as sketched in the text.

6.2.3. *Inappropriate utility function*

Example 6.4 (Paninski, 2005). *Suppose Y_x is a dichotomous random variable depending on two thresholds $\Theta_1 \in (0,1)$ and $\Theta_2 \in (1,2)$ through*

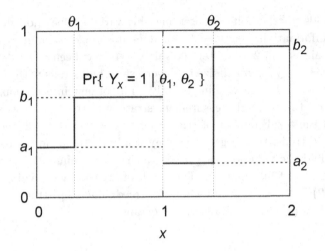

Fig. 6.5.   A two-threshold version of the model shown in Figure 6.2.

the conditional distribution

$$\Pr\{Y_x = 1 \mid \theta_1, \theta_2\} = \begin{cases} a_1, & x < \theta_1, \ x \in (0,1), \\ b_1, & x \geq \theta_1, \ x \in (0,1), \\ a_2, & x < \theta_2, \ x \in (1,2), \\ b_2, & x \geq \theta_2, \ x \in (1,2), \end{cases}$$

see Figure 6.5.

In this model, there are essentially two independent subproblems, the estimation of Θ_1 and the estimation of Θ_2. Indeed,

$$\mathrm{I}(Y_x; \Theta_1, \Theta_2 \mid \mathbf{y}) = \begin{cases} \mathrm{I}(Y_x; \Theta_1 \mid \mathbf{y}_1), & x \in (0,1), \\ \mathrm{I}(Y_x; \Theta_2 \mid \mathbf{y}_2), & x \in (1,2), \end{cases}$$

where $\mathbf{y}_1$ denotes the trial results placed on $(0,1)$ and $\mathbf{y}_2$ those placed on $(1,2)$, and so we have two independent instances of the single-threshold problem solved in Section 6.2.1. As the optimal gain (6.6) for each sub-problem only depends on the values of a_1, b_1 or a_2, b_2, one of three things will happen: either the inequality

$$\max_{x \in (1,2)} \mathrm{I}(Y_x; \Theta_1, \Theta_2 \mid \mathbf{y}) > \max_{x \in (0,1)} \mathrm{I}(Y_x; \Theta_1, \Theta_2 \mid \mathbf{y})$$

always holds, it always holds in the reverse direction, or equality always holds. In the first mentioned case, if one uses the joint entropy $\mathrm{H}(\Theta_1, \Theta_2 \mid \mathbf{y})$

as the loss function, then only the second threshold will ever be estimated. This is intuitively not the desired result.

However, even though Paninski (2005) presents this example as a failure of the greedy information maximization strategy, that is not the true cause of the problem as the greedy strategy is in fact optimal for minimizing the specified loss function. Instead, the true problem is inappropriateness of the loss function.

To avoid the problem, one might instead use the loss function

$$\max\{H(\Theta_1 \mid \mathbf{y}), H(\Theta_2 \mid \mathbf{y})\}.$$

In that case, both thresholds would be estimated to the same accuracy, but the proportion of trials spent on the more difficult-to-estimate threshold would be larger. This is intuitively the desired result.

6.3. Random Cost of Observation

In this section, we generalize the preceding discussion to the situation where the observation of Y_x is associated with some random cost $C_x \geq 0$, which given the value of Y_x, is independent of Θ and the results and costs of any other observations (Kujala, 2010):

The technical requirement that C_x depends on Θ only through Y_x is satisfied in particular if C_x is a component of Y_x. Thus, it leads to no loss of generality if the incurred costs are observable.

Costs of observation have been previously considered in the context of decision theory by DeGroot (1970) in several examples and there have been practical applications for such models especially in clinical trial design (e.g., Müller, Berry, Grieve, & Krams, 2006). Also, Paninski (2005) mentions a model where a certain price $C(x_{t-1}, x_t)$ has to be paid for each change of the state of the "observational apparatus" from x_{t-1} to x_t. What is common to all these examples is that the proposed objective function is based on some *difference* of the expected gain and the cost. However, with such an objective, one has to equate the units of gain with the units of cost. Our approach avoids this problem by using a different kind of placement strategy based on a *ratio* of gain and cost instead.

6.3.1. *Obtaining the best value for money*

Intuitively, we would like to make measurement more cost-effective by taking into account the fact that different trial placements may be associated with different costs. Two operational definitions of this goal were given in Kujala (2010):

(1) maximizing the expected amount of information given by an experiment that terminates when the total cost overruns a certain predetermined budget (for example, the cost could be defined as the random time taken by each trial in an experiment, and one might wish to maximize the expected total information gain over as many trials as can be completed in 15 minutes);

(2) minimizing the expected cost of an experiment that terminates when a predetermined amount of information has been obtained (for example, one might wish to minimize the expected time required to measure an unobservable variable to a certain predetermined level of accuracy).

Both definitions are reasonable, but the first one turns out to be more elegant in that it corresponds to the plain expected information maximization goal with an adaptive termination rule as discussed in Section 6.1.6. Hence, the optimal strategy under that goal is insensitive to the parametrization used to define the differential entropy measure of information. In contrast, simple counterexamples show that the second definition does not have this desirable property.

Remark 6.3. In the statement of the problem, we do not require the cost C_x to be observable. However, for the experiment to actually terminate when the budget is overrun, either the actual costs must be observable, or alternatively, any further trials could simply fail after the budget is overrun. In the latter case, the adaptive termination rule would be random. In either case, the actual costs are irrelevant to the final Bayesian estimates of Θ as they are assumed to depend on Θ only through the fully observable results Y_x.

Remark 6.4. Obviously exact maximization of the expected information gain under a given budget is generally intractable for the same reasons that the usual constant number of trials case is. However, even if the information gains and costs associated with each Y_x were known time-invariant constants, the problem of fitting the best value in a given constant budget would still be intractable — it is equivalent to the knapsack problem which is NP hard (see, Garey & Johnson, 1979). The heuristic we use is in fact similar to the heuristics used to find approximate solutions to the knapsack problem although we have the additional complications of randomness and the generally intractable sequential changes.

6.3.2. Heuristics

Here we repeat the discussion of heuristics in Kujala (2010).

Let us first define random variables denoting the gain and cost of the t-th observation:

$$G_t := U_t - U_{t-1} = u(Y_{X_1}, \ldots, Y_{X_t}) - u(Y_{X_1}, \ldots, Y_{X_{t-1}}),$$
$$C_t := C_{X_t}.$$

Assuming for a moment that the cost C_x is defined as the time taken by the observation of Y_x, one might think that choosing x so as to maximize the expected *rate* of information gain

$$\mathrm{E}\left(\frac{G_t}{C_t} \;\middle|\; \mathbf{y},\; X_t = x\right) \tag{6.8}$$

over the *duration* C_t of the next observation would be a good heuristic. However, even though the expected rate of gain is optimal over the next trial, that is in general no longer true over repeated trials. Over a large *constant* unit of time, repeated i.i.d. observations of Y_x are expected to result in each outcome (y_x, c_x) being observed for a total cumulative duration proportional to $p(y_x, c_x \mid \mathbf{y})c_x$. Thus, to estimate the average rate of gain obtainable from i.i.d. observations of Y_x, the expectation should be taken over the distribution

$$\frac{p(y_x, c_x \mid \mathbf{y})c_x}{\iint p(y_x, c_x \mid \mathbf{y})c_x \, dy_x \, dc_x}$$

instead. This leads to the objective function

$$\iint \left[\frac{u(\mathbf{y}, y_x) - u(\mathbf{y})}{c_x}\right] \frac{p(y_x, c_x \mid \mathbf{y})c_x}{\iint p(y_x, c_x \mid \mathbf{y})c_x \, dy_x \, dc_x} dy_x \, dc_x$$
$$= \frac{\mathrm{E}(G_t \mid \mathbf{y},\; X_t = x)}{\mathrm{E}(C_t \mid \mathbf{y},\; X_t = x)} \tag{6.9}$$

proposed in (Kujala, Richardson, & Lyytinen, 2010a). With the entropy loss function (6.1), this can be written in the convenient form

$$\frac{\mathrm{I}(Y_x; \Theta \mid \mathbf{y})}{\mathrm{E}(C_x \mid \mathbf{y})}. \tag{6.10}$$

Remark 6.5. Unlike the situation in the pure information maximization case, maximization of (6.9) does not correspond to maximization of the immediate expected utility. Thus, it is not the prototypical greedy algorithm, but it is still myopic in the sense that it expects the future sets of possible expected gains and costs to be similar to those of the current trial. In our view, this algorithm is the most natural generalization of the ubiquitous one-step greedy strategy to the situation where the costs of observation vary.

Remark 6.6. For the problem to be well-defined, the expected cost of a trial should always be positive. However, in some practically interesting situations there may be a positive probability that the actual cost is zero. (This is another reason why (6.8) is an inappropriate formulation — it would always be infinite in this case.) In fact, as long as the expectation is positive, there is no reason why we should not allow negative costs, too.

6.3.3. *Conditions for asymptotic optimality*

The following theorem implies that maximization of (6.9) is asymptotically optimal in the sense that it gives the asymptotically best gain-to-cost ratio if the set of the pairs of marginal distributions of $[(G_t, C_t) \mid \mathbf{y}, X_t = x]$ over all possible values of x do not change as each new outcome is added to the data $\mathbf{y}$. That is, the same marginal distributions of gain and cost are allowed to be associated with different x at different times as long as these x's at different times are in a one-to-one correspondence.

Theorem 6.1. *Suppose that the random variables G and C have finite expectations $\mathrm{E}(G)$ and $\mathrm{E}(C) \neq 0$. Then, almost surely (i.e., with probability 1)*

$$\lim_{n \to \infty} \frac{G_1 + \cdots + G_n}{C_1 + \cdots + C_n} = \frac{\mathrm{E}(G)}{\mathrm{E}(C)},$$

where G_k and C_k denote i.i.d. copies of G and C, respectively.

Under certain side conditions, optimality of the strategy can also be shown under the weaker assumption that the objective function always has the same maximum value α regardless of the past data:

Theorem 6.2. *Suppose that there exists a constant $\alpha > 0$ such that*

$$\max_{x \in X_t} \frac{\mathrm{E}(G_t \mid \mathbf{y}, X_t = x)}{\mathrm{E}(C_t \mid \mathbf{y}, X_t = x)} = \alpha \qquad (6.11)$$

for all possible sets **y** *of past observations. If X_t is defined as the maximizer of* (6.11) *and if for some $\sigma^2 < \infty$ and $\varepsilon > 0$,*

$$
\begin{cases}
\mathrm{Var}(G_t \mid Y_{X_1}, \ldots, Y_{X_{t-1}}) \leq \sigma^2, \\
\mathrm{Var}(C_t \mid Y_{X_1}, \ldots, Y_{X_{t-1}}) \leq \sigma^2, \\
\mathrm{E}(C_t \mid Y_{X_1}, \ldots, Y_{X_{t-1}}) \geq \varepsilon
\end{cases}
\tag{6.12}
$$

for all t, then the gain-to-cost ratio satisfies

$$
\lim_{t \to \infty} \frac{G_1 + \cdots + G_t}{C_1 + \cdots + C_t} \overset{a.s.}{=} \alpha.
$$

This is asymptotically optimal in the sense that for any other strategy that satisfies (6.12), *we have*

$$
\limsup_{t \to \infty} \frac{G_1 + \cdots + G_t}{C_1 + \cdots + C_t} \overset{a.s.}{\leq} \alpha.
$$

The proofs of these theorems can be found in Kujala (2010).

Of course, these conditions for asymptotic optimality are simple (in contrast with the theory of Paninski, 2005) and cannot be expected to hold in all practical models. We present them here simply to formalize the intuitive justification for the method that in the long run, only the *expected* values matter and the random variations in the gains and costs get averaged out. In the next section, we give a simple example where these conditions do apply.

6.3.4. *The simplified psychometric model*

We shall extend the model of Section 6.2.1 with random costs C_x associated with observing Y_x. Recall that in this model, for all $x \in \mathbb{R}$, $Y_x \in \{0, 1\}$ is a dichotomous random variable defined by

$$
\Pr\{Y_x = 1 \mid \theta\} = \begin{cases} a, & x < \theta, \\ b, & x \geq \theta \end{cases}
$$

for some $0 < a < b < 1$. Let us then assume that the distribution of the random cost C_x is fully determined by the value of Y_x, i.e., it does not depend directly on x. Then,

$$
\mathrm{E}(C_x \mid \theta) = \begin{cases} c_a, & x < \theta, \\ c_b, & x \geq \theta \end{cases}
$$

for some c_a and c_b, which we assume to be positive to avoid pathological cases.

6.3.4.1. *Optimal strategy under varying costs*

The expected information gain, as before, is given by

$$\begin{aligned}
I(Y_x; \Theta \mid \mathbf{y}) &= H(Y_x \mid \mathbf{y}) - E[H(Y_x \mid \Theta, \mathbf{y})] \\
&= h[a + (b - a)\Pr\{\Theta \le x \mid \mathbf{y}\}] \\
&\quad - [h(a) + (h(b) - h(a))\Pr\{\Theta \le x \mid \mathbf{y}\}]
\end{aligned}$$

and so the objective function is

$$\frac{I(Y_x; \Theta \mid \mathbf{y})}{E(C_x \mid \mathbf{y})} = \frac{h(a + rz) - (h_a + h_r z)}{c_a + c_r z},$$

where we denote $z := \Pr\{\Theta \le x \mid \mathbf{y}\}$, $r := b - a$, $h_a := h(a)$, $h_r := h(b) - h(a)$, and $c_r := c_b - c_a$. Thus, the objective function depends on the prior and the placement x only through z, which can attain any value in $[0, 1]$ provided that the prior distribution is absolutely continuous w.r.t. the Lebesgue measure. The function is zero for $z \in \{0, 1\}$ and positive for $z \in (0, 1)$. Therefore, the optimum can be found at a critical point of the objective function.

Differentiating w.r.t. z yields

$$\frac{(h'(a + rz)r - h_r)(c_a + c_r z) - (h(a + rz) - h_a - h_r z)c_r}{(c_a + c_r z)^2}.$$

The denominator is finite and positive and the numerator equals

$$\begin{aligned}
&c_r h_a - c_a h_r + rh'(a + rz)(c_a + c_r z) - c_r h(a + rz) \\
&= c_r h_a - c_a h_r + r[\log(1 - a - rz) - \log(a + rz)](c_a + \cancel{c_r z}) \\
&\quad + c_r[(a + \cancel{rz})\log(a + rz) + (1 - a - \cancel{rz})\log(1 - a - rz)] \\
&= c_r h_a - c_a h_r + (c_r a - c_a r)\log\omega + (c_a r - c_r a + c_r)\log(1 - \omega) \\
&= c_b h(a) - c_a h(b) + (c_b a - c_a b)\log\omega + (c_b(1 - a) - c_a(1 - b))\log(1 - \omega),
\end{aligned}$$

where we denote $\omega := a + rz = \Pr\{Y_x = 1 \mid \mathbf{y}\}$. It is easy to verify by differentiating and simplifying that the above expression is monotone for ω between a and b and therefore has a unique zero corresponding to the maximum value of the objective function.

Placing the next trial at this optimum is by Theorem 6.2 the asymptotically optimal strategy as it yields the same maximum value of the objective function at every trial regardless of the past data (and it is easy to show that the other assumptions of the theorem hold, too).

6.3.4.2. *Specific examples*

Assuming now $C_x = 1$, i.e., constant cost, we have $c_a = c_b = 1$ and the optimality condition becomes

$$h(a) - h(b) + (a - b)\log\omega + (b - a)\log(1 - \omega) = 0,$$

whence

$$\log\frac{1 - \omega}{\omega} = \frac{h(b) - h(a)}{b - a},$$

which yields the optimizer

$$\omega^* = \frac{1}{1 + \exp\left(\dfrac{h(b) - h(a)}{b - a}\right)}.$$

This result we have already seen in Section 6.2.1, although here we have adopted ω instead of $z = (\omega - a)/(b - a)$ as the parameter.

If we assume instead that $C_x = [Y_x = 0]$, i.e., each 0-result costs one unit, we have $c_a = 1 - a$ and $c_b = 1 - b$, and the equation becomes

$$\begin{aligned}
0 &= (1 - b)h(a) - (1 - a)h(b) \\
&\quad + ((1 - b)a - (1 - a)b)\log\omega \\
&\quad + ((1 - b)(1 - a) - (1 - a)(1 - b))\log(1 - \omega) \\
&= (1 - b)h(a) - (1 - a)h(b) + (a - b)\log\omega,
\end{aligned}$$

which yields the optimizer

$$\omega^* = \exp\left(-\frac{(1 - a)h(b) - (1 - b)h(a)}{b - a}\right).$$

This definition of cost was in fact used by Kujala et al. (2010a) in a child-friendly measurement formulation, which assumes a cost on each failure of a child due the fact that failures can lower motivation. If a child can only tolerate a certain number of failures, then this formulation should yield the maximum amount of information before that limit is reached.

6.3.4.3. *Non-adaptive and batch strategies*

The discussion of Section 6.2.1.2 generalizes directly. Denoting

$$I(Y_{x_1}, \ldots, Y_{x_n}; \Theta) = f(z) = -(Mz + v) \cdot \log(Mz + v) - u \cdot z - c$$

and

$$E(C_x) = g(z) = \sum_{k=1}^{n}(c_a + z_k c_r),$$

we have

$$\nabla\left(\frac{f}{g}\right)(z) = \frac{g(z)\nabla f(z) - f(z)\nabla g(z)}{g(z)^2}$$

and

$$H\left(\frac{f}{g}\right)(z) = \left[(\nabla g(z)\nabla f(z)^T + g(z)Hf(z) - \nabla f(z)\nabla g(z)^T - f(z)Hg(z))g(z)^2\right.$$

$$\left. - (g(z)\nabla f(z) - f(z)\nabla g(z))2g(z)\nabla g(z)^T\right]\frac{1}{g(z)^4}$$

$$= \frac{Hf(z)}{g(z)} - \frac{\nabla g(z)}{g(z)}\frac{\nabla f(z)^T}{g(z)} - \frac{\nabla f(z)}{g(z)}\frac{\nabla g(z)^T}{g(z)} + 2\frac{f(z)}{g(z)}\frac{\nabla g(z)}{g(z)}\frac{\nabla g(z)^T}{g(z)},$$

where $\nabla g(z) = (c_r, \ldots, c_r)$ and $Hg(z) = 0$. Although it is not obvious from the above expression, the Hessian of the objective function f/g was in practice always negative definite and so Newton's iteration worked fine here, too. However, we have no formal proof that the found optimum is in fact the global optimum when $n > 1$.

We do not go into any details of the single placement strategy $x_1 = \cdots = x_n$ here either.

6.3.4.4. *Comparison*

Figure 6.6 illustrates the optimal placements under the three strategies when each 0-result costs one unit.

Comparing the optimal adaptive placements in Figures 6.3 and 6.6 supports the intuitive characterization given in (Kujala et al., 2010a): while pure information maximization works much like binary search, roughly bisecting the uncertainty distribution at each step, the cost-aware variation instead chooses the placement at a certain higher percentile closer to the easier end.

The exact percentile of the optimal placement seems to depend mostly on the lapsing rate $1 - b$ and less on the guessing rate a. This dependence was to be expected: if there is going to be a large probability of careless mistakes anyway, then it does not pay off to make the trials very easy, and conversely, if careless mistakes are unlikely, then the easiest trials will be virtually free and the placements close to that end will yield the best value for money even though the gains over one trial are smaller.

The placements of the non-adaptive strategy are no longer close to evenly spaced. Instead, they cluster near the easier end.

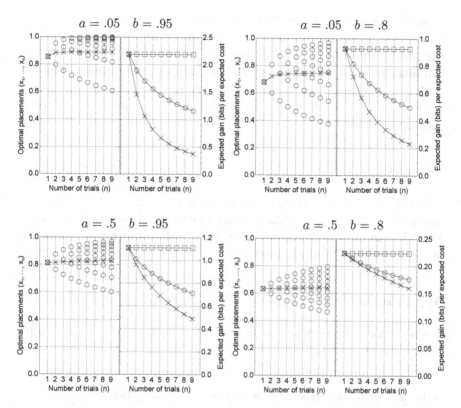

Fig. 6.6. Optimal placements and the corresponding optimal expected gain per expected cost ratios for the optimal adaptive strategy (squares), optimal non-adaptive strategy (circles), and optimal single-placement repeated observations strategy (crosses) under random cost of observation defined as $C_x = [Y_x = 0]$ (i.e., each 0-result costs one unit) in the model of Figures 6.2 and 6.3. The interpretation of the placement values is the same is in Figure 6.3.

6.3.5. *Discussion*

So far, we have only considered discrete cost variables explicitly. An obvious topic for future work is the calculation of the objective function (6.10) for some response time model with the cost C_x defined as the response time. In any experiments where the placement of a trial can affect its duration, this formulation can increase the efficiency per time unit over the pure information maximization greedy algorithm. In particular, in an n-choice task, the response times generally increase with n, and so a smaller value of n might turn out to be optimal even though it yields a higher guessing rate. However, the resulting time-efficiency also depends on any pre-stimulus delays, which should be included in the cost.

The cost-aware formulation may also be useful in adaptive memory retention experiments (see, e.g., Cavagnaro et al., 2010), where the delay after which the recall of a stimulus is tested is varied. Another topic for future work is generalization of the cost-aware strategy for a multi-step "lookahead" (see, e.g., King-Smith et al., 1994) to bring it closer to the globally optimal strategy.

Acknowledgments

This chapter is essentially an extended version of Kujala (2010). The author is grateful to Benja Fallenstein, Antti Penttinen, Rauli Ruohonen, and Matti Vihola for comments and discussions. This research was supported by the Academy of Finland grant 121855.

Appendix A. Auxiliary results

In this section, we present some lemmas and auxiliary results that were used in the main text.

A.1. Trade-off Between Guessing and Lapsing Rates

Often one can affect the guessing and lapsing rates by the design of the experiment. Obviously if both can be decreased, then the informativity of the experiment should increase. But what if the guessing rate can only be decreased at the cost of a higher lapsing rate, or vice versa? In this section, we give a definite answer under Lindley's (Lindley, 1956) method of comparing experiments, that is, we determine when one experiment is more informative than the other regardless of the prior information.

Lemma 6.1. *Let X be a real-valued random variable and let the functions $f, g : [a, b] \to \mathbb{R}$ be continuous and twice continuously differentiable on (a, b). If $\Pr\{X \in [a, b]\} = 1$ and $f''(x) \le g''(x)$ for all $x \in (a, b)$, then*

$$\mathrm{E}[f(X)] - f(\mathrm{E}[X]) \le \mathrm{E}[g(X)] - g(\mathrm{E}[X]).$$

The analogous result holds for "$<$" provided that X is not concentrated on a point.

Proof. Suppose $x_0 := \mathrm{E}[X] \in (a, b)$ (otherwise X is concentrated on either a or b and the result is trivial). Then, $f''(x) \leq g''(x)$ implies

$$f'(x) - f'(x_0) = \int_{x_0}^x f'' \leq \int_{x_0}^x g'' = g'(x) - g'(x_0)$$

for $x \in (x_0, b)$ and the inequality is reversed for $x \in (a, x_0)$. Integrating both sides again yields

$$\int_{x_0}^x [f'(t) - f'(x_0)]dt \leq \int_{x_0}^x [g'(t) - g'(x_0)]dt$$

for all $x \in (a, b)$, which implies

$$f(x) - f(x_0) - (x - x_0)f'(x_0) \leq g(x) - g(x_0) - (x - x_0)g'(x_0)$$

for all $x \in [a, b]$ (the inequality extends to the endpoints as every term is continuous in x). Substituting the random variable X for x and taking the expectation of both sides, we obtain

$$\mathrm{E}[f(X)] - f(x_0) - (\mathrm{E}[X] - x_0)f'(x_0) \leq \mathrm{E}[g(X)] - g(x_0) - (\mathrm{E}[X] - x_0)g'(x_0),$$

which implies the statement. The "$<$" version is a simple modification. $\square$

Theorem 6.3. *Let $Y_{a,b}$ be a dichotomous random variable depending on another random variable $\Theta \in [0, 1]$ through the conditional distribution*

$$\mathrm{Pr}\{Y_{a,b} = 1 \mid \theta\} = a + (b - a)\theta,$$

where $a, b \in [0, 1]$ and $a \neq b$. If both of the inequalities

$$\frac{a(1 - a)}{(b - a)^2} \leq \frac{a'(1 - a')}{(b' - a')^2}, \tag{A.1}$$

$$\frac{b(1 - b)}{(b - a)^2} \leq \frac{b'(1 - b')}{(b' - a')^2}, \tag{A.2}$$

are satisfied, then $\mathrm{I}(Y_{a',b'}; \Theta) \leq \mathrm{I}(Y_{a,b}; \Theta)$ regardless of the distribution of Θ. If, in addition, the inequality in (A.1) or (A.2) is strict, then $\mathrm{I}(Y_{a',b'}; \Theta) < \mathrm{I}(Y_{a,b}; \Theta)$ provided that Θ is not concentrated on a single point. If (A.1) holds with $<$ and (A.2) holds with $>$ (or vice versa), then, depending on the distribution of Θ, either one of $\mathrm{I}(Y_{a,b}; \Theta)$ and $\mathrm{I}(Y_{a',b'}; \Theta)$ can be strictly larger than the other.

Proof. The mutual information can be written as

$$
\begin{aligned}
\mathrm{I}(Y_{a,b}; \Theta) &= \mathrm{H}(Y_{a,b}) - \mathrm{E}[\mathrm{H}(Y_{a,b} \mid \Theta)] \\
&= h(a + (b - a)\,\mathrm{E}[\Theta]) - \mathrm{E}[h(a + (b - a)\Theta)] \\
&= h_{a,b}(\mathrm{E}[\Theta]) - \mathrm{E}[h_{a,b}(\Theta)],
\end{aligned}
$$

where

$$h(x) = -x \log x - (1 - x) \log(1 - x),$$
$$h_{a,b}(x) = h(a + (b - a)x).$$

As $h''(x) = -1/(x(1 - x))$, we obtain the expression

$$h''_{a,b}(x) = h''(a + (b - a)x)(b - a)^2$$
$$= -\frac{(b - a)^2}{[a + (b - a)x][1 - a - (b - a)x]},$$

which is continuous on $x \in (0, 1)$. Thus, Lemma 6.1 implies that the conclusion $I(Y_{a',b'}; \Theta) \leq I(Y_{a,b}; \Theta)$ follows if $h''_{a,b}(x) \leq h''_{a',b'}(x)$ for all $x \in (0, 1)$. But this inequality is equivalent to

$$\frac{[a + (b - a)x][1 - a - (b - a)x]}{(b - a)^2} \leq \frac{[a' + (b' - a')x][1 - a' - (b' - a')x]}{(b' - a')^2}$$

which is linear (the second order terms cancel) and therefore holds for all $x \in (0, 1)$ if and only if it holds at the end points $x \in \{0, 1\}$. This is precisely the condition given by (A.1) and (A.2) in the statement of the theorem. Furthermore, if the linear inequality is strict at either end point, then it is strict at every $x \in (0, 1)$ and Lemma 6.1 implies $I(Y_{a',b'}; \Theta) < I(Y_{a,b}; \Theta)$.

Finally, if $h''_{a,b}(0) < h''_{a',b'}(0)$ and $h''_{a,b}(1) > h''_{a',b'}(1)$ (or vice versa), then one can apply Lemma 6.1 to nonsingular distributions supported on sufficiently small ranges $[0, \varepsilon]$ and $[1 - \varepsilon, 1]$ to show that both $I(Y_{a,b}; \Theta) > I(Y_{a',b'}; \Theta)$ and $I(Y_{a,b}; \Theta) < I(Y_{a',b'}; \Theta)$ are possible. □

This general theorem gives immediately the following intuitively true corollary:

Corollary 6.1. *If $a' \neq b'$ and $[a', b'] \subset [a, b] \subset [0, 1]$, then $I(Y_{a,b}; \Theta) \geq I(Y_{a',b'}; \Theta)$.*

Proof. Given $b \in [b', 1]$, the inequality

$$\frac{a(1 - a)}{(b - a)^2} \leq \frac{a'(1 - a')}{(b' - a')^2}$$

obviously holds at $a = a'$. Differentiating the left side w.r.t. a yields

$$\frac{(1 - 2a)(b - a)^2 - a(1 - a)(-2)(b - a)}{(b - a)^4} = \frac{(b - a) + 2a(1 - b)}{(b - a)^3} \geq 0$$

and thus extends the inequality for all $a \in [0, a']$. The inequality (A.2) holds analogously. □

Using the general result, we can now prove the statement made in Example 6.2 on p. 137 in the main text. In this example, the result Y_n is distributed as

$$\Pr\{Y_n = 1 \mid \theta, m\} = \begin{cases} \theta + (1-\theta)\frac{1}{n}, & n \le m, \\ \frac{1}{n}, & n > m, \end{cases}$$

where $\Theta \in [0,1]$ is the unobservable variable we wish to estimate and $M \ge 2$ is an unobservable nuisance variable. Choosing the placement n can be seen as a trade-off between guessing and lapsing rates. Thus, using Theorem 6.3, we can find conditions under which $\mathrm{I}(Y_2; \Theta) > \mathrm{I}(Y_n; \Theta)$ regardless of the prior of Θ. Assuming that Θ and M are independent, we have

$$\Pr\{Y_n = 1 \mid \theta\} = \Pr\{n \le M\}\left[\theta + (1-\theta)\frac{1}{n}\right] + \Pr\{n > M\}\frac{1}{n}$$

$$= \frac{1}{n} + \Pr\{n \le M\}\frac{n-1}{n}\theta.$$

Using the notation of Theorem 6.3, we see that $[Y_n \mid \theta] \sim [Y_{a,b} \mid \theta]$ for $a = 1/n$ and $b - a = \Pr\{n \le M\}(n-1)/n$, and so $\mathrm{I}(Y_2; \Theta) > \mathrm{I}(Y_n; \Theta)$ if

$$1 = \frac{\frac{1}{2}(1 - \frac{1}{2})}{(1 - \frac{1}{2})^2} < \frac{\frac{1}{n}(1 - \frac{1}{n})}{\left(\Pr\{n \le M\}\frac{n-1}{n}\right)^2}$$

and

$$0 = \frac{1(1-1)}{(1 - \frac{1}{2})^2} \le \frac{[\ldots](1 - [\ldots])}{\left(\Pr\{n \le M\}\frac{n-1}{n}\right)^2}$$

both hold, which is equivalent to

$$\Pr\{n \le M\} < \frac{1}{\sqrt{n-1}}.$$

This is satisfied for all $n \ge 2$ in particular if we define the distribution of M by

$$p(m) \propto \begin{cases} 1/m, & m = 2, \ldots, 7, \\ 0, & \text{otherwise} \end{cases}$$

as in the example.

Appendix B. Technicalities

In the main text, we have avoided measure-theoretic technicalities and implicitly assumed certain regularity conditions. In this section we make these assumptions explicit. The same conditions must have been implicitly assumed in other works, too, as without them the mutual information $I(Y_x; \Theta)$ does not make sense and even Bayes's theorem does not generally hold. The material in this section is based on (Kujala, 2011). Proofs and further details can be found there.

B.1. Preliminaries

Let $(\Omega, \mathcal{F}, \Pr)$ be a probability space. A *random variable* is a measurable mapping $X : \Omega \to \mathsf{X}$ to some measurable space $(\mathsf{X}, \mathcal{X})$ (usually the real line $\mathbb{R}$ equipped with the Borel σ-algebra $\mathcal{B}(\mathbb{R})$). The *distribution* of the random variable is the measure $P_X(S) = \Pr(X^{-1}(S))$ induced on $\mathcal{X}$. If a transition measure $P_{X|Y} : \mathsf{Y} \times \mathcal{X} \to [0, \infty]$ satisfies $P_{X,Y}(S) = (P_{X|Y} \times P_Y)(S) := \int P_{X|Y}(y, S_y) dP_Y(y)$ for all $S \in \mathcal{X} \times \mathcal{Y}$, where we denote $S_y := \{ x : (x, y) \in S \}$, then it is called a *conditional distribution* of X given Y, or a *regular conditional probability* as Shiryaev (1996) calls it. In the following, we will use the shorthand $P_{X|y} := P_{X|Y}(y, \cdot)$.

If $P_X(S) = \int_S p_X(x) d\mu(x)$ for all $S \in \mathcal{X}$ for some measurable function $p_X : \mathsf{X} \to [0, \infty]$ and some measure $\mu : \mathcal{X} \to [0, \infty]$, then p_X is called a *density* of X w.r.t. μ. For brevity, we leave out the subscript of the density when it matches the arguments, i.e, instead of $p_X(x)$, we write simply $p(x)$. If a pair of random variables $(X, Y) : \Omega \to \mathsf{X} \times \mathsf{Y}$ has a joint density $p(x, y)$ w.r.t. the product $\mu \times \nu$ of arbitrary measures μ and ν (see, e.g., Mukherjea, 1972 for the definition), then Fubini's theorem implies that $p(x) = \int p(x, y) d\nu(y)$ and $p(y) = \int p(x, y) d\mu(x)$ are marginal densities w.r.t. μ and ν, respectively. If a conditional distribution $P_{X|Y}$ exists and satisfies $P_{X|y}(S) = \int_S p(x \mid y) d\mu(x)$ for all $S \in \mathcal{X}$, $y \in \mathsf{Y}$ for some measurable nonnegative function $(x, y) \mapsto p(x \mid y)$ and some measure μ, then $p(x \mid y)$ is called a *conditional density* of X given y. If a joint density $p(x, y)$ exists w.r.t. $\mu \times \nu$, then a conditional density can always be obtained as $p(x \mid y) := p(x, y)/p(y)$. (The value chosen for $p(y) = 0$ is immaterial as the conditional density is only determined $\mu \times P_Y$-a.e.)

B.2. Regularity Conditions for Bayesian Estimation

The following theorem gives a set of equivalent conditions under which we can avoid the potential problems of nonexistent distributions or densities.

Theorem 6.4. *Let $(X, Y) : \Omega \to \mathsf{X} \times \mathsf{Y}$ be a pair of random variables. Then, the following are equivalent:*

(1) X and Y have a joint density w.r.t. a product measure $\mu \times \nu$,
(2) $P_{X,Y} \ll P_X \times P_Y$,
(3) X has a conditional density $p(x \mid y)$ w.r.t. a σ-finite measure μ,
(4) X has a conditional distribution $P_{X|Y}$ such that $P_{X|y} \ll P_X$ for all y,
(5) X has a conditional distribution $P_{X|Y}$ and a marginal density $p(x)$ w.r.t. a measure μ such that $P_{X|y} \ll \mu$ for all y.

Obviously the same conditions with the roles of X and Y reversed are also equivalent. Furthermore,

(6) if the above conditions hold for X and Y, then they also hold for $X' = F(X)$ and $Y' = G(Y)$ where $F : \mathsf{X} \to \mathsf{X}'$ and $G : \mathsf{Y} \to \mathsf{Y}'$ are any measurable functions.

The conditions of Theorem 6.4 are precisely those under which Bayes's formula can be applied to a conditional density:

Theorem 6.5. *Let $(X, Y) : \Omega \to \mathsf{X} \times \mathsf{Y}$ be a pair of random variables and suppose that $p(y \mid x)$ is a conditional density of Y given X w.r.t. a measure ν. Then the following are equivalent:*

(a) X and Y satisfy the conditions of Theorem 6.4.
(b) If $p(x)$ is a marginal density of X w.r.t. a measure μ, then

$$p(x \mid y) := \frac{p(y \mid x)p(x)}{\int p(y \mid x)p(x)d\mu(x)}$$

is a conditional density of X given Y w.r.t. μ.

Remark 6.7. The conditions of Theorem 6.4 are also precisely those under which the Radon-Nikodým derivative in the measure-theoretic definition of mutual information

$$\mathrm{I}(X; Y) = \int dP_{X,Y} \log \frac{dP_{X,Y}}{d(P_X \times P_Y)}$$

exists (condition 2). In case $P_{X,Y}$ is singular w.r.t. $P_X \times P_Y$, Kolmogorov (1956) defines $\mathrm{I}(X; Y) = \infty$. Therefore, even though one might be able to work with conditional *distributions* directly, if one's utility function is the information gain, then Y_x and Θ in the main text must still satisfy these conditions for all x for the problem to be well-defined. Hence, there is no loss of generality in our formulations based on densities.

The conditions of Theorem 6.4 are mild, being satisfied whenever either X or Y is discrete as well as in most practical situations with continuous random variables. However, they preclude in particular the following example:

Example 6.5. Suppose that $X = Y \sim$ Uniform$[0, 1]$. The conditional distribution $P_{X|y}(S) = [y \in S]$ is singular w.r.t. $P_X = m_{[0,1]}$, where $m_{[0,1]}$ denotes the restriction of the Lebesgue measure to $[0, 1]$, and so condition 4 of Theorem 6.4 is not satisfied. The conditional density

$$p(x \mid y) = [x = y] := \begin{cases} 1, & x = y, \\ 0, & x \neq y \end{cases}$$

exists w.r.t. the counting measure, but this measure is not σ-finite and so this density does not satisfy condition 3. Even though the joint distribution can be written as

$$P_{X,Y}(S) = \int \left[\int_{S_y} [x = y] d\#(x) \right] dm_{[0,1]}(y),$$

where $\#$ is the counting measure, the integrand $[x = y]$ does not yield the joint density of condition 1 because the function $[x = y]$ is not integrable w.r.t. $\# \times m_{[0,1]}$ and so Fubini's theorem does not hold for the iterated integral.

B.3. Adaptive Sequential Estimation

If Y_x and Θ satisfy the conditions of Theorem 6.4 for all x, then one can apply Bayes's formula to any *finite* set $\mathbf{y} = \{y_{x_t}\}_{t=1}^T$ of results sequentially:

$$p(\theta \mid \mathbf{y}) \propto p(\theta)p(y_{x_1} \mid \theta) \cdots p(y_{x_T} \mid \theta).$$

This implies that $P_{\Theta|\mathbf{y}} \ll P_\Theta$ for all $\mathbf{y}$ (condition 4) and as this condition makes no reference to the distribution of $\mathbf{y}$, it follows that regardless of the decision function d, the whole-experiment outcome variable Y_d has a joint density with Θ provided that the experiment terminates with probability one (so that $\mathbf{y}$ is almost surely finite). Thus, the expected information gain $I(\Theta; Y_d)$ for the whole experiment is formally well-defined (although its value may still come out as ∞). However, if there is a positive probability that the experiment does not terminate, then it is possible that no joint density of Θ and Y_d exists, even for constant placements:

Example 6.6. Suppose that $X \sim \text{Uniform}[0, 1]$ and the random variables $Y_t \in \{0, 1\}$ for $t = 1, 2, \ldots$ are defined as a binary representation of X. Then, although the conditional density $p(x \mid y_1, \ldots, y_T)$ w.r.t. the Lebesgue measure is well-defined for any finite set of observations, the full sequence of results $Y := \{Y_t\}_{t=1}^{\infty}$ cannot have any joint density with X, because by condition 6 of Theorem 6.4, that would imply that also the transformed variable

$$Y' := F(Y) := \sum_{t=1}^{\infty} 2^{-t} Y_t$$

would have a joint density with $X = Y'$, which contradicts the negative result of Example 6.5.

References

Amzal, B., Bois, F. Y., Parent, E., & Robert, C. P. (2006). Bayesian optimal design via interacting particle systems. *Journal of the American Statistical Association, 101*, 773–785.

Brockwell, A., & Kadane, J. (2003). A gridding method for Bayesian sequential decision problems. *Journal of Computational & Graphical Statistics, 12*, 566–584.

Cavagnaro, D. R., Myung, J. I., Pitt, M. A., & Kujala, J. V. (2010). Adaptive design optimization: A mutual information-based approach to model discrimination in cognitive science. *Neural Computation, 22*, 887–905.

Cover, T. M., & Thomas, J. A. (2006). *Elements of Information Theory* (2nd ed.). NY: John Wiley & Sons.

DeGroot, M. H. (1970). *Optimal Statistical Decisions*. McGraw-Hill.

Garey, M. R., & Johnson, D. S. (1979). *Computers and Intractability: A Guide to the Theory of NP-Completeness*. W. H. Freeman.

King-Smith, P. E. (1984). Efficient threshold estimates from yes-no procedures using few (about 10) trials (Abstract). *American Journal of Optometry and Physiological Optics, 61*, 119P.

King-Smith, P. E., Grigsby, S. S., Vingrys, A. J., Benes, S. C., & Supowit, A. (1994). Efficient and unbiased modifications of the QUEST threshold method: Theory, simulations, experimental evaluation and practical implementation. *Vision Research, 34*, 885–912.

King-Smith, P. E., & Rose, D. (1997). Principles of an adaptive method for measuring the slope of the psychometric function. *Vision Research, 37*, 1595–1604.

Kolmogorov, A. N. (1956). On the Shannon theory of information transmission in the case of continuous signals. *IEEE Transactions on Information Theory, 2*, 102–108.

Kontsevich, L. L., & Tyler, C. W. (1999). Bayesian adaptive estimation of psychometric slope and threshold. *Vision Research, 39*, 2729–2737.

Kujala, J. V. (2004). On computation in statistical models with a psychophysical application (Doctoral dissertation). *Jyväskylä Studies in Computing, 46*. University of Jyväskylä.

Kujala, J. V. (2010). Obtaining the best value for money in adaptive sequential estimation. *Journal of Mathematical Psychology, 54*, 475–480.

Kujala, J. V. (2011). A remark on the assumptions of Bayes' theorem. *arXiv:1103.6136v1*.

Kujala, J. V., & Lukka, T. J. (2006). Bayesian adaptive estimation: The next dimension. *Journal of Mathematical Psychology, 50*, 369–389.

Kujala, J. V., Richardson, U., & Lyytinen, H. (2010a). A Bayesian-optimal principle for learner-friendly adaptation in learning games. *Journal of Mathematical Psychology, 54*, 247–255.

Kujala, J. V., Richardson, U., & Lyytinen, H. (2010b). Estimation and visualization of confusability matrices from adaptive measurement data. *Journal of Mathematical Psychology, 54*, 196–207.

Lesmes, L. A., Jeon, S. T., Lu, Z. L., & Dosher, B. A. (2006). Bayesian adaptive estimation of threshold versus contrast external noise functions: The quick TvC method. *Vision Research, 46*, 3160–3176.

Lindley, D. V. (1956). On a measure of the information provided by an experiment. *The Annals of Mathematical Statistics, 27*, 986–1005.

Mukherjea, A. (1972). A remark on Tonelli's theorem on integration in product spaces. *Pacific Journal of Mathematics, 42*, 177–185.

Müller, P., Berry, D. A., Grieve, A. P., & Krams, M. (2006). A Bayesian decision-theoretic dose-finding trial. *Decision Analysis, 3*, 197–207.

Müller, P., Sansó, B., & Iorio, M. D. (2004). Optimal Bayesian design by inhomogeneous Markov chain simulation. *Journal of the American Statistical Association, 99*, 788–798.

Myung, J. I., & Pitt, M. A. (2009). Optimal experimental design for model discrimination. *Psychological Review, 116*, 499–518.

Paninski, L. (2005). Asymptotic theory of information-theoretic experimental design. *Neural Computation, 17*, 1480–1507.

Shannon, C. E. (1948). A mathematical theory of communication. *The Bell System Technical Journal, 27*, 379–423, 623–656.

Shiryaev, A. (1996). *Probability* (2nd ed.). Springer.

Snoeren, P. R., & Puts, M. J. H. (1997). Multiple parameter estimation in an adaptive psychometric method: MUEST, an extension of the QUEST method. *Journal of Mathematical Psychology, 41*, 431–439.

Tanner, T. (2008). Generalized adaptive procedure for psychometric measurement. *Perception, 37 ECVP Abstract Supplement, 93*.

Verdinelli, I., & Kadane, J. B. (1992). Bayesian designs for maximizing information and outcome. *Journal of the American Statistical Association, 87*, 510–515.

Watson, A. B., & Pelli, D. G. (1983). QUEST: A Bayesian adaptive psychometric method. *Perception & Psychophysics, 33*, 113–120.

Chapter 7

Probabilistic Lattices: Theory with an Application to Decision Theory

Louis Narens

University of California Irvine

Standard probability theory is based on event spaces that are Boolean algebras of sets. Its main alternative in science—quantum probability theory—is based on non-Boolean event spaces consisting of closed subspaces of a Hilbert space. Quantum probability theory differs from standard probability theory in the following two ways: (i) Its probability functions are not additive over all pairs of disjoint elements of its event algebra; and (ii) it need not satisfy an equivalent of the distributivity property of Boolean algebras of sets, that is, it need not satisfy an equivalent of $A\cap(B\cup C) = (A\cap B)\cup(A\cap C)$. This chapter focuses on situations where, like in standard probability theory, probability functions are additive over all pairs of disjoint elements belonging to their event algebras. It is shown that this and technical considerations involving elements having probability 0 result in the algebras being greatly constrained, because they must (1) satisfy a property called "modularity," which is a generalization of the distributivity property, and (2) unless they are Boolean algebras, they cannot be uniquely complemented. An example of a non-Boolean probability theory with an algebra satisfying a generalization of unique complementation is used to formulate a new decision theory where the decision maker can change "modes", e.g., be in various emotional states while making a decision. The example's algebra is shown to be distributive. Its generalization of complementation—called "pseudo-complementation"—has the properties of the negation operation of a well-studied alternative to classical logic known in the literature as "intuitionistic logic." The example uses pseudo-complementation to allow for multiple interpretations of an event as the decision maker transitions through modes. A particular model of decision making is presented that (i) allows for modes, (ii) has an equation for utility that is very similar to the SEU decision model used throughout science, and (iii) is arguably normative for human decision makers, given that humans are emotional creatures.

7.1. Introduction and Overview

This chapter concerns alternative probability theories based on event spaces that are not Boolean algebras. Its emphasis is on understanding the limits of natural generalizations of standard probability theory and the application of one of these generalizations to rationality in human decisions.

The major alternative to classical probability theory in the literature is known collectively as "quantum logic." It is a collection of algebraic generalizations of J. von Neumann's approach in the 1930's to the foundations of quantum physics. In recent years, quantum logic has had increasing scientific application outside of physics.

In addition to quantum logic, an abundance of probability theories and logics have been produced, especially by philosophers. Most of these are not applicable in science, because they lack the kinds of algebraic structures to be useful in the manipulation of probabilistic quantities. One, however, appears to me to have scientific application. It began as the logic implicit in Brouwer's philosophy of mathematics. In this chapter, its propositional form is called a *Heyting lattice*. Heyting lattices have interpretations distinct from their use in Brouwer's philosophy. For example, Kolmogorov (1932) used Heyting lattices as a logic for describing mathematical constructions; Gödel (1933) employed them as a basis for modal logics that are useful for understanding proof theory of mathematical logic; Narens (2005) employed a variant of them as a basis for propositions that are either verifiable or refutable (i.e., *scientific propositions*); Narens (2007, 2009) employed a variant of them as a basis for formulating a concept "incompleteness" or "ambiguity" that people theoretically take into account in making probability judgments; and Narens (2011) and this chapter employ variants of Heyting lattices for formulating a version of "rationality" that take into account emotion and forms of psychological biases while making decisions.

Quantum and intuitionistic logics are generalizations of classical logic that allow for rich probabilistic calculi. This makes them obvious candidates for scientific application in situations requiring non-Boolean event spaces. Mathematical results of this chapter show that it is difficult to go beyond these logics—or minor variations of them—while at the same time retaining the basic algebraic properties of probability functions.

The organization of the chapter is as follows: First lattices are described. These are algebraic structures that are abstract algebraic generalizations of logics. A particular kind of lattice—called a *Boolean lattice*—corresponds to classical propositional logic and Boolean algebras of sets.

Next probability functions on lattices are described. Unlike Boolean lattices that have rich supplies of disjoint events, a general lattice need not have many "disjoint elements". This makes the usual definition of "probability function" unworkable, because its algebraic effectiveness relies on a rich supply of disjoint events. In this chapter, the usual definition of probability is generalized so that it applies to general lattices while at the same time being logically equivalent to the usual definition of probability function on Boolean lattices.

For various theoretical and technical reasons, the notion of probability function needs to be generalized a little further. This is because some lattices algebraically distinguish among some sets of probability 0, and, for various theoretical purposes, it is important to maintain this distinction for probability functions. This is done through the introduction of the notion of "positive infinitesimal probability" and by making appropriate modifications to the real number system and the definition of "probability".

The existence of a probability function adds a restriction to a lattice and therefore to a logic. This chapter investigates this restriction. The conclusion is that for lattices with a reasonable concept of "complementation" ("negation" in logic), the restriction is very constraining. There are essentially three kinds of "complemented" lattices with useful probability theories: (*i*) Boolean lattices where each element has a unique complement; (*ii*) lattices that share many properties with "quantum logics"; and (*iii*) lattices that have a particular kind of partial complement. This partial complement in lattice theory is called a *pseudo-complement* and in logic it is called *intuitionistic negation*. Lattices with pseudo-complements are a central focus of this chapter and they are shown to have a simple topological interpretation.

Their topological interpretation is used to formulate new concepts for decision theory. These lattice-topologically formulated concepts are then used to provide for a theory of rationality that is based on subjective experience instead of objective observation. This theory is then used to extend the rational Subjective Expected Utility (SEU) model of decision making to incorporate emotion and other psychological biases into decision making.

7.2. Lattices

This section provides some of the basic concepts and results of lattice theory used throughout this chapter.

Lattices are simple algebraic structures that apply to many mathematical situations. They are used in this chapter (i) to describe the minimal logical structure inherent in classical and non-classical propositional logics, and (ii) to capture a natural generalization of the probability calculus to non-classical propositional logics and event spaces that are not Boolean algebras.

Definition 7.1 (lattice). Let $\mathfrak{L} = \langle \mathcal{L}, \sqcup, \sqcap, 1, 0 \rangle$ be a *lattice algebra*, or just *lattice* for short, if and only if $\mathcal{L}$ is a nonempty set, $\sqcup$, called the *join operation*, and $\sqcap$, called the *meet operation*, are binary operations on $\mathcal{L}$, 1, called the *unit element* of $\mathcal{L}$, and 0, called the *zero element* of $\mathcal{L}$, are elements of $\mathcal{L}$, and the following eight conditions hold for all a, b, and c in $\mathcal{L}$:

(1) $1 \sqcap a = a$ and $1 \sqcup a = 1$.
(2) $0 \sqcup a = a$ and $0 \sqcap a = 0$.
(3) $a \sqcup (b \sqcup c) = (a \sqcup b) \sqcup c$.
(4) $a \sqcup b = b \sqcup a$.
(5) $a \sqcup (a \sqcap b) = a$.
(6) $a \sqcap (b \sqcap c) = (a \sqcap b) \sqcap c$.
(7) $a \sqcap b = b \sqcap a$.
(8) $a \sqcap (a \sqcup b) = a$.

In interpreting lattices, 1 plays the role of the sure event or "True", and 0 plays the role of the null event or "False". In general, as elements of a lattice, they should not be interpreted as the real numbers 1 and 0.

Birkhoff and von Neumann in their seminal article, *The Logic of Quantum Mechanics,* (Birkhoff & von Neumann, 1936) commented the following about the use of general lattices for generalizing classical logic:

> In any lattice $\mathfrak{L}$, [the identities in Conditions 3 to 8 in Definition 7.1] are true ... .
>
> Clearly [these identities] are well-known formal properties of *and* and *or* in ordinary logic. This gives an algebraic reason for admitting as a *postulate* (if necessary) the statement that a given calculus of propositions is a lattice. (p. 829)

Definition 7.2 (lattice $\leq$). Throughout this chapter, $\leq$, in addition to its ordinary use as a total ordering on number systems, will denote the following relation on a lattice: For all a and b in the domain of the lattice,

$$a \leq b \quad \text{iff} \quad a \sqcap b = a.$$

It is easy to show that $\leq$ is a partial ordering, that is, is a transitive and symmetric relation, and that for all x and y in the lattice, $x \sqcup y$ is the $\leq$-least upper bound (or sup) in the lattice of x and y, and $x \sqcap y$ is the $\leq$-greatest lower bound (or inf) in the lattice. Furthermore, 1 is the unique $\leq$-maximal element of the lattice and 0 is the unique $\leq$-minimal element of the lattice.

Throughout this chapter, a lattice $\mathfrak{L} = \langle \mathcal{L}, \sqcup, \sqcap, 1, 0 \rangle$ is often written as $\mathfrak{L} = \langle \mathcal{L}, \leq, \sqcup, \sqcap, 1, 0 \rangle$.

The following theorem is immediate from Definitions 7.2 and 7.1.

Theorem 7.1. *Suppose* $\mathfrak{L} = \langle \mathcal{L}, \leq, \sqcup, \sqcap, 1, 0 \rangle$ *is a lattice and a and x are arbitrary elements of $\mathcal{L}$ such that $a \leq x$. Then $x \sqcup a = x$.*

Proof. By Condition 5 of the definition of "lattice" (Definition 7.1),

$$x \sqcup a = x \sqcup (a \sqcap x) = x.$$

$\square$

Definition 7.3 (sublattice). $\mathfrak{L}_1 = \langle \mathcal{L}_1, \sqcup, \sqcap, u, z \rangle$ is said to be a *sublattice* of the lattice $\mathfrak{L} = \langle \mathcal{L}, \sqcup, \sqcap, 1, 0 \rangle$ if and only if $\mathfrak{L}_1$ is a lattice, $\mathcal{L}_1 \subseteq \mathcal{L}$, and for all a and b in $\mathcal{L}_1$, $a \sqcup b$ and $a \sqcap b$ are in $\mathcal{L}_1$. In this definition of "sublattice" it is not required that 1 is the unit element nor that 0 is the zero element of $\mathfrak{L}_1$—it is only required that $\mathfrak{L}_1$ has elements u and z that function respectively as 1 and 0 in the definition of "lattice" (Definition 7.1) so that in the lattice $\mathfrak{L}_1$, u is the unit element and z is the zero element.

Definition 7.4 (complemented and Boolean lattices). Let $\mathfrak{L} = \langle \mathcal{L}, \sqcup, \sqcap, 1, 0 \rangle$ be a lattice. Then $\mathfrak{L}$ is said to be *distributive* if and only if for all a, b, and c in $\mathcal{L}$,

$$a \sqcap (b \sqcup c) = (a \sqcap b) \sqcup (a \sqcap c).$$

An element b of $\mathcal{L}$ is said to be the *complement* of the element a of $\mathcal{L}$ if and only if

$$a \sqcap b = 0 \quad \& \quad a \sqcup b = 1.$$

The function $\frown$ on $\mathcal{L}$ is said to be a *complementation operation* of $\mathfrak{L}$ if and only if for each a in $\mathcal{L}$, $\frown a$ is a complement of a, and $\mathfrak{L}$ is said to be *complemented* if and only if it has a complementation operation. $\mathfrak{L}$ is said to be a *Boolean lattice* if and only if it is complemented and distributive.

Boolean algebras of sets are examples of Boolean lattices, with $\cup$ being the join operator, $\cap$ being the meet operator, and $-$ being the operation of set-theoretic complementation with respect to the unit element X. Throughout this chapter, when viewed as lattices, Boolean algebras of sets are often called "Boolean lattices of sets":

Definition 7.5 (Boolean lattice of sets). A lattice of the form $\mathfrak{X} = \langle \mathcal{X}, \subseteq, \cup, \cap, -, X, \varnothing \rangle$ is said to be a *Boolean lattice of sets* if and only if $\mathfrak{X}$ is a Boolean lattice.

Throughout this chapter, a generalization of a lattice complementation operation called a "negative operation" is used:

Definition 7.6 (negative operator). Let $\mathfrak{L} = \langle \mathcal{L}, \sqcup, \sqcap, 1, 0 \rangle$ be a lattice. Then $\vdash$ is said to be a *negative operator* on $\mathfrak{L}$, if and only if $\vdash$ is a function from $\mathcal{L}$ into $\mathcal{L}$ such that for all a in $\mathcal{L}$, $(\vdash a) \sqcap a = 0$. Note that if $\vdash$ is a negative operator on $\mathfrak{L}$ and $(\vdash a) \sqcup a = 1$, then $\vdash$ is also a complementation operator on $\mathfrak{L}$.

By convention, $\sqcup$ and $\sqcap$ dominate $\vdash$ in algebraic expressions. Thus $\vdash a \sqcup b$ stands for $(\vdash a) \sqcup b$ and $\vdash a \sqcap b$ stands for $(\vdash a) \sqcap b$.

By convention, the phrase "$\langle \mathcal{L}, \sqcup, \sqcap, \vdash, 1, 0 \rangle$ is a lattice" stands for "$\langle \mathcal{L}, \sqcup, \sqcap, 1, 0 \rangle$ is a lattice and $\vdash$ is a negative operator on $\langle \mathcal{L}, \sqcup, \sqcap, 1, 0 \rangle$." The analogous conventions hold for $\mathfrak{L} = \langle \mathcal{L}, \leq, \sqcup, \sqcap, \vdash, 1, 0 \rangle$ and $\mathfrak{L} = \langle \mathcal{L}, \leq, \sqcup, \sqcap, 1, 0 \rangle$.

Definition 7.7 (De Morgan's Laws). A lattice $\mathfrak{L} = \langle \mathcal{L}, \sqcup, \sqcap, 1, 0 \rangle$ is said to satisfy *De Morgan's Laws* if and only if it is complemented and for all a and b in $\mathcal{L}$,

$$\vdash (a \sqcap b) = (\vdash a) \sqcup (\vdash b), \tag{7.1}$$

and

$$\vdash (a \sqcup b) = (\vdash a) \sqcap (\vdash b). \tag{7.2}$$

Theorem 7.2. *Suppose $\mathfrak{L} = \langle \mathcal{L}, \leq, \sqcup, \sqcap, \vdash, 1, 0 \rangle$ is a lattice that satisfies De Morgan's Laws. Then if $a \leq b$, then $\vdash b \leq \vdash a$.*

Proof. Suppose $a \leq b$. Then

$$a = (a \sqcap b),$$

and therefore,

$$\neg\, a \;=\; \neg\,(a \sqcap b) \;=\; (\neg\, a) \sqcup (\neg\, b)\,,$$

and thus by Condition 8 of Definition 7.1,

$$(\neg\, b) \sqcap (\neg\, a) = (\neg\, b) \sqcap (\neg\, a \sqcup \neg\, b) = \neg\, b\,,$$

and therefore, $\neg\, b \le \neg\, a$. $\qquad\square$

The following theorem is well known:

Theorem 7.3. *Suppose* $\mathfrak{L} = \langle \mathcal{L}, \le, \sqcup, \sqcap, \neg, 1, 0 \rangle$ *is a Boolean lattice. Then* $\mathfrak{L}$ *satisfies De Morgan's Laws and for all a in $\mathcal{L}$, $a = \neg\neg\, a$.*

It is also well known that distributive (and therefore, Boolean) lattices satisfy the following additional form of "distributivity":

Theorem 7.4. *Suppose* $\mathfrak{L} = \langle \mathcal{L}, \le, \sqcup, \sqcap, \neg, 1, 0 \rangle$ *is a lattice. Then the following two statements are equivalent:*

(1) $\mathfrak{L}$ *is distributive.*
(2) ($\sqcup$-distributivity) For all a, b, c in $\mathcal{L}$,

$$a \sqcup (b \sqcap c) = (a \sqcup b) \sqcap (a \sqcup c)\,.$$

Boolean lattices are much more than complemented, they are *uniquely complemented:*

Theorem 7.5. *Suppose* $\mathfrak{L} = \langle \mathcal{L}, \le, \sqcup, \sqcap, \neg, 1, 0 \rangle$ *is a Boolean lattice and $\neg'$ is a complementation operation on $\mathfrak{L}$. Then $\neg\, = \neg'$.*

Proof. Suppose a is an arbitrary element of $\mathcal{L}$ and $b = \neg'\, a$. Because $\neg$ is also a complementation operation of $\mathfrak{L}$, it needs to only be shown that $\neg\, a = b$. Because $\mathfrak{L}$ is distributive and $a \sqcap b = 0$,

$$b = b \sqcap (a \sqcup \neg\, a) = (b \sqcap a) \sqcup (b \sqcap \neg\, a) = (b \sqcap \neg\, a)\,,$$

and thus $b \le \neg\, a$. Similarly, because $\mathfrak{L}$ is distributive and $a \sqcup b = 1$, it follows from Theorem 7.4 that

$$b = b \sqcup (a \sqcap \neg\, a) = (b \sqcup a) \sqcap (b \sqcup \neg\, a) = b \sqcup \neg\, a\,,$$

and therefore by Condition 8 of Definition 7.1,

$$\neg\, a \sqcap b = \neg\, a \sqcap (\neg\, a \sqcup b) = \neg\, a\,,$$

and thus $\neg\, a \le b$. This shows $\neg\, a = b$. $\qquad\square$

Birkhoff and von Neumann (1936) commented,

> Besides the (binary) operations of meet- and join-formation,
> there is a third (unary) operation which may be defined in partially
> ordered systems. This is the operation of complementation. ... In
> the case of closed linear subspaces of a Hilbert space (or Cartesian
> n-space), it corresponds to passage to the orthogonal complement
> ... one has the formal identities $[(a')' = a$ and the De Morgan Laws
> and their consequence, Theorem 7.2, if $a \leq b$, then $\vdash b \leq \vdash a]$.
> ... Up to now, we have only discussed formal features of logical
> structure which seem to be common to classical dynamics and the
> quantum theory. We now turn to the central difference between
> them—the *distributive identity* of propositional calculus: [distribu-
> tivity (Definition 7.4) and ⊔-distributivity (Theorem 7.4)] which is
> a law in classical, but not in quantum mechanics. ... Propositional
> calculi of classical mechanics are Boolean algebras. (p. 830)
>   ... Although closed linear subspaces of Hilbert space and Carte-
> sian n-space need not satisfy [the abovementioned distributive
> identity] relative to set-products and closed linear sums, the for-
> mal properties of these operations are not confined to [the proper-
> ties of general lattices and the De Morgan Laws]. ... The models
> for propositional calculi which have been considered in the pre-
> ceding sections are also interesting from the standpoint of pure
> logic. Their nature is determined by quasi-physical and technical
> reasoning, different from the introspective and philosophical con-
> siderations which have had to guide logicians hitherto. Hence it is
> interesting to compare the modifications which they introduce into
> Boolean algebra, with those which logicians on "intuitionist" and
> related ground have tried introducing.
>   The main difference seems to be that whereas logicians have
> usually assumed that [the properties given by the identity $(a')' = a$,
> the De Morgan Laws, and the consequence of De Morgan's Laws,
> if $a \leq b$, then $\vdash b \leq \vdash a$, stated in Theorem 7.2] of negation were
> the ones least able to withstand a critical analysis, the study of
> mechanics points to the *distributive identity* as the weakest link in
> the algebra of logic." (pp. 831-837)

Using "technical reasoning" about lattices, this chapter arrives at a different conclusion about negation than Birkhoff and von Neumann: When
a Boolean lattice appears to be inadequate for a scientific application in-

volving probabilistic ideas, the problem is likely due to the fact that the lattice necessarily has a *unique* complementation operation. In lattices, altering unique complementation to having multiple complementation operations leads necessarily (by theorems of this chapter) to a weakening of distributivity such as Birkhoff and von Neumann advocate. And the consideration of the other alternative—that is, consideration of lattices with a negative operation that is not a complementation operation—leads naturally (by theorems of this chapter) to the form of negation the intuitionists advocate. Lattices with such intuitionistic-like negation operations may apply to situations beyond mechanics, for example, this chapter applies it to decision theory.

The following well-known theorem of Stone (1936) characterizes Boolean lattices as Boolean lattices of sets.

Theorem 7.6 (Stone Representation Theorem). *Each Boolean lattice is isomorphic to a Boolean lattice of sets.*

The Stone Representation Theorem tells us that the only way to extend standard probability theory to algebraically different domains is to look beyond Boolean lattices. Birkhoff and von Neumann (1936) did this by weakening distributivity to the following condition:

Definition 7.8 (modular). A lattice $\mathcal{L}$ is said to be *modular* if and only if for all a, b, c in $\mathcal{L}$,

$$\text{if } a \leq b, \text{ then } (b \sqcap a) \sqcup (b \sqcap c) = b \sqcap (a \sqcup c).$$

Modular lattices were investigated by Dedekind in the late 19th Century. He also provided the following well-known characterization for them:

Theorem 7.7 (Dedekind Characterization Theorem). *A lattice is modular if and only if no sublattice is isomorphic to lattice N_5 in Figure 7.1.*

Using the Dedekind characterization theorem, Birkhoff provided the following characterization theorem for distributive lattices.

Theorem 7.8 (Birkhoff Characterization Theorem). *A lattice is distributive if and only if it has no sublattice isomorphic to either N_5 in Figure 7.1 or M_3 in Figure 7.2.*

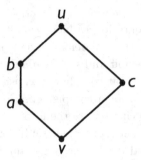

Fig. 7.1.   The lattice N_5. In this diagram, $v < a < b < u$; $v < c < u$; $a \sqcap c = b \sqcap c = v$; $a \sqcup c = b \sqcup c = u$, etc.

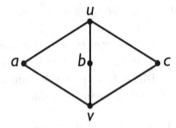

Fig. 7.2.   The lattice M_3. In this diagram, $v < a < u$; $v < b < u$; $v < c < u$; $a \sqcap b = a \sqcap c = b \sqcap c = v$; $a \sqcup b = a \sqcup c = b \sqcup c = u$, etc.

7.3. Extended Probability Functions

The concept of "probability function" generalizes to lattices as follows:

Definition 7.9 (probability function). Let $\mathfrak{L} = \langle \mathcal{L}, \leq, \sqcup, \sqcap, 1, 0 \rangle$ be a lattice. Then $\mathbb{P}$ is said to be a *probability function* on $\mathfrak{L}$ if and only if for all a, b, and c in $\mathcal{L}$,

- $\mathbb{P}(a)$ is a function from $\mathcal{L}$ into the closed interval of real numbers $[0,1]$,
- $\mathbb{P}(1) = 1$ and $\mathbb{P}(0) = 0$, and
- *(finite additivity)* if $b \sqcap c = 0$, then $\mathbb{P}(b \sqcup c) = \mathbb{P}(b) + \mathbb{P}(c)$.

While the probability concept of Definition 7.9 is very useful, it restricts in subtle ways the theoretical development of probability theory for lattices. This is because the concept of the real number system needs a piece of set theory for its description, and because of this, the concept of "probability function" has no purely algebraic description. The piece of set theory usually takes the form that the real numbers are Dedekind complete or the equivalent proposition that each nonempty, bounded set of real numbers has a real number as its least upper bound—that is, requiring concepts that cannot be formulated in terms of first-order logic. This makes the concept of probability function non-coordinate with first-order logic, thus obscuring some fundamental relationships between logic as a deductive system and probability as a means for measuring uncertainty. It also affects algebraic characterizations involving probabilities in other undesirable ways. For example, there are cases where $a < b$ but $\mathbb{P}(a) \not< \mathbb{P}(b)$. (This happens, for example, in Boolean lattice of sets if $a = 0$, $a \neq b$, and $\mathbb{P}(b) = 0$.)

For the purposes of this chapter, only very simple algebraic properties involving addition and multiplication of numbers are needed from a system of numbers containing both real and infinitesimal numbers. These properties consist of familiar identities and inequalities characteristic of the real numbers together with the assumption of the existence of a positive infinitesimal. A system with these identities and inequalities is called a "totally ordered field."

Definition 7.10 (totally ordered field). $\mathfrak{F} = \langle F, \leq, +, \cdot, 1, 0 \rangle$ is said to be a *totally ordered field* if and only if $\leq$ is a binary relation on F, $+$ and $\cdot$ are binary operations on F, 1 and 0 are elements of F, and the following four statements hold:

(1) $\leq$ is a total ordering on F.

(2) $\langle F, \leq, 0 \rangle$ satisfies the following five properties for all x, y, and z in F:
- $x + (y + z) = (x + y) + z$;
- $x + y = y + x$;
- $x + 0 = x$;
- there exists u (denoted by $-x$) such that $x + u = 0$;
- and $x \leq y$ iff $x + z \leq y + z$.

(3) Let $F^+ = \{x \mid 0 < x\}$. Then $\langle F^+, \cdot, 1 \rangle$ satisfies the following five properties for all x, y, and z in F^+:
- $x \cdot y$ is in F;
- $x \cdot (y \cdot z) = (x \cdot y) \cdot z$;
- $x \cdot y = y \cdot x$;
- $x \cdot 1 = x$;
- and $x \leq y$ iff $x \cdot z \leq y \cdot z$.

(4) $\langle F, \leq, +, \cdot, 1, 0 \rangle$ satisfies the following four properties for all x, y, and z in F:

- $1 \neq 0$;
- $x \cdot 0 = 0$;
- if $x \neq 0$, then there exists v (denoted by x^{-1}) such that $x \cdot v = 1$;
- and $x \cdot (y + z) = (x \cdot y) + (x \cdot z)$.

The real numbers with their usual addition and multiplication operations and their usual ordering form a totally ordered field. It is characterized up to isomorphism as a totally ordered field such that each of its nonempty bounded subsets A has a least upper bound that is an element of the field.

This chapter uses a slight extension of the concept of a probability function given in Definition 7.9. The extension allows for the assigning positive infinitesimal values to some non-zero elements. (Nikodym, 1960, introduced infinitesimal elements into modern probability theory through a related concept.)

Definition 7.11 (extended real numbers). $^*\mathfrak{R} = \langle {}^*\mathbb{R}, \leq, +, \cdot, 1, 0 \rangle$ is said to be an *extended system of real numbers,* or just the *extended reals* for short, if and only if the following two statements hold:

(1) $^*\mathfrak{R}$ is a totally ordered field.
(2) $^*\mathfrak{R}$ restricted to the set of real numbers $\mathbb{R}$ is the ordered field of real numbers.

$\langle {}^*\mathbb{R}, \leq, +, \cdot, 1, 0 \rangle$ is said to be a *proper* extended system of real numbers if and only if it is an extended system of real numbers and the following statement holds:

(3) $\langle {}^*\mathbb{R}, \leq, +, \cdot, 1, 0 \rangle$ has a *positive infinitesimal;* that is, it has an element $a \in {}^*\mathbb{R}$ such that for all positive b in $\mathbb{R}$, $0 < a < b$.

Let $\langle {}^*\mathbb{R}, \leq, +, \cdot, 1, 0 \rangle$ be a system of extended reals. Then, by definition,

$$^*[0, 1] = \{ x \mid x \in {}^*\mathbb{R} \ \ \& \ \ 0 \leq x \leq 1 \}.$$

Two elements r and s in $^*[0, 1]$ are said to be *infinitesimally close* if and only if $|r - s|$ is 0 or is a positive infinitesimal.

There are various methods in the literature for constructing or showing the existence of an extended system of real numbers. Chapter 3 of Narens (2007) provides a means of construction through ultrapowers.

It follows from Definition 7.11 that the system of real numbers is the special case of an extended system of real numbers where there is no positive infinitesimal.

The following definition extends probability theory to $\langle {}^*\mathbb{R}, \leq, +, \cdot, 1, 0 \rangle$.

Definition 7.12 (extended probability function).
Let $\mathfrak{L} = \langle \mathcal{L}, \leq, \sqcup, \sqcap, 1, 0 \rangle$ be a lattice. Then $\mathbb{P}$ is said to be an *extended probability function* on $\mathfrak{L}$ if and only if for all a, b, and c in $\mathcal{L}$,

- $\mathbb{P}$ is a function from $\mathcal{L}$ into the closed interval of the extended reals ${}^*[0,1]$,
- $\mathbb{P}(1) = 1$ and $\mathbb{P}(0) = 0$,
- *finite additivity:* if $b \sqcap c = 0$, then $\mathbb{P}(b \sqcup c) = \mathbb{P}(b) + \mathbb{P}(c)$, and
- *monotonicity:* if $a < b$ then $\mathbb{P}(a) < \mathbb{P}(b)$.

There are two important differences between the concepts of "extended probability function" and "probability function": (i) an extended probability function may have some non-real values, and (ii) an extended probability function $\mathbb{P}$ satisfies monotonicity, from which it follows that it also satisfies the condition,

$$\mathbb{P}(a) = 0 \ \text{iff} \ a = 0.$$

In contrast, some probability functions $\mathbb{Q}$ have elements b in their domain such that $b \neq 0$ and $\mathbb{Q}(b) = 0$. The later situation has the unfortunate consequence that there are elements that have a possibility of occurring while at the same time their probability of occurrence is the same as the probability of the occurrence of the impossible element 0. The existence of positive infinitesimal probabilities together with monotonicity eliminate this kind of circumstance.

An extended probability function $\mathbb{P}$ can be made into probability function ${}^\circ\mathbb{P}$ as follows: Let

$${}^\circ\mathbb{P}(a) = \text{the real number that is infinitesimally close to } \mathbb{P}(a).$$

(Because the real number system satisfies the least upper bound condition, it is easy to show that there is exactly one real number infinitesimally close to $\mathbb{P}(a)$.) Then ${}^\circ\mathbb{P}$ is a probability function whose values are infinitesimally close to $\mathbb{P}$.

7.4. Properties of Lattices with Extended Probability Functions

7.4.1. *Probabilistic lattices*

This section explores how the existence of an extended probability function on a lattice $\mathfrak{L}$ restricts the algebraic form of $\mathfrak{L}$—or equivalently, restricts the laws of a logic.

Definition 7.13 (probabilistic lattice). Let $\mathfrak{L} = \langle \mathcal{L}, \leq, \sqcup, \sqcap, 1, 0 \rangle$ be a lattice. Then $\mathfrak{L}$ is said to be *probabilistic* if and only if there exists an extended probability function on $\mathfrak{L}$.

The following theorem shows that Boolean algebras are examples of probabilistic lattices.

Theorem 7.9. *Each Boolean lattice is a probabilistic lattice.*

Proof. By the Stone Representation Theorem (Theorem 7.6), each Boolean lattice $\mathfrak{L}$ is isomorphic to a Boolean lattice of sets. Luxemburg (1962) shows that each Boolean lattice of sets has an extended probability function. Then it easily follows by isomorphism that $\mathfrak{L}$ has an extended probability function.   □

Definition 7.14 (section). Let $\mathfrak{L} = \langle \mathcal{L}, \leq, \sqcup, \sqcap, 1, 0 \rangle$ be a lattice and a be an element of $\mathcal{L}$.
  Then by definition,

$$[0, a] = \{x \mid x \in \mathcal{L} \ \& \ x \leq a\}.$$

$[0, a]$ is called the *section (determined by a)* of $\mathfrak{L}$. It easily follows from the definition of "lattice" (Definition 7.1) that $[0, a]$ is a sublattice of $\mathfrak{L}$.
  $\mathfrak{L}$ is said to be *section complemented* if and only if for each b in $\mathcal{L}$, the lattice $[0, b]$ is complemented.
  Note that because $\mathfrak{L} = [0, 1]$, it follows that if $\mathfrak{L}$ is section complemented then $\mathfrak{L}$ is complemented.

  Section complementation is important for the theory of probabilistic lattices because it allows for the existence of conditional extended probabilities. Suppose $\mathbb{P}$ is an extended probability function on a lattice $\mathfrak{L}$ and $[0, a]$ is a section of $\mathfrak{L}$. Then $[0, a]$ is a lattice. Define $\mathbb{P}_{[0,a]}$ on $[0, a]$ as follows:

for all b in $[0, a]$,

$$\mathbb{P}_{[0,a]}(b) = \frac{\mathbb{P}(b)}{\mathbb{P}(a)}.$$

It easily follows that $\mathbb{P}_{[0,a]}$ is an extended probability function on $[0, a]$.

A form of complementation stronger than section complementation is relative complementation.

Definition 7.15 (relative complement). Let $\mathfrak{L} = \langle \mathcal{L}, \leq, \sqcup, \sqcap, 1, 0 \rangle$ be a lattice and a and b be arbitrary elements of $\mathcal{L}$ such that $a \leq b$. Then by definition,

$$[a, b] = \{ x \mid x \in \mathcal{L} \ \& \ a \leq x \leq b \}.$$

It is immediate that $[a, b]$ is a sublattice of $\mathfrak{L}$, and $[a, b]$ is said to be *relatively complemented* if and only if it is complemented as a lattice. For each x in $[a, b]$, an element y in $[a, b]$ such that

$$a = x \sqcap y \ \& \ b = x \sqcup y$$

is called the *relative complement of x with respect to* $[a, b]$.

Theorem 7.10. *Let* $\mathfrak{L} = \langle \mathcal{L}, \sqcup, \sqcap, 1, 0 \rangle$ *be a complemented modular lattice and* a, b, x, *and* t *be elements of* $\mathcal{L}$ *such that* $a \leq x \leq b$ *and* t *is a complement of* x. *Then*

$$y = (a \sqcup t) \sqcap b$$

is the relative complement of x *in the lattice* $[a, b]$.

Proof. Because $a \leq x$ and $t \sqcap x = 0$, it follows by the modularity of $\mathfrak{L}$ that

$$x \sqcap y = (a \sqcup t) \sqcap b \sqcap x = (a \sqcup t) \sqcap x = (a \sqcap x) \sqcup (t \sqcap x) \quad (7.3)$$
$$= a \sqcup (t \sqcap x) = a \sqcup 0 = a.$$

Because $a \leq x \leq b$ and $x \sqcup t = 1$, it then follows from $x \sqcup a = x$ (Theorem 7.1) and the modularity of $\mathfrak{L}$ that

$$x \sqcup y = x \sqcup [(a \sqcup t) \sqcap b] = x \sqcup [(a \sqcap b) \sqcup (t \sqcap b)] = x \sqcup [a \sqcup (t \sqcap b)] (7.4)$$
$$= (x \sqcup a) \sqcup (t \sqcap b) = x \sqcup (t \sqcap b) = (x \sqcap b) \sqcup (t \sqcap b)$$
$$= (x \sqcup t) \sqcap b = 1 \sqcap b = b.$$

It follows from (7.3) and (7.4) that y is the relative complement of x in the lattice $[a, b]$. □

The following is the converse of Theorem 7.10. The proof follows Szász (1963).

Theorem 7.11. *Let* $\mathfrak{L} = \langle \mathcal{L}, \sqcup, \sqcap, 1, 0 \rangle$ *be a complemented modular lattice and a, b, and x be elements of $\mathcal{L}$ such that $a \leq x \leq b$. Then for each relative complement y of x in $[a, b]$ there exists a complement t of x in $\mathfrak{L}$ such that*

$$y = (a \sqcup t) \sqcap b.$$

Proof (Szász, 1963, p. 112). Let y be an arbitrary relative complement of x in $[a, b]$. Let

- v be a relative complement of a in $[0, y]$,
- u be a relative complement of b in $[y, 1]$,
- and t the relative complement of y in $[v, u]$.

Then by the definition of "relative complement," the following equations hold:

$$v \leq t \leq u \tag{7.5}$$

$$x \sqcap y = a \quad \text{and} \quad x \sqcup y = b, \tag{7.6}$$

$$a \sqcap v = 0 \quad \text{and} \quad a \sqcup v = y, \tag{7.7}$$

$$b \sqcap u = y \quad \text{and} \quad b \sqcup u = 1, \tag{7.8}$$

$$y \sqcap t = v \quad \text{and} \quad y \sqcup t = u. \tag{7.9}$$

The following argument shows that t is a complement of x in $\mathfrak{L}$: By assumption, $x = x \sqcap b$. By (7.5), (7.8), (7.6), (7.9), and (7.7),

$$x \sqcap t = (x \sqcap b) \sqcap (u \sqcap t) = x \sqcap (b \sqcap u) \sqcap t$$
$$= x \sqcap y \sqcap t = (x \sqcap y) \sqcap (y \sqcap t) = a \sqcap v = 0.$$

And similarly, using $x \sqcup a = x$ (Theorem 7.1),

$$x \sqcup t = (x \sqcup a) \sqcup (v \sqcup t) = x \sqcup (a \sqcup v) \sqcup t$$
$$= x \sqcup y \sqcup t = (x \sqcup y) \sqcup (y \sqcup t) = b \sqcup u = 1.$$

t also satisfies,

$$y = (a \sqcup t) \sqcap b,$$

because by (7.5), (7.7), (7.9), and (7.8),

$$(a \sqcup t) \sqcap b = (a \sqcup (v \sqcup t)) \sqcap b = [(a \sqcup v) \sqcup t)] \sqcap b$$
$$= (y \sqcup t) \sqcap b = u \sqcap b = y.$$

$\square$

It is useful to have a statement of Theorem 7.10 for the special case of sections.

Theorem 7.12. *Suppose* $\mathfrak{L} = \langle \mathcal{L}, \leq, \sqcup, \sqcap, \ulcorner, 1, 0 \rangle$ *is a complemented modular lattice. Then* $\mathfrak{L}$ *is section complemented, and for each* b *in* $\mathcal{L}$ *and each* x *in* $[0, b]$,

$$y = \ulcorner x \sqcap b$$

is a complement of x *in the lattice* $[0, b]$.

Proof. By Theorem 7.10, $\mathfrak{L}$ is relatively complemented. It then follows from the definition of "relatively complemented" that $\mathfrak{L}$ is section complemented. The theorem then follows from the statement of Theorem 7.10 with $a = 0$ and $t = \ulcorner x$. $\square$

Theorem 7.13. *Suppose* $\mathfrak{L} = \langle \mathcal{L}, \sqcup, \sqcap, 1, 0 \rangle$ *is section complemented and* $\mathbb{P}$ *is an extended probability function on* $\mathfrak{L}$. *Then for all* a *and* b *in* $\mathcal{L}$,

$$\mathbb{P}(a) + \mathbb{P}(b) = \mathbb{P}(a \sqcup b) + \mathbb{P}(a \sqcap b).$$

Proof. Let a and b be arbitrary elements of $\mathcal{L}$. Because $0 \leq a \sqcap b \leq b$ and $\mathfrak{L}$ is section complemented, let x in the lattice $[0, b]$ be such that

$$(a \sqcap b) \sqcap x = 0 \quad \& \quad (a \sqcap b) \sqcup x = b.$$

Then, because $\mathbb{P}$ is an extended probability function,

$$\mathbb{P}(a \sqcap b) + \mathbb{P}(x) = \mathbb{P}(b),$$

that is,

$$\mathbb{P}(a \sqcap b) = \mathbb{P}(b) - \mathbb{P}(x). \tag{7.10}$$

Because $x \leq b$,

$$a \sqcap x = a \sqcap (b \sqcap x) = (a \sqcap b) \sqcap x = 0. \tag{7.11}$$

Because $a = a \sqcup (a \sqcap b)$ (Theorem 7.1), it then follows by the choice of x,

$$a \sqcup x = [a \sqcup (a \sqcap b)] \sqcup x = a \sqcup [(a \sqcap b) \sqcup x] = a \sqcup b. \tag{7.12}$$

Because $\mathbb{P}$ is an extended probability function, it then follows from (7.11) and (7.12) that

$$\mathbb{P}(a \sqcup b) = \mathbb{P}(a \sqcup x) = \mathbb{P}(a) + \mathbb{P}(x). \tag{7.13}$$

Adding (7.10) and (7.13) then yields,

$$\mathbb{P}(a \sqcup b) + \mathbb{P}(a \sqcap b) = \mathbb{P}(a) + \mathbb{P}(b).$$

$\square$

Theorem 7.14. *Suppose* $\mathfrak{L} = \langle \mathcal{L}, \sqcup, \sqcap, 1, 0 \rangle$ *is probabilistic. Then the following two statements are equivalent:*

(1) $\mathfrak{L}$ *is section complemented.*
(2) $\mathfrak{L}$ *is complemented and modular.*

Proof. Because $\mathfrak{L}$ is probabilistic, let $\mathbb{P}$ be an extended probability function on $\mathfrak{L}$.

Suppose Statement 1. Then, because $\mathfrak{L}$ is section complemented, it is complemented. Thus to show Statement 2, it only needs to be shown that $\mathfrak{L}$ is modular. By Theorem 7.7, modularity immediately follows if $\mathfrak{L}$ has no N_5 sublattice. Suppose $\mathfrak{L}$ has a N_5 sublattice. A contradiction will be shown. Let a, b, and c be elements of the N_5 sublattice such that

$$a < b, \ b \sqcap c = a \sqcap c < a, \ \& \ b < b \sqcup c = a \sqcup c.$$

(See, for example, Figure 7.1 that was previously presented.) Then by the definition of "extended probability function,"

$$\mathbb{P}(a) < \mathbb{P}(b), \tag{7.14}$$

and by Theorem 7.13,

$$\mathbb{P}(a) + \mathbb{P}(c) = \mathbb{P}(a \sqcup c) + \mathbb{P}(a \sqcap c) = \mathbb{P}(b \sqcup c) + \mathbb{P}(b \sqcap c) = \mathbb{P}(b) + \mathbb{P}(c). \tag{7.15}$$

Equations (7.14) and (7.15) are contradictory.

Suppose Statement 2. Then Statement 1 follows from Theorem 7.12. $\square$

Theorem 7.15. *Suppose* $\mathfrak{L} = \langle \mathcal{L}, \leq, \sqcup, \sqcap, \frown, 1, 0 \rangle$ *is a lattice. Then the following two statements are equivalent:*

(1) $\mathfrak{L}$ *is uniquely complemented, section complemented, and probabilistic.*
(2) $\mathfrak{L}$ *is Boolean.*

Proof. Suppose Statement 1. Because $\mathfrak{L}$ is complemented, to show that $\mathfrak{L}$ is Boolean, it needs only to be shown that $\mathfrak{L}$ is distributive. By Theorem 7.14, $\mathfrak{L}$ is a complemented modular lattice. Therefore, to show distributivity, it is sufficient by Theorems 7.7 and 7.8 to show that $\mathfrak{L}$ has no M_3 sublattice. This is done by contradiction.

Suppose $\mathfrak{L}$ had a M_3 sublattice, say with distinct elements a, b, c, u, and v, where

$$u = a \sqcup b = a \sqcup c = b \sqcup c \ \& \ v = a \sqcap b = a \sqcap c = b \sqcap c,$$

that is, a, b, and c, are distinct and both b and c are relative complements of a in $[v, u]$. (For example, see Figure 7.2 that was previously presented.) Because $\mathfrak{L}$ is uniquely complemented, it follows from Theorems 7.10 and 7.11 that all relative complements of a in $[v, u]$ have the form $v \sqcup (\mathbin{\vdash} a \sqcap u)$, and thus it must be the case that $b = v \sqcup (\mathbin{\vdash} a \sqcap u) = c$, which is impossible, because by hypothesis $b \neq c$.

Suppose Statement 2. Then because $\mathfrak{L}$ is Boolean, it is complemented and modular, and therefore by Lemma 7.12, it is section complemented. It is probabilistic by Theorem 7.9. Thus to show Statement 1, it needs only to be shown that $\mathfrak{L}$ is uniquely complemented. In order to do this suppose x and y are arbitrary elements of $\mathcal{L}$ such that

$$x \sqcup y = 1 \quad \& \quad x \sqcap y = 0.$$

Because $\mathbin{\vdash}$ is a complementation operation of $\mathfrak{L}$, it needs to only be shown that $\mathbin{\vdash} x = y$. Because $\mathfrak{L}$ is distributive and $x \sqcap y = 0$,

$$y = y \sqcap (x \sqcup \mathbin{\vdash} x) = (y \sqcap x) \sqcup (y \sqcap \mathbin{\vdash} x) = (y \sqcap \mathbin{\vdash} x),$$

and thus $y \leq \mathbin{\vdash} x$. Similarly, because $\mathfrak{L}$ is distributive and $x \sqcap y = 0$,

$$x = x \sqcap (y \sqcup \mathbin{\vdash} y) = (x \sqcap y) \sqcup (x \sqcup \mathbin{\vdash} y) = x \sqcup \mathbin{\vdash} y,$$

which by Theorems 7.3 and 7.2 yield $\mathbin{\vdash} x \leq \mathbin{\vdash}\mathbin{\vdash} y$, and thus by Theorem 7.3, $\mathbin{\vdash} x \leq y$. Therefore, $\mathbin{\vdash} x = y$, because $y \leq \mathbin{\vdash} x$ and $\mathbin{\vdash} x \leq y$. □

A pragmatic consequence of Theorem 7.15 is that if one wants to study generalizations of probability theory with a probability function defined on all elements of a non-Boolean lattice, then one should look to lattices that are either (i) non-complemented or (ii) have two or more complementation operations or (iii) are not section complemented. (iii) weakens the probability theory in that the concept of "conditional probability" becomes problematic. While lattices with (ii) appear in quantum mechanics, they do not use, for scientific purposes, a probability function that is defined on all elements of a lattice; instead they use a "partial probability function" that is only guaranteed to be additive for disjoint elements that satisfies a property called "orthogonality". The next section studies an approach to probability theory that uses (i).

7.5. Pseudo-Complemented Lattices

This section weakens the property of complementation of Boolean lattices by providing distributive lattices with two new negative operations. One of these—*relative pseudo-complementation*—has been much studied in the foundations of mathematics. It corresponds to the implication operator of intuitionistic logic. The second—*pseudo-complementation*—is a much weaker condition. It is definable in terms of the first, and corresponds to the negation operator of intuitionistic logic. The characteristic difference between a Boolean lattice and a pseudo-complemented lattice is that in a Boolean lattice, $(\vdash a) \sqcup a = 1$ for all a in the lattice's domain, whereas in pseudo-complemented lattice, there may be some b in its domain such that $(\vdash b) \sqcup b \neq 1$. This section uses distributive lattices with an operation of pseudo-complementation as event spaces for extended probability functions.

Definition 7.16 (relative pseudo-complementation). Let $\mathfrak{L} = \langle \mathcal{L}, \leq, \sqcup, \sqcap, 1, 0 \rangle$ be a lattice and a, b, and c be arbitrary elements of $\mathcal{L}$. Then the following definitions hold:

(1) c is said to be a *pseudo-complement of a relative to b*, in symbols, $c = (a \Rightarrow b)$, if and only if for all x in $\mathcal{L}$,

$$x \leq c \ \text{ iff } \ a \sqcap x \leq b.$$

(2) $\mathfrak{L}$ is said to be *relatively pseudo-complemented* if and only if for all a and b in $\mathfrak{L}$, the relative complement of a relative b exists. If $\mathfrak{L}$ is relatively pseudo-complemented, the operation $\Rightarrow$ is called the *operation of relative pseudo-complementation* for $\mathfrak{L}$.

(3) $\mathfrak{L}$ is said to be a *Heyting lattice* if and only if it is relatively pseudo-complemented.

Definition 7.17 (pseudo-complementation). Let $\mathfrak{L} = \langle \mathcal{L}, \leq, \sqcup, \sqcap, 1, 0 \rangle$ be a lattice and a and c be arbitrary elements of $\mathcal{L}$. Then the following definitions hold:

(1) c is said to be the *pseudo-complement* of a if and only if for each x in $\mathcal{L}$,

$$x \leq c \ \text{ iff } \ a \sqcap x = 0.$$

(2) $\mathfrak{L}$ is said to be *pseudo-complemented* if and only if each element of $\mathcal{L}$ has a pseudo-complement.

Note by Definitions 7.16 and 7.17 that c is the pseudo-complement of a if and only if c is pseudo-complement of a relative to 0, that is, $c = (a \Rightarrow 0)$.

Throughout this chapter the following conventions are used:

- If ⌐ is an operation on $\mathfrak{L}$ that assigns to each element of $\mathcal{L}$ its pseudo-complement, then the notation $\langle \mathcal{L}, \leq, \sqcup, \sqcap, \neg, 1, 0 \rangle$ will often be used in place of $\langle \mathcal{L}, \leq, \sqcup, \sqcap, 1, 0 \rangle$ in describing pseudo-complemented lattices.
- Similarly, if $\Rightarrow$ is a binary operation of relative pseudo-complementation on $\mathfrak{L}$, then the notation $\langle \mathcal{L}, \leq, \sqcup, \sqcap, \neg, \Rightarrow, 1, 0 \rangle$ will often be used in place of $\langle \mathcal{L}, \leq, \sqcup, \sqcap, \Rightarrow, 1, 0 \rangle$ and $\langle \mathcal{L}, \leq, \sqcup, \sqcap, \neg, 1, 0 \rangle$ in describing relative pseudo-complemented lattices.

The next theorem establishes the uniqueness of the operations of pseudo- and relative pseudo-complementation.

Theorem 7.16. *Suppose* $\mathfrak{L} = \langle \mathcal{L}, \leq, \sqcup, \sqcap, 1, 0 \rangle$ *is a lattice and* a, b, *and* c *are arbitrary elements of* $\mathcal{L}$. *Then the following two statements hold:*

(1) If the pseudo-complement of a *relative to* b *exists, then it is unique.*

(2) If the pseudo-complement of c *exists, then it is unique.*

***Proof*.** Immediate consequences of Definitions 7.16 and 7.17. □

Relative pseudo-complementation is a much stronger condition that pseudo-complementation. The next theorem shows that it implies distributivity.

Theorem 7.17. *Let* $\mathfrak{L} = \langle \mathcal{L}, \leq, \sqcup, \sqcap, \neg, \Rightarrow, 1, 0 \rangle$ *be a Heyting lattice. Then* $\mathfrak{L}$ *is distributive.*

***Proof*.** Suppose a, b, and c are arbitrary elements of $\mathcal{L}$. Because

$$a \sqcap b \leq (a \sqcap b) \sqcup (a \sqcap c) \quad \& \quad a \sqcap c \leq (a \sqcap b) \sqcup (a \sqcap c), \qquad (7.16)$$

it follows from Definition 7.16 that

$$b \leq a \Rightarrow [(a \sqcap b) \sqcup (a \sqcap c)] \quad \& \quad c \leq a \Rightarrow [(a \sqcap b) \sqcup (a \sqcap c)].$$

Thus,

$$b \sqcup c \leq a \Rightarrow (a \sqcap b) \sqcup (a \sqcap c),$$

and therefore, by (7.16) and Definition 7.16,

$$a \sqcap (b \sqcup c) \leq (a \sqcap b) \sqcup (a \sqcap c). \qquad (7.17)$$

By Theorem 7.1 and properties of $\leq$,

$$a \sqcap b \leq a \sqcap (b \sqcup c) \quad \& \quad a \sqcap c \leq a \sqcap (b \sqcup c),$$

from which it follows that

$$(a \sqcap b) \sqcup (a \sqcap c) \leq a \sqcap (b \sqcup c). \qquad (7.18)$$

By (7.17) and (7.18),

$$a \sqcap (b \sqcup c) = (a \sqcap b) \sqcup (a \sqcap c).$$

$\square$

Theorem 7.18. *Suppose* $\mathfrak{L} = \langle \mathcal{L}, \leq, \sqcup, \sqcap, \vdash, 1, 0 \rangle$ *is a pseudo-complement lattice. Then the following two statements hold:*

(1) $\vdash 0 = 1$ & $\vdash 1 = 0$.
(2) For all a and b in $\mathcal{L}$, if $a \leq b$ then $\vdash b \leq \vdash a$.

Proof. 1. Because for all a in $\mathcal{L}$, $a \leq 1$ and $a \sqcap 0 = 0$, it follows from Definition 7.17 that $\vdash 0 = 1$. Because for all a in $\mathcal{L}$, $1 \sqcap a = 0$ if and only if $a = 0$, it follows from Definition 7.17 that $\vdash 1 = 0$.

2. Suppose a and b are arbitrary elements of $\mathcal{L}$ and $a \leq b$. Then,

$$a \sqcap \vdash b = (a \sqcap b) \sqcap \vdash b = a \sqcap (b \sqcap \vdash b) = a \sqcap 0 = 0.$$

Thus by Definition 7.17, $\vdash b \leq \vdash a$. $\square$

The next theorem shows that a Boolean lattice with its version of the classical implication operation formally meets the criterion of being a relatively pseudo-complement lattice.

Theorem 7.19. *Suppose* $\mathfrak{L} = \langle \mathcal{L}, \leq, \sqcup, \sqcap, \vdash, 1, 0 \rangle$ *is a Boolean lattice. Define* $\Rightarrow$ *on $\mathcal{L}$ as follows: For all a and b in $\mathcal{L}$,*

$$a \Rightarrow b = \vdash a \sqcup b.$$

Then $\Rightarrow$ *is the relative pseudo-complementation operation on $\mathfrak{L}$ and $\vdash$ is the pseudo-complementation operation on $\mathfrak{L}$.*

Proof. Let a and b be arbitrary elements of $\mathcal{L}$. It will be shown that $\Rightarrow$ satisfies the definition of a relative pseudo-complementation operator (Definition 7.16). Note that

$$a \sqcap (\vdash a \sqcup b) = (a \sqcap \vdash a) \sqcup (a \sqcap b) = a \sqcap b.$$

Suppose that x is an arbitrary element of $\mathcal{L}$ such that $a \sqcap x \leq b$. Then

$$a \sqcap x \leq a \sqcap b.$$

Because $x \sqcap \ulcorner a \le \ulcorner a$, it follows that

$$(x \sqcap \ulcorner a) \sqcup (a \sqcap b) \le \ulcorner a \sqcup (a \sqcap b).$$

Thus

$$x = x \sqcap 1 = x \sqcap (\ulcorner a \sqcup a) = (x \sqcap \ulcorner a) \sqcup (a \sqcap x) \le (x \sqcap \ulcorner a) \sqcup (a \sqcap b) \le \ulcorner a \sqcup (a \sqcap b),$$

that is,

$$x \le \ulcorner a \sqcup b.$$

Thus it has been shown that

$$\text{if } a \sqcap x \le b \text{ then } x \le \ulcorner a \sqcup b. \tag{7.19}$$

Suppose y is an arbitrary element of $\mathcal{L}$ such that $y \le \ulcorner a \sqcup b$. Then

$$a \sqcap y \le a \sqcap (\ulcorner a \sqcup b) = (a \sqcap \ulcorner a) \sqcup (a \sqcap b) = a \sqcap b \le b,$$

that is,

$$a \sqcap y \le b.$$

Thus it has been shown that

$$\text{if } y \le \ulcorner a \sqcup b \text{ then } a \sqcap y \le b. \tag{7.20}$$

Because x and y in (7.19) and (7.20) were chosen to be arbitrary elements of $\mathcal{L}$, it follows from (7.19) and (7.20) and Definition 7.16 that $\ulcorner a \sqcup b$ the operation $\Rightarrow$ of relative pseudo-complementation for $\mathfrak{L}$.

Because $(a \Rightarrow 0) = \ulcorner a \sqcup 0 = \ulcorner a$, $\ulcorner$ is the pseudo-complementation operator of $\mathfrak{L}$. $\qquad\square$

The following theorem summarizes the relationships between Boolean, Heyting, and pseudo-complement lattices.

Theorem 7.20. *The following statements hold:*

(1) If $\mathfrak{L}$ is a Boolean lattice, then it is a Heyting lattice.

(2) If $\mathfrak{L}$ is a Heyting lattice, then it is a pseudo-complemented distributive lattice.

(3) There exists $\mathfrak{L}$ that is a Heyting lattice but not a Boolean lattice.

Proof. 1. Statement 1 immediately follows from Theorem 7.19.

2. Statement 2 immediately follows from Theorem 7.17 and the definition of "pseudo-complement" (Definition 7.17).

3. Let A and B be the open squares,

$$A = \{(x, y) \mid (-1 < x < 0) \ \& \ (0 < y < 1)\},$$
$$B = \{(x, y) \mid (0 < x < 1) \ \& \ (0 < y < 1)\},$$

and L be the open rectangle,

$$L = A \cup B \cup \{(0, y) \mid 0 < y < 1\},$$

(Figure 7.3). Let $\mathcal{L} = \{L, \varnothing, A, B, A \cup B\}$. The following shows that

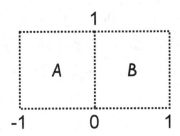

Fig. 7.3.   Example of a Heyting lattice

$\Rightarrow$ is the operation of relative pseudo-complementation on the lattice $\langle \mathcal{L},$ $\subseteq, \cup, \cap, L, \varnothing \rangle$: For each x and y in $\mathcal{L}$, if $x \subseteq y$, then $(x \Rightarrow y) = L$, because $x \cap L = x \subseteq y$. By inspection, $(A \Rightarrow B) = \varnothing$, and similarly $(B \Rightarrow A) = \varnothing$. By inspection, $[(A \cup B) \Rightarrow A] = A$, and similarly, $[(A \cup B) \Rightarrow B] = B$. Also by inspection, $[(A \cup B) \Rightarrow \varnothing] = \varnothing$. By inspection, for $x = \varnothing, A, B, A \cup B, L$, $(L \Rightarrow x) = x$. This takes care of all of the possibilities of $u \Rightarrow v$ for all u and v in $\mathcal{L}$. The above facts show that $\mathfrak{L} = \langle \mathcal{L}, \subseteq, \cup, \cap, \neg, \Rightarrow, L, \varnothing \rangle$ is a Heyting lattice, where for all x in $\mathcal{L}$, $\neg x = (x \Rightarrow \varnothing)$. $\mathfrak{L}$ is not Boolean, because $A \cup B$ does not have a complement in $\mathfrak{L}$.   □

The proof Statement 3 of Theorem 7.20 uses open sets from the Euclidean plane to provide an example of a Heyting lattice and therefore a pseudo-complemented lattice. The following shows that this can be done in a much more general manner.

Definition 7.18 (lattice of open sets). Let $\mathcal{T}$ be a collection of sets that forms a topology with universal set X. Let $\mathcal{X}$ be a subset of $\mathcal{T}$ such that $\mathfrak{X} = \langle \mathcal{X}, \cup, \cap, X, \varnothing \rangle$ is a lattice. Then $\mathfrak{X}$ is called a *lattice of open sets (from the topology $\mathcal{T}$)*.

Definition 7.19 (pseudo-complemented lattice of open sets). Let $\mathfrak{X} = \langle \mathcal{X}, \cup, \cap, X, \varnothing \rangle$ be a lattice of open sets from the topology $\mathcal{T}$. Let $\vdash$ be the following function from $\mathcal{X}$ into $\mathcal{T}$: For each A in $\mathcal{X}$, $\vdash A =$ the interior of the set-theoretic complement of A, $X - A$. Then

$$\mathfrak{X} = \langle \mathcal{X}, \cup, \cap, \vdash, X, \varnothing \rangle$$

is said to be a *pseudo-complemented* lattice of open sets if and only if $\vdash A$ is in $\mathcal{X}$ for each A in $\mathcal{X}$.

The following is an immediate consequence of Definition 7.18.

Theorem 7.21. *Let* $\mathfrak{X} = \langle \mathcal{X}, \cup, \cap, \vdash, X, \varnothing \rangle$ *be a pseudo-complemented lattice of open sets from the topology* $\mathcal{T}$. *Then the following two statements hold:*

(1) $\mathfrak{X}$ *is a pseudo-complemented lattice.*
(2) If $\mathcal{X} = \mathcal{T}$, *then* $\mathfrak{X}$ *is a Heyting lattice.*

Stone (1937) showed that each distributive lattice was isomorphic to a lattice of sets that is a lattice of the form $\langle \mathcal{X}, \cup, \cap, X, \varnothing \rangle$, where $\mathcal{X}$ is a set of subsets of the nonempty set X, and that each Heyting lattice was isomorphic to a lattice of open sets from a topology $\mathcal{T}$, $\langle \mathcal{X}, \cup, \cap, X, \varnothing \rangle$, where $\mathcal{X} = \mathcal{T}$. The proof of the following theorem uses his result about distributive lattices to extend his topological result about Heyting lattices to pseudo-complemented lattices.

Theorem 7.22. *Each pseudo-complemented distributive lattice is isomorphic to a pseudo-complemented lattice of open sets.*

Proof. Theorem 8.8 of Narens (2007). □

Definition 7.20 (extended lattice probability function). Let $\mathfrak{L} = \langle \mathcal{L}, \leq, \sqcup, \sqcap, 1, 0 \rangle$ be a lattice. Then $\mathbb{P}$ is said to be an *extended lattice probability function* on $\mathfrak{L}$ if and only if for all a, b, and c in $\mathcal{L}$,

- $\mathbb{P}$ is a function from $\mathcal{L}$ into the closed interval of the extended reals $^\star[0,1]$,
- $\mathbb{P}(1) = 1$ and $\mathbb{P}(0) = 0$,
- *finite lattice additivity:* $\mathbb{P}(b \sqcup c) = \mathbb{P}(b) + \mathbb{P}(c) - \mathbb{P}(b \sqcap c)$, and
- *monotonicity:* if $a < b$ then $\mathbb{P}(a) < \mathbb{P}(b)$.

If an extended lattice probability function is into the real interval $[0,1]$, it is called a *lattice probability function*.

Theorem 7.23. *Each pseudo-complemented distributive lattice* $\mathfrak{L} =$ $\langle \mathcal{L}, \sqcup, \sqcap, \ulcorner, 1, 0 \rangle$ *has an extended lattice probability function.*

Proof.   By Theorem 7.22, let $\mathfrak{X} = \langle \mathcal{X}, \cup, \cap, \ulcorner, X, \varnothing \rangle$ be a pseudo-complemented lattice of open sets that is isomorphic to $\mathfrak{L}$. Let $\mathcal{Y}$ be the set of all subsets of X. Then

$$\mathfrak{Y} = \langle \mathcal{Y}, \cup, \cap, -, X, \varnothing \rangle$$

is a Boolean lattice of sets, where $-$ is the operation set-theoretic complementation with respect to X. By Theorem 7.9, let $\mathbb{P}$ be an extended probability function on $\mathfrak{L}$. Then by Theorem 7.13, $\mathbb{P}$ is an extended lattice probability function. Let $\mathbb{P}'$ be the restriction of $\mathbb{P}$ to $\mathcal{X}$. Then $\mathbb{P}'$ is an extended lattice probability function on $\mathfrak{X}$. It is then easy to use the isomorphism of $\mathfrak{X}$ and $\mathfrak{L}$ to map $\mathbb{P}'$ onto a function $\mathbb{Q}$ on $\mathcal{L}$ that is an extended lattice probability function on $\mathfrak{L}$. $\square$

The following two theorems show the similarities and differences between Boolean lattices and pseudo-complemented distributive lattices. The first shows that operation of pseudo-complementation in distributive lattices shares several of well-known properties of the operation of complementation in Boolean lattices. The second shows the failure in pseudo-complemented distributive lattices of some important properties of the complementation operation of Boolean lattices.

Theorem 7.24. *Suppose* $\mathfrak{L} = \langle \mathcal{L}, \leq, \sqcup, \sqcap, \ulcorner, 1, 0 \rangle$ *is a pseudo-complemented distributive lattice. Then the following eight statements hold for all a and b in* $\mathcal{L}$:

(1) $\ulcorner 1 = 0$ *and* $\ulcorner 0 = 1$.
(2) If $b \leq a$, *then* $\ulcorner a \leq \ulcorner b$.
(3) If $a \sqcap b = 0$, *then* $b \leq \ulcorner a$.
(4) $a \sqcap \ulcorner a = 0$.
(5) $a \leq \ulcorner\ulcorner a$.
(6) $\ulcorner a = \ulcorner\ulcorner\ulcorner a$.
(7) $\ulcorner a \sqcup \ulcorner b \leq \ulcorner (a \sqcap b)$.
(8) $\ulcorner (a \sqcup b) = \ulcorner a \sqcap \ulcorner b$.

Proof. (1) and (2) follow from Theorem 7.18.

(3) and (4) are immediate consequences of the definition of "pseudo-complementation."

(5). By (4), $\vdash a \sqcap a = 0$. Then by (3), $a \leq \vdash\vdash a$.

(6). By (5), $a \leq \vdash\vdash a$. Then by (2), $\vdash\vdash\vdash a \leq \vdash a$. By (5),

$$\vdash a \leq \vdash\vdash (\vdash a) = \vdash\vdash\vdash a.$$

The above shows that $\vdash a = \vdash\vdash\vdash a$.

(7). By $a \sqcap b \leq a$ & $a \sqcap b \leq b$ and (2),

$$\vdash a \leq \vdash (a \sqcap b) \quad \& \quad \vdash b \leq \vdash (a \sqcap b).$$

Therefore $\vdash a \sqcup \vdash b \leq \vdash (a \sqcap b)$.

(8). By Theorem 7.22, it may be assumed that $\mathfrak{L}$ is a pseudo-complemented lattice of open sets. Then using simple topological considerations, it is easy to verify that the interior of the set-theoretic complement of the union of two open sets is the same as the intersection of the interiors of their set-theoretic complements, that is, it is easy to verify that (8) holds. $\square$

Theorem 7.25. *There exists a pseudo-complemented lattice* $\mathfrak{L} = \langle \mathcal{L}, \subseteq, \cup, \cap, \vdash, L, \varnothing \rangle$ *such that*

(1) for some A in $\mathcal{L}$, $A \cup \vdash A \neq L$;

(2) for some A in $\mathcal{L}$, $\vdash\vdash A \neq A$; and

(3) for some A and B in $\mathcal{L}$, $\vdash (A \cap B) \neq \vdash A \cup \vdash B$.

Proof. Let L be the set of real numbers and $\mathcal{L}$ be the set of open subsets of the reals determined by the $\leq$ ordering on the real numbers. Let $C = $ the open interval $(0, \infty)$ and $D = $ the open interval $(-\infty, 0)$. For each Y in $\mathcal{L}$, let $\vdash Y = $ the interior of the set-theoretic complement of Y, $L - Y$. Then it is easy to verify that $\mathfrak{L} = \langle \mathcal{L}, \subseteq, \cup, \cap, \vdash, L, \varnothing \rangle$ is a pseudo-complemented lattice. It is also easy to verify that (1) follows if $A = C$, (2) if $A = C \cup D$, and (3) if $A = C$ & $B = D$. $\square$

7.6. An Application to Decision Theory

7.6.1. *Introduction*

This section examines a new form of decision theory designed to incorporate emotion, bias, and other modes that influence human decision makers. It is based on the idea that the decision maker's subjective interpretation of events can vary with mode in a manner that influences her decision. This is modeled by having the subjective interpretations of events take place in an pseudo-complemented lattice of open sets instead of a Boolean algebra. This allows for an event to have multiple subjective interpretations, depending on mode, of the same event while at the same time retaining critical coherences necessary for "rationality." A more detailed account of this form of decision theory can be found in Narens (2011).

7.6.2. *Subjective and objective coherences*

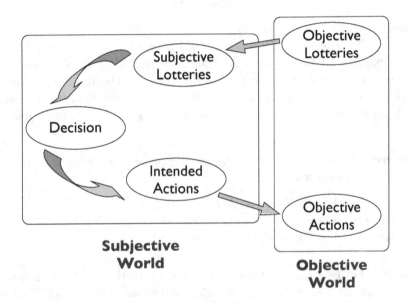

Fig. 7.4.   Decision Making

Figure 7.4 illustrates the subjective and objective (observable) parts of decision making involving lotteries. The objective world consists of observable behavior of the decision maker, for example, the decision maker's ranking of lotteries. The subjective world is a theoretical framework for how decisions get made. *It is not assumed that the act of interpreting objective lotteries into subjective ones produces an isomorphism.* This is because it is assumed that in making a decision, the decision maker can enter in various "modes." These modes result from factors exogenous to content of the lotteries, e.g., differences in the way they are stated or displayed, or from factors endogenous to the lotteries, e.g., one lottery contains a catastrophic outcome for the decision maker, while another does not. In fact, if lotteries L_1 and L_2 contain a common event A, then it may be the case that the interpretation for a mode M, $\mathcal{I}_M(A)$, may differ from the interpretation for a mode N, $\mathcal{I}_N(A)$. For example, if in L_1 the event A is paired with an outcome that is catastrophic for the decision maker in L_1 and induces fear in her, she may give it a different interpretation than for the lottery L_2 that has an outcome that when paired with A does not induce fear in her. The decision maker uses her interpretations of lotteries to make a decision about which action she intends to take in the objective world. It is assumed that the decision maker carries out the intended action in the objective world, e.g., she chooses the lottery she intends, or she ranks lotteries as she intends. This chapter focuses on the question of when such decisions are "rational" or "normative".

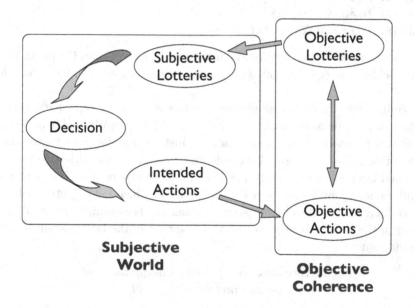

Fig. 7.5. Objective Coherence

"Coherence" is a concept used in decision theory to justify concepts of "rationality." The double arrows in Figure 7.5 represent how coherence is usually employed. Note that it applies to only things in the objective world. For this reason, it is called here "objective coherence" to distinguish it from other coherence concepts developed in this chapter that focus on subjective matters.

An example when objective coherence cannot be achieved is a Dutch Book: An individual, called a *bookie*, sets prices for lotteries so that for each lottery she is indifferent between buying or selling the lottery at its price. Then a *Dutch Book* can be made against the bookie if her prices can be arbitraged—that is, and if an arbitrageur can buy lotteries from her and sell lotteries to her at their prices in a manner so that arbitrageur will always make a profit, no matter what event occurs. The bookie prices for the lotteries is said to be rational if no Dutch Book can be made against them. The following is an example of a set of prices that yield a Dutch Book against the bookie:

The events Y, B, R are pairwise disjoint, $X = Y \cup B \cup R$ is the sure event, and the bookie offers the following bets:

(1) $(\$1, Y; \$0, X - Y) = \$\frac{1}{4}$,

(2) $(\$1, B; \$0, X - B) = \$\frac{1}{4}$,

(3) $(\$1, Y \cup B; \$0, X - [Y \cup B]) = \$\frac{2}{3}$.

A Dutch book is made against the bookie by buying (1) and (2) (cost: $\$\frac{1}{2}$) and selling (3) (generating: $\$\frac{2}{3}$), guaranteeing a profit no matter which event occurs.

Subjective coherence is similar to objective coherence except that it takes place in the subjective world instead of the objective world and is between subjective lotteries and intended actions instead of objective lotteries and objective actions. In Figure 7.6 the double arrows between subjective lotteries and intended actions corresponds to subjective coherence. The lotteries and monetary prices given in (1), (2) and (3), which fail objective coherence, may nevertheless be subjectively coherent. For example, in the above case of the bookie, it need not be the case that in the bookie's subjective world that

(the interpretation of Y) $\cup$ (the interpretation of B)
$=$ the interpretation of $(Y \cup B)$,

because she can be in one mode when interpreting Y and when interpreting B and a different mode when interpreting $Y \cup B$.

7.6.3. *Multimode utility theory: General principles*

Multimode Utility Theory—MUT for short—is an approach to modeling decision making where the decision maker can vary in modes, e.g., emotional states, in the evaluation of a lottery or can change modes with lotteries. MUT contains many subtheories. The one that is the focus of this chapter

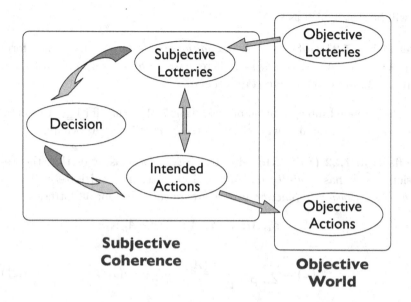

Fig. 7.6. Subjective Coherence

is called *Descriptive Subjective Expected Utility*—or *DSEU* for short. It is a theory of decision making where the decision maker is *subjectively* rational. When the decision maker has only one mode, DSEU degenerates into a variant of the standard SEU model of (objectively) rational decision making.

MUT is based on the following four principles:

(i) The decision maker has various modes that influence her decisions. $\mathcal{M}$ represents the set of modes.

(ii) The subjective interpretation of an event can vary with mode.

(iii) Each lottery is interpreted subjectively as a lottery.

(iv) The meanings of subjective interpretations of the same event are semantically related.

Semantic relatedness in (iv) is discussed later in the chapter.

Subjective Expected Utility—SEU or the SEU model—was formulated in Savage (1954). Savage used unconditional probability as part of his foundation. Luce and Krantz (1971) formulated a variant of SEU using conditional probability. This chapter uses the variant of Luce and Krantz. In this chapter, this variant is generally referred to as "Conditional SEU", or "(Conditional) SEU", or just "SEU". When Savage's formulation is needed, it will be distinguish as such.

Definition 7.21 (lottery). *A (conditional) lottery* has the form $(a_1, A_1; \ldots; a_i, A_i; \ldots; a_n, A_n)$, where $a_1, \ldots, a_n$ are *outcomes* and $A_1, \ldots, A_n$ are disjoint *uncertain events.*

The intended interpretation of Definition 7.21 is that if $\bigcup_{i=1}^{n} A_i$ occurs, then the decision maker will receive outcome a_i if event A_i occurs.

Definition 7.22 (SEU Model). The *SEU Model* assumes that the decision maker has a *utility function* u over outcomes and lotteries and a *subjective probability function* P over events such that for all lotteries

$$L = (a_1, A_1; \ldots; a_i, A_i; \ldots; a_n, A_n),$$

$$u(L) = \sum_{i=1}^{n} \frac{\mathsf{P}(A_i)}{\mathsf{P}(A_1) + \cdots + \mathsf{P}(A_n)} \cdot u(a_i). \qquad (7.21)$$

Equation 7.21 is called the *SEU Model* for L.

Note that in (7.21),

$$\frac{\mathsf{P}(A_i)}{\mathsf{P}(A_1) + \cdots + \mathsf{P}(A_n)}$$

is the subjective *conditional probability* of A_i occurring given that $\bigcup_{i=1}^{n} A_i$ occurs.

Because MUT allows for multiple interpretations of the same event, it is convenient to follow a modeling idea introduced by Tversky and Koehler (1994) and have subjective evaluations and utilities be defined in terms of descriptions of lotteries instead of directly in terms of lotteries.

Definition 7.23 (descriptions). A typical description of the lottery

$$L = (a_1, A_1; \ldots; a_i, A_i; \ldots; a_m, A_m)$$

has the form

$$\mathbf{L} = (\mathbf{a}_1, \mathbf{A}_1; \ldots; \mathbf{a}_i, \mathbf{A}_i; \ldots; \mathbf{a}_m, \mathbf{A}_m).$$

L is called the *objective interpretation* of $\mathbf{L}$ and is denoted by $\mathcal{I}_O(\mathbf{L})$. Similarly, the *objective interpretations* of $\mathbf{A}_i$ and $\mathbf{a}_i$ are, respectively, A_i and a_i, and are denoted by, respectively, $\mathcal{I}_O(\mathbf{A}_i)$ and $\mathcal{I}_O(\mathbf{a}_i)$.

Definition 7.24 (Modes and Interpretations). $\mathcal{M}$ denotes the set of states of the decision maker. These states are called the *modes* of the decision maker. A special mode is sometimes needed. This mode, O, is called the *state of objectivity* or the *objective mode*, and it may or may not be a state of the decision maker, that is, may or may not be in $\mathcal{M}$.

When the decision maker is in mode M, $\mathcal{I}_M(\mathbf{A})$ denotes her interpretation of $\mathbf{A}$. Similarly for $\mathcal{I}_M(\mathbf{a})$ and $\mathcal{I}_M(\mathbf{L})$.

7.6.4. *DSEU model*

MUT modeling is a general way of taking emotion, biases, and contexts into account for decision making. One subset of MUT models, called *DSEU models,* generalize SEU modeling in a manner that stresses subjective rationality while being able to provide alternative modeling for a number of phenomena in the literature. It also provides the means to model phenomena where utility and probability interact in situations like in the following experiment:

Kissing Your Favorite Movie Star Experiment. Rottenstreich and Hsee (2001) conducted the following experiment. A participant was placed in one of two conditions, a certainty condition or low probability condition. Certainty condition participants were asked to imagine that they could receive either "the opportunity to meet and kiss your favorite movie star" or $50 in cash. Low probability condition participants were asked to imagine that they could take part in either a lottery offering a 1% chance of winning "the opportunity to meet and kiss your favorite movie star" or a lottery offering a 1% chance of winning $50 in cash. The results were: *Certainty condition:* 70% of participants preferred the cash over the kiss. *Low-probability condition:* 65% of participants preferred the kiss lottery over the cash lottery.

This data not only violates SEU but the large class of utility models that assume a fundamental axiom of measurement theory called "conjoint independence" holding between probabilities and utilities of positive outcomes. (See Chapter 6 of Krantz, Luce, Suppes, & Tversky, 1971, for formal definitions and discussion of conjoint independence, which they call "independence".) This is how Rottenstreich and Hsee (2001) interpret the result of their experiment:

> Weighting functions will be more S-shaped for lotteries involving affect-rich than affect-poor outcomes. That is, people will be more sensitive to departures from impossibility and certainty but less sensitive to intermediate probability variations for affect-rich outcomes. We corroborated

this prediction by observing probability-outcome interactions: An affect-poor prize was preferred over an affect-rich prize under certainty, but the direction of preference reversed under low probability. We suggest that the assumption of probability-outcome independence, adopted by both expected-utility and prospect theory, may hold across outcomes of different monetary values, but not different affective values. (p. 185)

It is claimed by many that rational decision making for lotteries demand the SEU model to hold. The bases for such claims are various axiomatizations of SEU, e.g., those of Savage (1954) and Luce and Krantz (1971). The intent of DSEU modeling is to produce models that are structurally very close to SEU, are "subjectively rational", and allow for phenomena such as the result of the Kissing Your Favorite Movie Star experiment.

Convention 7.1. Throughout the rest of this chapter, it is assumed that the interpretations of descriptions of events are into a lattice of open sets $\mathfrak{X} = \langle \mathcal{X}, \subseteq, \cup, \cap, \neg, X, \varnothing \rangle$ (Definition 7.18) and that P is a lattice probability function (Definition 7.20) on $\mathfrak{X}$. Elements of $\mathcal{X}$ are often called *interpreted events*.

Definition 7.25 (DSEU expected utility formula). By definition, u is the decision maker's utility function over outcomes and lotteries. The *DSEU expected utility formula* is said to hold if and only if for all lotteries L,

$$u(L) = \sum_{i=1}^{m} \frac{\mathsf{P}[\mathcal{I}_{M_i}(\mathbf{A}_i)]u(a_i)}{\mathsf{P}[\mathcal{I}_{M_1}(\mathbf{A}_1)] + \cdots + \mathsf{P}[\mathcal{I}_{M_m}(\mathbf{A}_m)]}, \qquad (7.22)$$

where for $i = 1, \ldots, m$, M_i is the mode of the decision maker during the evaluation of the lottery term description $a_i, \mathbf{A}_i$ as part of her computation the utility of L, *and there is a function F such that $M_i = F(a_i, \mathbf{A}_i)$.* Note that in (7.22), $u(a_i)$ (which, by a previous definition, is $u[\mathcal{I}_O(a_i)]$), does not depend on the mode of the decision maker. This feature is called *outcome objectivity*.

Definition 7.26 (semantical coherence). Let A be a lottery event and M and N be modes. Then $\mathcal{I}_M(\mathbf{A})$ and $\mathcal{I}_N(\mathbf{A})$ are said to be *semantically related* if and only if

$$\neg \neg \mathcal{I}_M(\mathbf{A}) = \neg \neg \mathcal{I}_N(\mathbf{A}).$$

Semantic coherence is said to hold if and only if for all lottery events B, C, and D and all modes V and W in $\mathcal{M}$,

(i) $B \cap C = \varnothing$ iff $\mathcal{I}_V(\mathbf{B}) \cap \mathcal{I}_V(\mathbf{C}) = \varnothing$, and

(ii) $\mathcal{I}_V(\mathbf{D})$ is semantically related to $\mathcal{I}_W(\mathbf{D})$.

Note that it follows from properties of $\frown$ that if $\mathcal{I}_M(\mathbf{A})$ and $\mathcal{I}_N(\mathbf{A})$ are semantically related, then for all elements B in $\mathcal{X}$ and all modes W in $\mathcal{M}$,

$$\mathcal{I}_W(\mathbf{B}) \cap \mathcal{I}_M(\mathbf{A}) = \varnothing \text{ iff } \mathcal{I}_W(\mathbf{B}) \cap \mathcal{I}_N(\mathbf{A}) = \varnothing . \qquad (7.23)$$

Also, it follows from semantic coherence and (7.23) that the subjective interpretation of a lottery is a lottery.

Definition 7.27 (DSEU model). The *DSEU model* is said to hold if and only if the DSEU expected utility formula and semantic coherence hold.

In summary, the DSEU model is characterized by the following properties:

- *Decision weights:* P is a lattice probability function on a pseudo-complemented lattice of open sets.
- *Outcome objectivity:* For all modes M and all outcomes a, $\mathcal{I}_M(\mathbf{a}) = \mathcal{I}_O(\mathbf{a}) = a$.
- *DSEU Utility Formula:*

$$u(L) = \sum_{i=1}^{m} \frac{\mathsf{P}[\mathcal{I}_{M_i}(\mathbf{A}_i)]u(a_i)}{\mathsf{P}[\mathcal{I}_{M_1}(\mathbf{A}_1)] + \cdots + \mathsf{P}[\mathcal{I}_{M_m}(\mathbf{A}_m)]} .$$

- *Dependency of mode on description:* $M_i = F(\mathbf{a}_i, \mathbf{A}_i)$.
- *Semantic coherence:* For all lottery events A and B and all modes M and N,

 (i) $\mathcal{I}_M(\mathbf{A})$ is semantically related to $\mathcal{I}_N(\mathbf{A})$, and

 (ii) $A \cap B = \varnothing$ iff $\mathcal{I}_M(\mathbf{A}) \cap \mathcal{I}_M(\mathbf{B})$.

Testing DSEU is very much like the *qualitative* testing of SEU: One presents pairs of lotteries to the participant and ask which she prefers, and tests whether or not her preference behavior is consistent with core preference properties derived from SEU's utility equation,

$$u(L) = \sum_{i=1}^{m} \frac{\mathsf{P}[\mathcal{I}_{M_i}(\mathbf{A}_i)]u(a_i)}{\mathsf{P}[\mathcal{I}_{M_1}(\mathbf{A}_1)] + \cdots + \mathsf{P}[\mathcal{I}_{M_m}(\mathbf{A}_m)]} .$$

It is important that the function F in the equation,

$$M_i = F(\mathbf{a}_i, \mathbf{A}_i) ,$$

is specified *prior* to conducting the test; otherwise there are too many ways of fitting the data. In testing, the equation $M_i = F(\mathbf{a}_i, \mathbf{A}_i)$ should be taken as a theory about what gives rise to modes. The case $M_i = F(\mathbf{a}_i)$ for various F captures various extensions of SEU in the literature.

Luce (1991) provided a theory of rank- and sign-dependent linear utility models for binary gambles (i.e., binary lotteries with monetary outcomes). The following is the formula for the sign-dependent case he derives from qualitative assumptions for the utility of a gamble:

$$u[(\$a, A; \$b, B)] = \begin{cases} u(\$a)w^+(A) + u(\$b)(1 - w^+(B)) & \text{if } \$a > \$0,\ \$b > \$0, \\ u(\$a)w^-(A) + u(\$b)(1 - w^-(B)) & \text{if } \$a < \$0,\ \$b < \$0, \\ u(\$a)w_1(A) + u(\$b)(1 - w_1(B)) & \text{if } \$a > \$0,\ \$b \leq \$0, \\ u(\$a)w_2(A) + u(\$b)(1 - w_2(B)) & \text{if } \$a \leq \$0,\ \$b > \$0. \end{cases}$$
$$(7.24)$$

The model described in (7.24) is for descriptive modeling, and, for this reason, w^+, w^-, w_1, and w_2 are called "weighting functions," because they need not be additive over disjoint events.

The following is a DSEU model of a closely related situation. For each description $\mathbf{c}, \mathbf{A}$ of a term of a gamble, suppose DSEU and

$$F(\mathbf{c}, \mathbf{A}) = F(\mathbf{c}) = \begin{cases} + & \text{if } \$c > \$0, \\ - & \text{if } \$c \leq \$0. \end{cases}$$

Then for each binary gamble $(a, A; b, B)$,

if $\$a > \$0$ and $\$b > \$0$, then
$$u[(a, A; b, B)] = \frac{\mathsf{P}[\mathcal{I}_+(\mathbf{A})]u(a)}{\mathsf{P}[\mathcal{I}_+(\mathbf{A})] + \mathsf{P}[\mathcal{I}_+(\mathbf{B})]} + \frac{\mathsf{P}[\mathcal{I}_+(\mathbf{B})]u(b)}{\mathsf{P}[\mathcal{I}_+(\mathbf{A})] + \mathsf{P}[\mathcal{I}_+(\mathbf{B})]};$$

if $\$a < \$0$ and $\$b < \$0$, then
$$u[(a, A; b, B)] = \frac{\mathsf{P}[\mathcal{I}_-(\mathbf{A})]u(a)}{\mathsf{P}[\mathcal{I}_-(\mathbf{A})] + \mathsf{P}[\mathcal{I}_-(\mathbf{B})]} + \frac{\mathsf{P}[\mathcal{I}_-(\mathbf{B})]u(b)}{\mathsf{P}[\mathcal{I}_-(\mathbf{A})] + \mathsf{P}[\mathcal{I}_-(\mathbf{B})]};$$

if $\$a > \$0$ and $\$b \leq \$0$, then
$$u[(a, A; b, B)] = \frac{\mathsf{P}[\mathcal{I}_+(\mathbf{A})]u(a)}{\mathsf{P}[\mathcal{I}_+(\mathbf{A})] + \mathsf{P}[\mathcal{I}_-(\mathbf{B})]} + \frac{\mathsf{P}[\mathcal{I}_-(\mathbf{B})]u(b)}{\mathsf{P}[\mathcal{I}_+(\mathbf{A})] + \mathsf{P}[\mathcal{I}_-(\mathbf{B})]};$$

and if $\$a \leq \$0$ and $\$b > \$0$, then
$$u[(a, A; b, B)] = \frac{\mathsf{P}[\mathcal{I}_-(\mathbf{A})]u(a)}{\mathsf{P}[\mathcal{I}_-(\mathbf{A})] + \mathsf{P}[\mathcal{I}_+(\mathbf{B})]} + \frac{\mathsf{P}[\mathcal{I}_+(\mathbf{B})]u(b)}{\mathsf{P}[\mathcal{I}_-(\mathbf{A})] + \mathsf{P}[\mathcal{I}_+(\mathbf{B})]}.$$

The DSEU and Luce's models are very similar. The principal difference is that DSEU requires much more additivity than Luce's model, because P is required to be a lattice probability function. This makes the DSEU model much more rational or normative-like than the Luce model, and thus a better model for matters such as public policy decisions, where loss aversion is included as a priority. On the other hand, for laboratory studies of gambling with relatively small sums of money, Luce's model is likely to describe data better than DSEU, because of well-known failures of probabilistic additivity in related studies.

The above DSEU model, which has two lattice probability functions, does not distinguish between some pairs of gambles having both positive and negative outcomes in the manner of the Luce model that has four probability functions. This distinction, if needed, can be incorporated into a multimode extension of the DSEU model that has four modes.

The following are two other examples from the literature that have similar DSEU models:

Lopes (1987) looks at the impacts of hope and fear on decision making. Interpreted in terms of DSEU, a large gain puts the decision maker in a hope mode that increases the subjective probability associated with the gain, and a large loss puts the decision maker in a fear mode that increases the subjective probability associated with the loss. Note that in this case a mode M is determined by a formula of the form $M = F(\mathbf{a})$.

Chichilnisky (2009) provided a theory of how catastrophes impact decision making. Interpreted in terms of DSEU the key qualitative phenomena are the following:

A description of a catastrophic outcome $\mathbf{a}$ puts the decision maker in a fear mode which increases her subjective probability of the outcome. Note that in this case a mode M is determined by a formula of the form $M = F(\mathbf{a})$.

A description of a catastrophic event $\mathbf{A}$ puts the decision maker in a fear mode which increases her subjective probability of the outcome. In this case a mode M is determined by a formula of the form $M = F(\mathbf{A})$.

7.6.5. *Rationality and DSEU*

SEU, exemplified by Savage's axioms, has become the gold standard for rationality, especially in economics. This has been questioned by some economists: Gilboa, Postlewaite and Schmeidler (2009) write:

> We reject the view that rationality is a clear-cut, binary notion that can be defined by a simple set of rules or axioms. There are various ingredients to rational choice. Some are of internal coherence, as captured by Savage's axioms. Others have to do with external coherence with data and scientific reasoning. The question we should ask is not whether a particular decision is rational or not, but rather, whether a particular decision is more rational than another. And we should be prepared to have conflicts between the different demands of rationality. When such conflicts arise, compromises are called for. Sometimes we may relax our

demands of internal consistency; at other times we may lower our standards of justifications for choices. But the quest for a single set of rules that will universally define the rational choice is misguided.

The "rationality" inherent in DSEU is demonstrated in the following theorem. With a very reasonable assumption[1] about P, the following theorem is shown in Narens (2011):

Theorem 7.26. *A DSEU model's subjective part,*

$$\langle \mathcal{S} = \{\mathcal{I}_M(\boldsymbol{A} \mid M \in \mathcal{M} \ \& \ A \ a \ lottery \ event\},$$

$$P \ restricted \ to \ \mathcal{S}, \{u(L) \mid L \ is \ a \ lottery\},$$

$$\mathcal{OUT} = \{\mathcal{I}_O(\boldsymbol{a}) \mid a \ is \ a \ lottery \ outcome\}, \{u(a) \mid a \ is \ in \ \mathcal{OUT}\}\rangle,$$

is isomorphically imbeddable in a (Conditional) SEU model.

Proof.   Theorem 1 of Narens (2011).   □

This theorem implies that any irrationality in the subjective part of a DSEU model can be transferred to a SEU model via the isomorphic imbedding. However, the SEU model cannot have any irrationality. Therefore, the DSEU model also does not have any *subjective* irrationality, because it is a subpart of a rational system.

7.6.6. *Summary and conclusion for MUT and DSEU*

Three major changes into the formalism of the SEU theory of decision making has been introduced:

- The introduction of modes of the decision maker. This allows for a wider range of mental activity for the decision maker while making decisions.
- The use of descriptions of events as a primitive concept. This allows for events to have multiple interpretations by allowing the interpretation of the description to vary with mode.

---

[1]The assumption is: If $\lesssim$ is defined on the set of interpreted events $\mathcal{X}$ by,

$$A \lesssim B \ \text{iff} \ \mathbb{P}(A) \leq \mathbb{P}(B),$$

and if $\mathbb{Q}$ is a lattice probability function such that for all A and B in $\mathcal{X}$,

$$A \lesssim B \ \text{iff} \ \mathbb{Q}(A) \leq \mathbb{Q}(B),$$

then $\mathbb{P} = \mathbb{Q}$. Using proof techniques from the theory of measurement, it is easy to show that this condition holds for most situations of economic interest. It is assumed here so that a theorem from the literature that uses it can be cited. I believe it is likely that this assumption is not needed to obtain Theorem 7.26.

- The change of the event space from a Boolean algebra to a pseudo-complemented lattice of open sets. This provides a mechanism for allowing subjective probabilities to vary with modes while maintaining coherence within and across lotteries under change in mode. The coherence feature guarantees that lotteries will remain lotteries under change of mode and will have the interpretations of the same event description and be coherently related to one another.

Narens (2011) comments the following about these changes and the importance of subjective rationality for decision making:

> While these three changes are major innovations with respect to SEU, they essentially leave intact the ideas involved in SEU's formula for computing the utility of lotteries. This allows for the formulation of the DSEU model that maintains a highly rational flavor in the tradition of SEU while allowing the decision maker to fluctuate among modes during her calculations of utilities. ...
>
> Humans are emotional. And while some psychologists may designate certain human emotional states as "neutral", these "neutral" states do not appear to have much to do with rational or normative decision making. DSEU shows how shifting affect can impact subjective probabilities while yielding coherent decision making. I conclude from this result that it is possible to include affect into normative decision making—or put differently—to redefine normative decision making to include affect. In my opinion, such an inclusion allows for a better decision theory for humans, both from normative and descriptive perspectives, than those, like SEU, that ignore affect.
>
> Humans are also subjective. In traditional decision theories, subjectivity is treated like a crazed person in need of a straight jacket. In SEU, the straight jacket takes the form of a rigid structure that only allows subjectivity for the importance of outcomes and the size of probabilities. In contrast, MUT sees instead a colorful individual in need of the kind of clarity provided by, for example, DSEU.

Acknowledgments

The research for this chapter was supported by the AFOSR grant FA9550-08-1-0389.

References

Birkhoff, G., & von Neumann, J. (1936). The logic of quantum mechanics. *Annals of Mathematics, 37*, 823–843.

Chichilnisky, G. (2009). The topology of fear, *Journal of Mathematical Economics, 45*, 807–816.

de Finetti, B. (1937). La prévision: Ses lois logiques, ses sources subjectives. *Annales de l'Institut Henri Poincaré, 7*, 1–68.

Gilboa, I., Postlewaite, A., & Schmeidler, D. (2009). Is it always rational to satisfy Savage's axioms? *Economics and Philosophy, 25*, 285–296.

Gödel, K. (1933). Eine Interpretation des intuitionistischen Aussagenkalküls. *Ergebnisse eines Mathematischen Kolloquiums, 4*, 39–40. English translation in J. Hintikka (Ed.), *The Philosophy of Mathematics*. Oxford, 1969.

Kolmogorov, A. (1932). Zur Deutung der intuitionistischen Logik. *Mathematische Zeitschrift, 35*, 58–65. English translation in P. Mancosu (Ed.), *From Brouwer to Hilbert : The Debate on the Foundations of Mathematics in the 1920s*. Oxford University Press, 1998.

Krantz, D. H., Luce, R. D., Suppes, P., & Tversky, A. (1971). *Foundations of Measurement, Vol. 1: Additive and Polynomial Representation*. NY: Academic Press.

Lopes, L. L. (1987). Between hope and fear: The psychology of risk. In L. Berkowitz (Ed.), *Advances in Experimental Social Psychology*. NY: Academic Press.

Luce, R. D. (1991). Rank- and sign-dependent linear utility models for binary gambles. *Journal of Economic Theory, 53*, 75–100.

Luce, R. D., & Krantz, D. H. (1971). Conditional expected utility. *Econometrica, 39*, 253–271.

Luxemburg, W. A. J. (1962). Two applications of the method of construction by ultrapowers to analysis. *Bulletin of the American Mathematical Society, 68*, 416–419.

Narens, L. (2005). A theory of belief for scientific refutations. *Synthese, 145*, 397–423.

Narens, L. (2007). *Theories of Probability: An Examination of Logical and Qualitative Foundations*. London: New World Scientific.

Narens, L. (2009). A foundation for support theory based on a non-boolean event space. *Journal of Mathematical Psychology, 53*, 399–407.

Narens, L. (2011). Multimode utility theory. Unpublished manuscript.

Nikodym, O. (1960). Sur le mesure non-archimedienne effective sur une tribu de Boole arbitraire. *Comptes Rendus Mathmatique. Acadmie des*

Sciences. Paris, 251, 2113–2115.

Rottenstreich, Y., & Hsee, C. K., (2001). Money, kisses, and electric shocks: On the affective psychology of risk. *Psychological Science, 12*, 185–190.

Savage, L. J. (1954). *The Foundations of Statistics*. NY: John Wiley & Sons.

Stone, M. H. (1936). The theory of representations for boolean algebras. *Transactions of the American Mathematical Society, 40*, 37–111.

Stone, M. H. (1937). Topological representations of distributive lattices and Brouwerian logics. *Časopis Pro Pestovaní Matematiky A Fysiky, 67*, 1–25.

Szász, G. (1963). *Introduction to Lattice Theory* (3rd ed.). NY: Academic Press.

Tversky, A., & Koehler, D. (1994). Support theory: A nonextensional representation of subjective probability. *Psychological Review, 101*, 547–567.

von Mises, R. (1936). *Wahrscheinlichkeit, Statistik un Wahrheit* (2nd ed.). English translation, *Probability, Statistics and Truth*. Dover, 1981.

Chapter 8

Presumption of Equality as a Requirement of Fairness

Wlodek Rabinowicz

Lund University

Presumption of Equality enjoins that individuals be treated equally in the absence of discriminating information. My objective in this chapter is to make this principle more precise, viewing it as a norm of fairness, in order to determine why and under what conditions it should be obeyed.

Presumption norms are procedural constraints, but their justification might come from the expected outcomes of the procedures they regulate. This outcome-oriented approach to fairness is pursued in the chapter. The suggestion is that in the absence of information that would discriminate between the individuals, equal treatment minimizes the expected unfairness in the outcome. Another suggestion is that, under these circumstances, equal treatment also minimizes maximal possible unfairness, i.e., it is minimally unfair if 'worst comes to worst'. Whether these suggestions are correct depends on the properties of the underlying unfairness measure.

8.1. Introduction

This chapter examines Presumption of Equality (PE), which enjoins us to treat different individuals equally if we cannot discriminate between them on the basis of the available information. I will view this principle as a requirement of fairness – more specifically, as a procedural principle whose goal is to promote *fairness in outcome*. The objective is to make PE so understood more precise and to determine why and under what conditions it should be obeyed.

Why, then, should PE be obeyed? A natural answer is that, in the absence of relevant discriminating information, treating some individuals better than others is *arbitrary*, which is a bad thing. There is certainly some truth in this explanation. But while arbitrariness considerations are important, they are not always decisive. In some cases in which the discriminating information is absent, unequal treatment might still be right, despite its arbitrariness. To illustrate, suppose two individuals compete for two scholarships, one of which is more attractive than the other. Your task is to make the decision, but the information you have is limited. While you know that both candidates are deserving, you have no clue which of them, if any, has stronger merits. Suppose you have no opportunity to gather further information. Since the scholarships are not equally attractive, to give one to one individual and the other to the other is to treat them un-equally, which is arbitrary and to that extent unsatisfactory. At the same time, one *could* treat them equally by withholding scholarships from both. Such 'levelling down', however, would be grossly unfair given that both of them deserve a scholarship. Avoidance of arbitrariness is thus not all that matters. To justify equal treatment, in cases in which such treatment *can* be justified, we need to rely on other considerations.

At this point, I expect an objection: Why not decide who gets what scholarship by a toss of a coin? This would give each individual a fifty-fifty chance of getting the more attractive scholarship, but even the loser would not come away empty-handed: She would receive the other scholarship instead. Arguably, such a lottery is itself a form of equal treatment, since it gives each person equal chances. By tossing a coin, we avoid arbitrariness but at the same time see to it that both persons are awarded, which they deserve.

It is true that, in the case at hand, drawing lots or tossing a coin is the obvious thing to do. Avoidance of arbitrariness *is* important. I would deny, however, that equal lottery on unequal treatments is on a par with equal treatment. On an outcome-oriented approach to fairness, which is adopted in this chapter, this is not so. The outcome of the lottery will still be unequal. And inequality in outcome may well matter from the point of view of fairness.

Here is the suggestion I instead want to examine: Principles of fairness can be constraints on procedures or constraints on outcomes. Presumption norms such as PE constrain procedures, but procedural constraints can often be justified in terms of the expected outcomes of the procedures that obey these constraints. This is the path that will be pursued in my chapter: The suggestion is that equal treatment should be chosen because, and to the extent that, it *minimizes the expected unfairness* in outcome. When the available information does not discriminate between the individuals concerned, expected unfairness will normally be at its lowest if individu-als are treated equally. Normally, but not always. As will be seen, the scholarship example provides an exception to this rule.

It has been suggested that the core reason behind presumption norms is to be found in the differential costs of potential errors.[1] Thus, for example, presumption of innocence in criminal law is justified by the greater moral cost of punishing an innocent person as compared with that of letting a guilty person go free. Louis Katzner applied this idea to the choice between presumption of equality and its opposite, presumption of inequality:

> The only possible basis for opting for one of them rather than the other is which state of affairs one would rather see – that in which some of those who are similar are treated differently or that in which some of those who are different are treated similarly. (Katzner 1973, p. 92)[2]

My approach to PE is different. As will be seen, for this principle to hold, the moral cost of treating equals unequally need not be greater than that of the equal treatment of unequals.

Some presumptions might have more to do with the differential probability of errors than with their differential costs: What is being presumed is deemed to be sufficiently probable to function as a default assumption.[3] But my argument does not assume that equality in deserts is probabilistically privileged in this way. Indeed, PE might well be justified even in those cases in which it is very *im*probable, or perhaps even excluded, that the individuals are equally deserving.

While the justification I offer does not appeal to the differences in the costs of errors or in the probabilities of errors, it does appeal to the differences in the *expected* costs of errors. The suggestion is that in the absence of discriminating information we should treat individuals as if they were equal because the expected moral cost of error is minimized in this way: Equal treatment minimizes expected unfairness. To get an idea why it is so, note that unfairness may be seen as a kind of distance between the

---

[1] See Ullmann-Margalit (1983), p. 159: "It is the justification of presumptions in normative terms which touches what I take to be the core of the concept of presumption. [...] this normative type of consideration has to do with the acceptability of error." (p. 159) Making a presumption is grounded in "certain evaluative considerations which are primarily concerned with the differential acceptability of the relevant sorts of expected errors: the fact that one sort of error is judged to be, in the long run and all things considered, preferred on grounds of moral values or social goals to the alternative sort(s) constitutes an overriding reason for the decision underlying the presumption rule." (p. 162)

[2] Quoted in Ullmann-Margalit (1983).

[3] Cf. Ullmann-Margalit (1983), p. 157: "with presumption rules relating to presumptions that accord with the normal balance of probability the chance of an error [...] is reduced."

way individuals are treated and the way they deserve to be treated. A treatment's expected unfairness is on this view its expected distance from the fair treatment. We can thus think of the set of possible treatments as a set of points forming a spatial area. One of these points is the fair treatment but, in the absence of information, we don't know which point it is. Consequently, we do not know how large is the distance between the treatment we choose and the fair treatment. Now, the conjecture is that equal treatment lies in the center of this spatial area. At the same time, in the absence of information that discriminates between individuals, no direction in the area is more privileged than other directions. Therefore, by positioning ourselves in the center we should minimize our *expected* distance from the fair treatment, i.e. keep the expected unfairness in outcome at a minimum.

I want to examine under what conditions on the unfairness measure equal treatment will in fact have this feature of centrality. I will also inquire what happens if we instead choose a 'minimax' approach, i.e. opt for a treatment that is least unfair if 'worst comes to worst'. In other words, I will also examine under the conditions on the unfairness measure under which equal treatment minimizes maximal possible unfairness. Intuitively, centrality should guarantee this result as well.

8.2. Individuals and Treatments

The model to be used is highly abstract and allows of different interpretations. Its main components are a non-empty finite set $\mathbf{I} = \{i_1, \ldots, i_n\}$ of *individuals* and a non-empty set $\mathbf{T} = \{a, b, c, \ldots\}$ of possible *treatments* of individuals in $\mathbf{I}$. Every treatment a in $\mathbf{T}$ is assumed to be a vector $(a_1, \ldots, a_n)$, where, for every k $(1 \leq k \leq n)$, a_k is the way in which individual i_k is treated in a. We shall sometimes use the notation $a(i_k)$ for a_k. Treatment a is *equal* iff $a_1 = \ldots = a_n$. Other components of the model will be introduced later.

The three interpretations of the model that follow are themselves relatively abstract. Each can in turn be instantiated in many different ways.

Interpretation 1: Cake-divisions
A 'cake' is a homogeneous object or resource that is to be divided, without remainder, among the individuals in $\mathbf{I}$. A treatment a is a vector of real numbers, $(a_1, \ldots, a_n)$, with each a_k being the share of the cake assigned by a to individual i_k. The shares are all non-negative and together they

sum up to one. **T** is the set of all possible vectors of this kind. The equal treatment, $(1/n, ..., 1/n)$, divides the cake equally among the members of **I**.

Representing cake-divisions in this way means that we view them as *types* rather than tokens. Thus, to illustrate, if a cake is divided in pieces of equal size, it does not matter who gets which piece, since the cake is homogeneous. There is therefore no reason to make this distinction in the model. This is a general feature of our approach. Treatments are interpreted as types that specify the relevant characteristics of their tokens. As a result, any two treatments in the model are supposed to be relevantly different from each other.

Interpretation 2: Rankings
On this interpretation, **T** is the set of all possible rankings of the individuals in **I**. That a treatment a ranks i above j means that i is treated better than j in a. A tie in a between i and j means that they are treated equally well. The ranking interpretation of treatments is appropriate when *ordinal* differences between the individuals are all that matters from the point of view of fairness, i.e., when fairness only requires that the more deserving individuals should be better treated and that the equally deserving individuals should be treated equally well.

A ranking may be represented as an assignment of ordinal numbers to individuals, with 1 being the highest level in the ranking, 2 being the second highest level, etc. The assignment of levels starts from the highest one and continues downwards. Thus, equal treatment is the ranking in which every individual is assigned the highest level: $(1, ..., 1)$. (For another logically equivalent representation of rankings see below, Section 8.3.)

Interpretation 3: Indivisible Goods
Suppose that G is a set of indivisible objects that are to be distributed, with or without remainder, among the individuals in **I**. $a(i)$ is the subset of G that treatment a assigns to an individual i. For some i, $a(i)$ may be empty, and for distinct i and j, $a(i)$ and $a(j)$ are disjoint, i.e. have no elements in common. Some objects in G may be withheld from the distribution. The scholarship case provides an example. There, G consists of two scholarships, the more attractive one, A, and the less attractive B, and possible treatments amount to different partial or total distributions of G

among the two individuals involved. Thus, for example, $a = (\{A\}, \{B\})$ is
the assignment in which A goes to i and B goes to j, while $b = (\{B\}, \{A\})$
assigns B to i and A to j. The equal treatment is the distribution $(\emptyset, \emptyset)$ in
which scholarships are withheld from both idividuals, i.e., each of them is
assigned the empty set.

For simplicity, I exclude decision problems in which there is no equal
treatment or in which several treatments are equal. (The latter restriction
would be violated, for example, in Cake-Division, if we allowed divisions
in which part of the cake remains undistributed. The number of equal
treatments would then increase from one to infinity.) I will also assume
that **T** is closed under permutations on individuals. Thus, we impose two
conditions on the set of treatments:

A1. For every permutation f on **I** and every a in **T**, **T** contains some b
such that for every i in **I**, $f(i)$ is treated in b as i is treated in a. That
is, for every i, $b(f(i)) = a(i)$.

A2. There is a unique element of **T**, call it **e**, such that **e** is an equal
treatment.

A2 is a substantial restriction. So is A1, which is a kind of completeness
requirement on **T**. If **T** is the set of actually *available* treatments, this set
sometimes might be too small for A1 to be satisfied. Here, however, I will
ignore this difficulty and assume that **T** is 'roomy' enough.

There are two ways of looking at **T**. If **T** is seen as the set of *conceivable*
treatments, it is plausible to suppose that **T** is large enough to satisfy such
conditions as A1. This way of looking at **T** is appropriate if we think of the
elements of **T** as the possible ways in which the individuals might *deserve*
to be treated. But if **T** instead is interpreted as the set of ways in which
we can treat the individuals, i.e., as the set of *available* treatments, A1 is
not that plausible. Still, in the context of our discussion, the simplifying
assumption that the set of available treatments is as large as the set of
conceivable treatments is innocuous: Remember that we want to know
whether equal treatment minimizes expected unfairness (and/or maximal
possible unfairness), as compared with available alternatives, in the absence
of information that discriminates between the individuals. If this conjecture
turns out to hold when the set of available treatments is large, it will
obviously still hold if that set is diminished.

Given A1, every permutation f on **I** induces the corresponding permutation on **T** that for every a in **T** assigns some b in **T** in which for every

individual i, $f(i)$ is treated in the same way as i is treated in a. I will refer to the union of f and the permutation that f induces on **T** as an *automorphism* and use symbols p, p', etc., to stand for different automorphisms. Intuitively, then, an automorphism is a simultaneous permutation of individuals into individuals and of treatments into treatments, in which the former permutation induces the latter.

Definition 8.1. An *automorphism*, p, is a simultaneous permutation of **I** and of **T** such that for all i in **I** and all a in **T**, $p(a)(p(i)) = a(i)$.

Corollary of A1: Every permutation on **I** is included in exactly one automorphism.

This notion of an automorphism will come in handy below.

Here follow some examples of automorphisms. Suppose that **I** consists of three individuals, i_1, i_2, i_3, and let **T** be the set of cake-divisions among the members of **I**. One automorphism would then permute i_1 into i_2, i_2 into i_3 and i_3 into i_1. This would effect the corresponding permutation on cake-divisions. For example, $(0, 2/3, 1/3)$ would be permuted into $(1/3, 0, 2/3)$. Analogously, if **T** is the set of rankings of i_1, i_2, and i_3, the automorphism that permutes i_1 into i_2, i_2 into i_3 and i_3 into i_1 involves the corresponding permutation on rankings. For example, it permutes the ranking with i_1 on top, followed by i_2 and i_3, in that order, into the ranking with i_2 on top, followed by i_3 and i_1.

It is easy to see that only equal treatment, **e**, stays invariant under all automorphisms: For all p, $p(\mathbf{e}) = \mathbf{e}$, and for all $a \in \mathbf{T}$, if $a \neq \mathbf{e}$, then for p, $p(a) \neq a$.

We now define a relation between treatments that is going to be important in what follows:

Definition 8.2 (Structural Identity). A treatment a is *structurally identical* to a treatment b iff there exists some automorphism p such that $p(a) = b$.

Intuitively, this relation obtains between two treatments if we can get one from the other just by reshuffling individuals, while otherwise keeping the treatment unchanged. Structural identity is an equivalence relation: it is reflexive, symmetric, and transitive.[4] We can therefore partition **T** into *structures*, S, S', etc., which are equivalence classes of treatments with

---

[4]This follows because it is true by definition that the set of automorphisms contains the identity automorphism and is closed under inverses and relative products.

respect to the relation of structural identity. As an example, suppose that **T** is the set of cake-divisions among three individuals, i_1, i_2, and i_3. Consider a cake-division $a = (1, 0, 0)$. Its structure consists of three treatments:

$$(1, 0, 0), (0, 1, 0) \text{ and } (0, 0, 1).$$

On the other hand, the structure of $b = (1/2, 1/3, 1/6)$ consists of six treatments. In b, each of the three individuals gets a different share and there are six ways in which we can assign these three different shares to three individuals.

For any treatment a in **T**, let S_a be the structure of a. The number of treatments in a structure may vary but is always finite given that **I** is finite: If **I** contains n individuals, the number of treatments in a structure is at most equal to $n!$, which is the number of possible permutations on **I**. Different automorphisms correspond to different permutations on **I** and the number of treatments in a structure cannot exceed the number of automorphisms. But it can be smaller: In some cases, several permutations on **I** induce automorphisms transforming a into the same treatment, which decreases the size of S_a. This will be the case whenever two or more individuals are treated equally in a. Thus, for example, there are two permutations on individuals that give rise to automorphisms transforming $(1, 0, 0)$ into $(0, 1, 0)$. Each of them assigns i_2 to i_1, but they differ in their assignments to i_2 and i_3. Since the latter two individuals are treated equally in $(1, 0, 0)$, in both cases $(1, 0, 0)$ is transformed into the same treatment, $(0, 1, 0)$.

At one extreme, all the individuals are treated in the same way in **e**. Therefore, **e**'s structure contains only **e** itself. At the other extreme, if all the individuals are treated differently in a, any two distinct automorphisms will transform a into different treatments, which means that the number of treatments in a's structure will equal the number of automorphisms.

8.3. Unfairness Measure

Before I introduce the last component of the model, an unfairness measure, let me first make a further simplifying assumption: In the situations to be considered, the agent knows that *there is exactly one (perfectly) fair treatment in* **T**, i.e. exactly one treatment in which everyone gets what he or she deserves. That there is at least one such treatment in **T** is plausible if we think of **T** as the set of conceivable ways in which the individuals might deserve to be treated. (See the discussion of A1 in the preceding section.) But that there is no more than one such way is a non-trivial constraint on the model. In some cases, it would need to be given up.

The unfairness measure **d** is based on this assumption. **d** is a function from pairs of treatments to real numbers, with the following interpretation: **d**(a, b) specifies the degree of unfairness of a on a hypothetical supposition that it is b that is the (perfectly) fair treatment. This degree of unfairness can be seen as the *distance* from a to b: To the extent a's unfairness is greater, a is farther away from b. On this interpretation, **d** can be assumed to satisfy the standard conditions on a distance measure:

(D0) **d**(a, b) ≥ 0;   *(Non-negativity)*

(D1) **d**(a, b) = 0 iff a = b;   *(Minimality)*

(D2) **d**(a, b) = **d**(b, a);   *(Symmetry)*

(D3) **d**(a, b) + **d**(b, c) ≥ **d**(a, c).   *(Triangle Inequality)*

Interpreting **d** as distance gives the model a geometric flavour. The pair (**T**, **d**) is then a *metric space*: a set of points with a distance measure defined on it. The set **I** of individuals may be seen as the set of *dimensions* of that space: A point $a = (a_1, \ldots, a_n)$ is defined by its coordinates on the different dimensions i_k in **I**.

As will be seen in the next section, interpreting unfairness as distance goes well beyond what is needed for our purposes. In particular, it need not be assumed that the unfairness measure is symmetric or that it satisfies the triangle inequality. Especially since some of these superfluous assumptions, such as symmetry, have quite notable implications.[5] Nevertheless, this geometric interpretation is not implausible and, in addition, it makes the model more intuitive and easier to grasp.

One further very natural condition on **d** is Impartiality, which requires **d** to be invariant under automorphisms:

> *Impartiality*: For all automorphisms p and all a, b in **T**, **d**(p(a), p(b)) = **d**(a, b).

According to this condition, if one permutes the individuals in two treatments in the same way, the distance between the treatments does not change. This means that the unfairness measure pays no attention to personal identities. Thus, for example, giving all of the cake to the individual who only deserves a small share is equally unfair independently of who it

---

[5]Thus, symmetry implies that for every a in **T**, **d**(e, a) = **d**(a, e), which means that equal treatment of unequals is just as unfair as the correspondingly unequal treatment of equals. This means that, in the presence of symmetry, justification of PE cannot be traced back to the differential moral costs of errors (cf. Section 8.1 above).

is who gets this unfair advantage.

How is **d** to be understood on the different interpretations of our model? Consider Cake-Divisions first. It seems plausible that for each individual i, the distance between two cake-divisions should be an increasing function of the (absolute) difference between the shares of the cake they give to i. That is, the distance between treatments a and b should be an increasing function of $|a_1 - b_1|, \ldots, |a_n - b_n|$. The simplest function of this kind is the sum: $|a_1 - b_1| + \ldots + |a_n - b_n|$. This kind of measure is sometimes called *city-block* distance. If we instead go for the sum of the *squared* differences and then take the square root of that sum, we get Euclidean distance. City-block and the Euclidean measure are two instances of the class of *Minkowski distance* functions. **d** belongs to this class iff, for some $k \geq 1$, and for all a, b in **T**, $\mathbf{d}(a, b) =$ the k-th root of the sum $|a_1 - b_1|^k + \ldots + |a_n - b_n|^k$. If $k = 1$, the so-defined **d** is the city-block distance; if $k = 2$, it is the Euclidean distance. The higher k is, the more disproportionate influence is given, by exponentiation, to the larger differences $|a_i - b_i|$, as compared with smaller differences. Only if $k = 1$, all the differences are given influence proportionate to size.

How is the distance measure to be understood for rankings? Again, there are several possibilities, but the proposal due to Kemeny and Snell seems especially plausible. A ranking might be seen as a set of ordered pairs of individuals: A pair (i, j) belongs to a ranking a iff a ranks i at least as highly as it ranks j. We fully specify a given ranking by providing a list of such ordered pairs. Now, the distance between two rankings, a and b, can be measured by the number of pairs that belong to either a or b but not to both of these rankings. It is easy to show that this definition satisfies the standard conditions on a distance measure (D0 - D3 above).[6]

For the case of Indivisible Goods we do not have a plausible definition of distance that is *generally* applicable. Different situations that exemplify the general structure of Indivisible Goods require different specifications of the unfairness measure. Let us therefore focus on the scholarship example we have started with. The example involves two individuals, i and j, and two scholarships, one more attractive, A, and the other, B, less so. Let us

---

[6]Cf. Kemeny & Snell 1962, chapter on *Preference Ranking: An Axiomatic Approach.* The authors show that this measure is the only distance function on rankings that satisfies the following condition:

If a ranking b lies 'between' rankings a and c, in the sense that it is included in their union and includes their intersection, then $\mathbf{d}(a, b) + \mathbf{d}(b, c) = \mathbf{d}(a, c)$.

That is, if b lies between a and c in the specified sense, then, intuitively, b lies on the shortest line connecting a and c.

suppose that one of the individuals (though unknown which one) is more deserving than the other.[7]

There is no need to fully specify a suitable unfairness measure **d**. I will only assume that the following holds: the distance between alternative treatments in which each individual gets a scholarship, i.e., the distance between $a = (\{A\}, \{B\})$ and $b = (\{B\}, \{A\})$, is shorter than the distance to each of these treatments from the equal treatment $\mathbf{e} = (\emptyset, \emptyset)$, in which none of the individuals gets anything. To put it formally,

$$\mathbf{d}(\mathbf{e}, a) > \mathbf{d}(b, a) = \mathbf{d}(a, b) < \mathbf{d}(\mathbf{e}, b).$$

Intuitive motivation: If a is fair or if b is fair, both individuals deserve a scholarship. But then withholding the scholarship from both is even more unfair than giving the somewhat more attractive scholarship to the less deserving individual. Also, since **d** is symmetric, $\mathbf{d}(b, a) = \mathbf{d}(a, b)$.

8.4. Information Measure and Expected Unfairness

Apart from the fixed components, **I**, **T**, **e**, and **d**, our model contains one variable component: a probability distribution P on **T**, which reflects the agent's information about the case at hand. For every a in **T**, $P(a)$ stands for the agent's probability for a being the (perfectly) fair treatment.[8]

Since, as we have assumed, the agent knows that there is exactly one fair treatment in **T**, we take it that the P-values for different treatments sum up to one. There is a difficulty here, though. On some interpretations, such as Cake-Divisions, the number of possible treatments in **T** is infinite. In such cases, the sum of P-values for different treatments might be lower than one and might even be zero if the probability is distributed uniformly over **T**. In such a uniform distribution, each treatment gets the probability zero (unless we allow infinitesimals as probability values) and the sum of zeros is zero. This difficulty could be dealt with by replacing summation with integration, but to keep the calculations at the elementary level I will assume that there exists a finite subset of **T** such that the agent is certain that the fair treatment belongs to that subset. Then P-values of different treatments will sum up to one, as desired, and, for any subset Y of **T**, we

---

[7]This assumption is made in order to guarantee that the fair treatment belongs to **T**. If both candidates were equally deserving, none of the available treatments would be fair.
[8]Fixed components are marked in bold, while the variable component is italicized. We take P to be variable, since we want to examine whether equal treatment ought to be chosen for every P that does not discriminate between the individuals.

can define $P(Y)$, the probability that Y contains the fair treatment, as the sum of P-values assigned to the elements of Y:

$$P(Y) = \Sigma_{a \in Y} P(a).$$

To say that the available information does not discriminate between the individuals in $\mathbf{I}$ must mean that structurally identical treatments are assigned the same P-values. Thus, we are led to the following definition:

Definition 8.3. P *does not discriminate between the individuals* iff for all structurally identical a and b in $\mathbf{T}$, $P(a) = P(b)$.[9]

Two treatments are structurally identical iff there is an automorphism that transforms one into the other. Thus, P does not discriminate between the individuals iff it is *invariant under automorphisms*: for every p and a, $P(a) = P(p(a))$. Note also that if P does not discriminate between individuals, then for every a, $P(a)$ equals the probability of a's structure S_a, divided by the number of treatments that belong to this structure: $P(a) = P(S_a)/\text{card}(S_a)$.

It is easy to define the *expected unfairness* of a treatment a with respect to a given probability function P as the P-weighted sum of its distances to different possible treatments. For every treatment b, the distance from a to b is weighted with the probability $P(b)$ of b being *the* fair treatment.

Definition 8.4 (Expected unfairness).

$$\text{ExpUnf}_P(a) = \Sigma_{b \in T} P(b) \, \mathbf{d}(a, b).$$

For the expected value to be a meaningful notion, it is enough if the underlying value function is unique up to positive affine transformations, i.e., up to the choice of unit and zero. Representing unfairness as *distance* implies that the only thing that is left for an arbitrary decision is the unit of measurement. The zero-point for distance is not arbitrary: That each point's distance to itself, and only to itself, equals zero is a defining feature of a distance measure.

Even apart from the fixity of the zero-point, it should be clear that interpreting the unfairness measure $\mathbf{d}$ as a distance function is much more than we need in order to give meaning to the notion of expected unfairness. Neither symmetry nor triangle inequality are needed for this purpose. But,

---

[9]Our finiteness constraint on P does not hinder P from being indiscriminative in this way. The reason is that every structure is finite if $\mathbf{I}$ is finite.

as suggested above, treating unfairness as distance is not implausible and it makes the model easier to grasp.

8.5. Expected Unfairness and Equal Treatment

Consider the following hypothesis:

> *ExpUnf-minimization*: For every P that does not discriminate between the individuals and for every treatment a in $\mathbf{T}$, $\mathrm{ExpUnf}_P(\mathbf{e}) \leq \mathrm{ExpUnf}_P(a)$.

In other words, on this hypothesis, equal treatment minimizes expected unfairness in the absence of discriminating information. This would explain why PE should be accepted.

We want to know under what circumstances the hypothesis is going to hold. More precisely, we want to know what condition on the *unfairness measure* would make ExpUnf-minimization valid.

If Y is a finite set of treatments, let $\bar{\mathbf{d}}(a, Y)$ stand for a's *average distance* to the treatments in Y:

$$\bar{\mathbf{d}}(a, Y) = \Sigma_{b \in Y} \, \mathbf{d}(a, b)/\mathrm{card}(Y).$$

The following condition on $\mathbf{d}$ can be shown to be both *necessary and sufficient* for ExpUnf-minimization:

> *Structure Condition:* For every structure $S \subseteq \mathbf{T}$ and every $a \in \mathbf{T}$, $\bar{\mathbf{d}}(\mathbf{e}, S) \leq \bar{\mathbf{d}}(a, S)$.

The condition states that, for every structure S, equal treatment has a minimal average distance to S, as compared with other treatments.

Theorem 8.1 (Sufficiency).

> *Structure Condition* $\Rightarrow$ *ExpUnf-minimization*.

Proof.

> *Claim:* If P does not discriminate between the individuals, then for every $a \in \mathbf{T}$,

$$\mathrm{ExpUnf}_P(a) = \Sigma_{S \subseteq T} \, P(S) \, \bar{\mathbf{d}}(a, S).$$

That is, in the absence of discriminating information, a's expected unfairness is a weighted sum of its average distances to different structures, with weights being the probabilities of these structures. Here is the proof of the

Claim:

$ExpUnf_P(a)$

$=\Sigma_{b \in T} P(b)\, \mathbf{d}(a, b)$ [by the definition of ExpUnf]

$=\Sigma_{S \subseteq T}\, \Sigma_{b \in S} P(b)\, \mathbf{d}(a, b)$ [since $\mathbf{T}$ can be partitioned

                                               into structures]

$=\Sigma_{S \subseteq T}\, \Sigma_{b \in S}\, (P(S)/\mathrm{card}(S))\, \mathbf{d}(a, b)$ [since P is indiscriminative]

$=\Sigma_{S \subseteq T} P(S)\, (\Sigma_{b \in S}\, \mathbf{d}(a, b)/\mathrm{card}(S))$ [by algebra]

$=\Sigma_{S \subseteq T}\, P(S)\, \bar{\mathbf{d}}(a, S)$ [by the definition of $\bar{\mathbf{d}}$]

Given the Structure Condition, the average distance from **e** to a structure S never exceeds the corresponding distance from any a to S. Consequently, the Claim implies that $\mathrm{ExpUnf}_P(\mathbf{e}) \leq \mathrm{ExpUnf}_P(a)$. $\square$

We now want to prove that the Structure Condition is *necessary* for ExpUnf-minimization:

Theorem 8.2 (Necessity).

         Structure Condition $\Leftarrow$ ExpUnf-minimization.

Proof. We need to show that if the Structure Condition is violated by our model, i.e., if for some structure S and treatment a, $\bar{\mathbf{d}}(a, S) < \bar{\mathbf{d}}(\mathbf{e}, S)$, then ExpUnf-minimization is violated as well: there exists a probability function P that does not discriminate between the individuals and is such that, with respect to P, the expected unfairness of a is lower than the expected unfairness of **e**. To construct a P like this, we simply let it be the uniform probability distribution on S. $\square$

To forestall possible misunderstandings, it should be pointed out that for a *particular* P that does not discriminate between the individuals, **e** might minimize expected unfairness with respect to that P even if the underlying unfairness measure **d** happens to violate Structure Condition. However, the Structure Condition is necessary if **e** is to minimize expected unfairness for *all* possible P that do not discriminate between the individuals, as required by the hypothesis of ExpUnf-minimization.

Is the Structure Condition satisfied by the different interpretations of our model? I think it is fair to say that this condition *usually* holds. It can be shown to hold for all Minkowski-distance measures on cake-divisions.[10]

---

[10]For the proof, see Rabinowicz (2008), Appendix A. In that appendix, I consider a more general interpretation on which $\mathbf{T}$ consists of all possible real-number assignments to individuals, i.e., not only those in which the assigned numbers are non-negative and add up to 1 (as in Cake-Divisions). On this interpretation, there are non-denumerably

It can also be shown to hold for the Kemeny-Snell distance measure on rankings.[11] On the other hand, this condition is violated in the scholarship example. There, as we remember, the distance between treatments $a = (\{A\}, \{B\})$ and $b = (\{B\}, \{A\})$ is shorter than the distance to each of them from the equal treatment $\mathbf{e} = (\emptyset, \emptyset)$. Since the set $\{a, b\}$ is a structure, it immediately follows that the average distance from a to this structure is shorter (in fact, more than twice as short) than the corresponding average distance from $\mathbf{e}$ to the structure in question:

$$\bar{\mathbf{d}}(a, \{a, b\}) = (\mathbf{d}(a, a) + \mathbf{d}(a, b))/2 = \mathbf{d}(a, b)/2$$
$$< (\mathbf{d}(\mathbf{e}, \mathbf{a}) + \mathbf{d}(\mathbf{e}, \mathbf{b}))/2 = \bar{\mathbf{d}}\,(\mathbf{e}, \{a, b\}).$$

Since the Structure Condition is violated in this case, it follows that there exists a probability function P that does not discriminate between the individuals and with respect to which $\mathbf{e}$'s expected unfairness exceeds the expected unfairness of a: One such P is the uniform probability distribution on $\{a, b\}$. If we are certain that the fair treatment is either a or b, with each of these treatments being an equally likely candidate to the title, treating the individuals equally by withholding the scholarships from both will not minimize expected unfairness.

As we have seen in this section, the Structure Condition is both sufficient and necessary if equal treatment is to minimize expected unfairness. But this condition is neither especially transparent nor intuitive. It has a feel of a constraint that itself should be derivable from some more basic and simple conditions. What these conditions might be is not clear to me, however. One of them would probably be Impartiality mentioned in Section 8.3, which implies that for each structure, all its elements are equi-distant from the equal treatment. But we obviously need other conditions as well.[12] The conjecture is that they would guarantee, together with Impartiality, that for every structure, the treatments in that structure form vertices of

---

many equal treatments in **T**. The following condition is shown to be satisfied by every Minkowki-distance measure on such a set **T**:

Generalized Structure Condition: For every $a \in \mathbf{T}$, there exists an equal treatment $\mathbf{e}_a \in \mathbf{T}$ such that for every structure $S \subseteq \mathbf{T}$, $\bar{\mathbf{d}}(\mathbf{e}_a, S) \leq \bar{\mathbf{d}}(a, S)$.

For Minkowski spaces, $\mathbf{e}_a$ is obtained from a by averaging: For every individual i, $\mathbf{e}_a(i) = (a_1 + \ldots + a_n)/n$.

When **T** is restricted to the set of Cake-Divisions, in which real values assigned to the different individuals add up to 1, $\mathbf{e}_a$ will coincide with $\mathbf{e}$, for every cake-division a. Therefore, for this restricted set of treatments, the simple Structure Condition will hold.

[11] For the proof, see Rabinowicz (2008), Appendix B.

[12] Impartiality is satisfied in the scholarship example despite the fact that this example violates the Structure Condition.

a regular geometric figure that has equal treatment in its center, thereby implying the Structure Condition. Finding these underlying conditions is, as I see it, a major outstanding problem that is left for further inquiry.

8.6. Minimax

Given the Structure Condition, equal treatment minimizes expected unfairness in the absence of discriminating information. However, for some people, this might not be a decisive consideration. They might feel that the proper course of action is not to minimize expected disvalue but rather to minimize the maximal potential disvalue, i.e. to minimize unfairness in *the worst possible case* that has a non-zero probability. What is the position of equal treatment from this 'minimax' perspective?

A treatment a's *maximal possible unfairness* with respect to a probability distribution P can be defined as a's maximal distance to a positively P-valued treatment:

Definition 8.5 (Maximal possible unfairness).

$$\text{MaxUnf}_P(a) = \max\{\mathbf{d}(a, b)\colon b \in \mathbf{T} \ \& \ P(b) > 0\}.^{13}$$

With this notion in hand, we can consider the following hypothesis:

> *MaxUnf-minimization:* For all P that do not discriminate between the individuals, and all a in $\mathbf{T}$, $\text{MaxUnf}_P(\mathbf{e}) \leq \text{MaxUnf}_P(a)$.

If this hypothesis holds, it would provide another potential reason for accepting PE.

As is easily seen, the following condition on $\mathbf{d}$ is both necessary and sufficient for the validity of MaxUnf-minimization:

> *Minimax Condition:* For every a in $\mathbf{T}$ and every structure $S \subseteq \mathbf{T}$,
>
> $$\max\{\mathbf{d}(\mathbf{e}, b)\colon b \in S\} \leq \max\{\mathbf{d}(a, b)\colon b \in S\}.$$

According to the Minimax Condition, equal treatment minimizes maximal distance to every structure, as compared with other treatments in $\mathbf{T}$.

---

[13]If the set $\{\mathbf{d}(a, b)\colon b \in \mathbf{T} \ \& \ P(b) > 0\}$ is infinitely large, it might lack a maximum (or even an upper bound). However, this problem will not arise as long as we hold on to our simplifying assumption that the number of treatments in $\mathbf{T}$ with positive P-values is finite.

Theorem 8.3 (Sufficiency).

Minimax Condition $\Rightarrow$ MaxUnf-minimization.

Proof. Let c be any treatment such that $P(c) > 0$ and $\mathrm{MaxUnf}_P(\mathbf{e}) = \mathbf{d}(\mathbf{e}, c)$. Let S_c be the structure of c. By the Minimax Condition, it holds for every a in $\mathbf{T}$ that

$$\max\{\mathbf{d}(\mathbf{e}, b)\colon b \in S_c\} \leq \max\{\mathbf{d}(a, b)\colon b \in S_c\}.$$

But then, for some $c' \in S_c$, $\mathbf{d}(a, c') = \max\{\mathbf{d}(a, b)\colon b \in S_c\} \geq \max\{\mathbf{d}(\mathbf{e}, b)\colon b \in S_c\} = \mathbf{d}(\mathbf{e}, c)$. Since $P(c) > 0$ and P does not discriminate between the individuals, $P(c') = P(c) > 0$. Consequently, $\mathrm{MaxUnf}_P(a) \geq \mathbf{d}(a, c')$. Since $\mathbf{d}(a, c') \geq \mathbf{d}(\mathbf{e}, c) = \mathrm{MaxUnf}_P(\mathbf{e})$, it follows that $\mathrm{MaxUnf}_P(a) \geq \mathrm{MaxUnf}_P(\mathbf{e})$. $\qquad\square$

Theorem 8.4 (Necessity).

Minimax Condition $\Leftarrow$ MI-minimization.

Proof. Suppose that for some a and S, Minimax Condition is violated:

$$\max\{\mathbf{d}(\mathbf{e}, b)\colon b \in S\} > \max\{\mathbf{d}(a, b)\colon b \in S\}.$$

Let P be the uniform probability distribution on S. Then P does not discriminate between the individuals, but $\mathrm{MaxUnf}_P(\mathbf{e}) = \max\{\mathbf{d}(\mathbf{e}, b)\colon b \in S\} > \max\{\mathbf{d}(a, b)\colon b \in S\} = \mathrm{MaxUnf}_P(a)$. Which means that MaxUnf-minimization is violated as well. $\qquad\square$

If the distance measure satisfies the Structure Condition and Impartiality, there is no need to impose the Minimax Condition as an independent constraint. It can be shown that

Theorem 8.5.

Structure Condition & Impartiality $\Rightarrow$ Minimax Condition.

Proof. Impartiality implies that for every a in $\mathbf{T}$ and every automorphism p, $\mathbf{d}(\mathbf{e}, a) = \mathbf{d}(p(\mathbf{e}), p(a)) = \mathbf{d}(\mathbf{e}, p(a))$. Since a and b are structurally identical iff $b = p(a)$ for some automorphism p, and since $\mathbf{e}$ is invariant under automorphisms, the following must hold given Impartiality:

For all structurally identical treatments a, b in $\mathbf{T}$, $\mathbf{d}(\mathbf{e}, a) = \mathbf{d}(\mathbf{e}, b)$.

Thus, for every structure S, **e** is equi-distant from every treatment in S. Now, consider any treatment a in **T**. By the Structure Condition, $\bar{d}(e, S) \leq \bar{d}(a, S)$. Therefore, since **e**'s distance to different treatments in S is constant, the maximal distance from **e** to the elements of S cannot exceed the maximal distance from a to the elements of S. Minimax Condition follows.   □

Thus, given the impartiality of the distance measure, equal treatment will automatically minimize not only expected unfairness but also maximal possible unfairness, if that measure satisfies the Structure Condition.[14] There is therefore no need for an independent worry about minimax considerations. However, an interesting question is whether there are any plausible interpretations of our model in which the distance measure satisfies Minimax but violates the Structure Condition. On such interpretations, it will be possible to have probability distributions that do not discriminate between individuals and with respect to which **e** does not minimize expected unfairness, even though on all non-discriminative probability distributions **e** will minimize maximal possible unfairness. The issue whether to opt for equal treatment will on such interpretations sometimes depend on whether we adhere to the minimization of expected disvalue or to minimaxing. I do not know, however, of any plausible interpretation of this kind.

8.7. Extensions

The model we have presented rests on a series of simplifying assumptions. While it always is a good idea to start out with a simple formal framework, this makes our approach unrealistic in several respects. It is therefore natural to consider possible extensions of the modeling. Here are some rather obvious questions to ask:

(i) What if the set of treatments contains *more than one equal treatment*?

In some cases of this kind there might not exist any equal treatment that minimizes expected unfairness or maximal possible unfairness as compared with all other treatments in **T**. What one might hope for, though, is that

---

[14]This means, in particular, that equal treatment minimizes maximal possible unfairness in rankings with Kemeny-Snell distance and in Cake-Divisions with Minkowski distance, but does *not* minimize it in the scholarship example. If $P(\{A\}, \{B\}) = P(\{B\}, \{A\}) = 1/2$, the equal treatment $(\emptyset, \emptyset)$ will be a bad choice for a minimaxer.

for every treatment a in **T** there is always *some* equal treatment that is at least as satisfactory as a in terms of minimization of expected unfairness and/or minimization of maximal possible unfairness. This would mean that treating individuals unequally is never preferable to all forms of equal treatment. More precisely, if we just focus on minimization of expected unfairness, what one might hope for is the following:

> *Generalized ExpUnf-minimization:* For every treatment a in **T**, **T** contains *some* equal treatment $\mathbf{e}_a$ such that for every P that does not discriminate between individuals, $\mathrm{ExpUnf}_P(\mathbf{e}_a) \leq \mathrm{ExpUnf}_P(a)$.

It is easy to prove that the condition on the distance measure that is both necessary and sufficient for Generalized ExpUnf-minimization is a generalized version of the Structure Condition:

> *Generalized Structure Condition:* For every $a \in \mathbf{T}$, there exists an equal treatment $\mathbf{e}_a \in \mathbf{T}$ such that for every structure $S \subseteq \mathbf{T}$,
> $$\bar{\mathrm{d}}(\mathbf{e}_a,\, S) \leq \bar{\mathrm{d}}\,(a,\, S).^{15}$$

A natural question is whether this Generalized Structure Condition can be derived from some set of more intuitive and basic assumptions about the distance measure.

(ii) What if the available information is indiscriminative between the individuals *within a subgroup* $X \subseteq \mathbf{I}$, but not necessarily outside that subgroup?

Let me introduce some definitions:

Definition 8.6. If $X \subseteq \mathbf{I}$, p is an X-*automorphism* iff p is an automorphism that permutes X onto X and for all i in **I** that do not belong to X, $p(i) = i$.

Definition 8.7. A probability distribution P on **T** *does not discriminate between the individuals in* $X \subseteq \mathbf{I}$ iff P is invariant under X-automorphisms, i.e., iff for all X-automorphisms p and all $a \in \mathbf{T}$, $P(p(a)) = P(a)$.

Definition 8.8. If $X \subseteq \mathbf{I}$ and $a, b \in \mathbf{T}$, a and b are X-*structurally identical* iff there is some X-automorphism p such that $p(a) = b$.

Definition 8.9. a is an X-equal treatment iff for all i, j in X, $a(i) = a(j)$.

If we just focus on the minimization of expected unfairness, we might be interested in the following hypothesis:

---

[15] For a discussion of Generalized Structure Condition, see footnote 10 above.

Subgroup-Generalized ExpUnf-minimization: For every $X \subseteq \mathbf{I}$ and every $a \in \mathbf{T}$, $\mathbf{T}$ contains *some* X-equal treatment $\mathbf{e}_{a,X}$ such that for every P that does not discriminate between individuals in X, $\mathrm{ExpUnf}_P(\mathbf{e}_{a,X}) \leq \mathrm{ExpUnf}_P(a)$.

This hypothesis provides reasons for equal treatment of those individuals between which our information does not discriminate.

Since X-structural identity is an equivalence relation, just like the ordinary structural identity, $\mathbf{T}$ can be partitioned into X-*structures* – equivalence classes with respect to X-structural identity. The condition on the distance measure that is necessary and sufficient for Subgroup-Generalized ExpUnf-minimization is a further generalization of the Generalized Structure Condition:

Subgroup-Generalized Structure Condition: For every $X \subseteq \mathbf{I}$ and every $a \in \mathbf{T}$, there exists an X-equal treatment $\mathbf{e}_{a,X} \in \mathbf{T}$ such that for every X-structure $S \subseteq \mathbf{T}$, $\bar{\mathbf{d}}\,(\mathbf{e}_{a,X}, S) \leq \bar{\mathbf{d}}\,(a, S)$.[16]

It is easy to see that this condition entails the Generalized Structure Condition as a special case (for $X = \mathbf{I}$), but it would be interesting to know how to derive it from some more basic conditions on a distance measure.

(iii) What if there might be *several (perfectly) fair treatments*, and not just one?

We would then need to work with a different unfairness measure:

$\mathbf{d}(a, Y)$ – the degree of unfairness of a on the hypothetical assumption that Y is the set of all fair treatments in $\mathbf{T}$.

We take it that $\mathbf{d}(a, Y)$ is defined only if Y is non-empty and finite. (Finiteness is assumed for the sake of simplicity.) An obvious requirement on this measure is that $\mathbf{d}(a, Y) = 0$ iff $a \in Y$. In other words, the degree of unfairness of a is zero if and only if a is one of the (perfectly) fair treatments. In fact, $\mathbf{d}(a, Y)$ could simply be defined in terms of distance between treatments, as the minimal distance from a to the elements of Y.[17]

---

[16]In the case of Cake-Divisions, we may conjecture that $\mathbf{e}_{a,X}$ is obtainable from a as follows: (i) for every i in X, $\mathbf{e}_{a,X}$ assigns to i the average of the values assigned by a to the members of X; while (ii) for every i outside X, $\mathbf{e}_{a,X}(i) = a(i)$. It is less clear, however, how to construct $\mathbf{e}_{a,X}$ in the case of rankings. The individuals in X should of course be ranked equally in $\mathbf{e}_{a,X}$. But how should these individuals be ranked vis-à-vis the individuals that do not belong to X? There is no obvious answer to this question.

[17]That is, we could let $\mathbf{d}(a, Y)$ be $\min\{b \in Y : \mathbf{d}(a, b)\}$, with $\mathbf{d}(a, b)$ now interpreted as $\mathbf{d}(a, \{b\})$. If Y were allowed to be infinite, $\mathbf{d}(a, Y)$ could be instead be identified with the greatest lower bound of $\{\mathbf{d}(a, b) : b \in Y\}$.

We would also need a different measure of information. The new measure would have to be a probability distribution on sets of treatments rather than on individual treatments:

$P(Y)$ – the probability that Y is the set of all fair treatments in **T**.

For such a probability measure P, the notion of non-discrimination would have to be appropriately re-defined:

P does not discriminate among the individuals iff for all automorphisms $p, P(p(Y)) = P(Y)$, where $p(Y) = \{b: \exists a \in Y\, p(a) = b\}$.

And we would need to correspondingly re-define the notions of expected unfairness and maximal expected unfairness:

$\text{ExpUnf}_P(a)$: $\Sigma_{Y \subseteq T}\ P(Y)\ \mathbf{d}(a, Y)$,
$\text{MaxUnf}_P(a) = \max\{\mathbf{d}(a, Y): Y \subseteq \mathbf{T}\ \&\ P(Y) > 0\}$.

The obvious question is: What conditions on the re-defined unfairness measure will then guarantee that, in the absence of discriminating information, equal treatment will minimize expected unfairness and/or maximal expected unfairness? Will some condition analogous to our Structure Condition do the job?[18]

These are just some of the follow-up questions that could be raised. But their examination would require another paper.

Acknowledgments

This is a significantly revised and streamlined version of Rabinowicz (2010). I am much indebted to the participants of the Purdue Winer Memorial Lectures, West Lafayette, Indiana, October 2010, for helpful comments and suggestions.

---

[18]Here is a natural candidate for such a condition:

If Σ is a family of subsets of **T** such that, for some finite $Y \subseteq \mathbf{T}$, $\Sigma = \{Z \subseteq \mathbf{T}:$ for some automorphism $p, p(Y) = Z\}$, then $\bar{\mathbf{d}}(\mathbf{e}, \Sigma) \leq \bar{\mathbf{d}}(a, \Sigma)$ for all $a \in \mathbf{T}$.

If some automorphism permutes Y into Z, then sets Y and Z can be said to be structurally identical. The set Σ, which is the set of images of Y under automorphisms, can therefore be seen as the *structure* of Y. Thus, the suggested condition says the **e** minimizes the average distance to the structure of every finite set of treatments.

References

Katzner, L. I. (1973). Presumptions of reason and presumptions of justice. *The Journal of Philosophy, 70,* 89-100.

Kemeny, J. G., & Snell, J. L. (1962). Preference rankings. In J. G. Kemeny and J. L. Snell (Eds.), *Mathematical Models in the Social Sciences.* Cambridge, MA: MIT Press.

Rabinowicz, W. (2008). Presumption of equality. In M. L. Jönsson (Ed.), *Proceedings of the 2008 Lund-Rutgers Conference.* Lund: Lund Philosophy Reports.

Rabinowicz, W. (2010). If in doubt, treat'em equally – A case study in the application of formal methods to ethics. In T. Czarnecki, K. Kijania-Placek, O. Poller, and J. Wolenski (Eds.), *The Analytical Way – Proceedings of the 6th European Congress of Analytic Philosophy.* London: College Publications.

Ullmann-Margalit, E. (1983). On presumption. *The Journal of Philosophy, 80,* 143-163.

Chapter 9

Ternary Paired Comparisons Induced by Semi- or Interval Order Preferences

Michel Regenwetter and Clintin P. Davis-Stober

University of Illinois at Urbana-Champaign and University of Missouri

Loomes and Sugden (1995) discussed how to incorporate a probabilistic element into decision theories. We follow up on their discussion of random preference specifications for *transitivity of preference*, including a well-known necessary condition for these models, the *triangle inequality*. This condition applies traditionally to binary forced choice, whereas we consider ternary paired comparisons, in which respondents are permitted to express indifference between two offered choice alternatives. We review a general method for fully specifying necessary and sufficient conditions for certain random preference models. We provide such specifications for the random preference models whose *core theories* are *semiorders* or *interval orders*, in the case of four choice alternatives.

9.1. Introduction

"Semiorders" and "interval orders" (Luce, 1956; Fishburn, 1985; Pirlot & Vincke, 1997) are mathematical representations of preference that capture the notion that we only have preferences among choice alternatives that are 'sufficiently different' from each other. "Semiorders" model preferences as consistent with a fixed threshold of discrimination among choice alternatives. More generally, "interval orders" permit the threshold of discrimination to depend on the objects being compared. "Semiorders" and "interval orders" have the interesting feature that strict preference (i.e., beyond threshold preference) is transitive, whereas indifference may be intransitive. For instance, A may not be discriminable from B, B may not be discriminable from C, but A may be discriminable from C, and as a consequence, a person might be indifferent between A and B, indifferent between B and C, but not indifferent between A and C.

Probabilistic choice models offer arguably the most canonical representation of variable or uncertain preferences. We build directly on Regenwetter

Fig. 9.1. Hasse diagram of a linear order on four objects. The 24 permutations of the
4 nodes yield 24 different strict linear orders on $\mathcal{C} = \{A, B, C, D\}$.

and Davis-Stober (2008), Regenwetter, Dana, and Davis-Stober (2010), and
Regenwetter, Dana, and Davis-Stober (2011), where we discussed a broad
range of issues with probabilistic models of transitive preferences, based on
prior important work by, e.g., Loomes and Sugden (1995), Luce (1995,1997),
and Hey (2005). In the terminology of Loomes and Sugden (1995) we study
"random preference models" (which we also often call "mixture models" to
highlight that we make no distributional assumptions) whose "core theories" are "semiorders" or "interval orders." To put it more formally, we
discuss probabilistic choice models that are mathematically equivalent to

the "semiorder polytope" or "interval order polytope" for "ternary paired comparison probabilities."

Throughout, the relevant empirical data are paired comparisons among gambles with the option of expressing indifference between alternatives ("ternary paired comparisons"). We keep the background discussion to a minimum. But, to keep the chapter self-sufficient, we repeat a few known definitions and results.

9.2. Random Preference Models of Semiorders and Interval Orders

Since any laboratory experiment on decision making or preference will use a limited set of choice options, we assume that the set C of choice options (relevant to the experiment) is finite.

Definition 9.1. A *(binary) relation* R on a finite set of choice alternatives C is a set of the form $R \subseteq C \times C$. We also write xRy to denote $(x, y) \in R$, and say that a person with preference relation R prefers x to y. Let $I_C = \{(c, c) | \ c \in C\}$. If R and S are two relations, let $RS = \{(z, y) \in C \times C | \ \exists x, zRx, xSy\}$. If $R \subseteq C \times C$, let $R^{-1} = \{(x, y) \in C \times C | \ yRx\}$ and let $\bar{R} = \{(x, y) \in C \times C | \ (x, y) \notin R\}$. A relation R on C is

$$
\begin{aligned}
complete & \quad \text{if } R \cup R^{-1} \cup I_C = C \times C, \\
asymmetric & \quad \text{if } R \cap R^{-1} = \varnothing, \\
transitive & \quad \text{if } RR \subseteq R, \\
negatively\ transitive & \quad \text{if } \bar{R}\,\bar{R} \subseteq \bar{R}.
\end{aligned}
$$

A *strict partial order* is an asymmetric and transitive binary relation. An *interval order* is a strict partial order R with the property that

$$R \overline{R^{-1}} R \subseteq R.$$

A *semiorder* is an interval order R with the property that

$$R R \overline{R^{-1}} \subseteq R.$$

Denote by

$$
\begin{aligned}
\mathcal{SO}(C) & \quad \text{the collection of semiorders on } C, \\
\mathcal{IO}(C) & \quad \text{the collection of interval orders on } C.
\end{aligned}
$$

A *strict weak order* is an asymmetric and negatively transitive binary relation. A *strict linear order* is a transitive, asymmetric, and complete binary

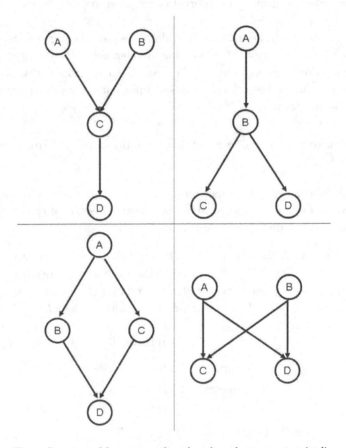

Fig. 9.2. Hasse diagrams of four types of weak orders that are not strict linear orders. By permuting the nodes one can construct 12 strict weak orders like the one on the upper left (exchanging A and B leads to an equivalent Hasse diagram that represents the same strict weak order), 12 strict weak orders like the one on the upper right, 12 strict weak orders like the one on the lower left, and 6 strict weak orders like the one on the lower right.

relation. By *indifference* we mean a lack of preference. A person with preference relation R is *indifferent* between x and y if this person neither prefers x to y nor y to x, i.e., whenever $x\bar{R}y$ and $y\bar{R}x$.

Strict linear orders, strict weak orders, interval orders, semiorders, and strict partial orders are decreasingly restrictive models of transitive preference. Among these, as we will see in more detail, interval orders, semiorders, and strict partial orders permit indifference to be intransitive.

Consider the directed graph that features a directed edge from node X to node Y if and only if $(X, Y) \in R$. A *Hasse diagram* of a binary preference relation R is a simplified representation of that directed graph where all directed edges implied by transitivity are omitted to ease readability. Also, the Hasse diagram is oriented so that all directed edges point downwards. Figures 9.1 through 9.6 provide Hasse diagrams of the different types of binary relations for $C = \{A, B, C, D\}$.

Figure 9.1 shows the Hasse diagram of a strict linear order. The collection of all $4! = 24$ strict linear orders on $C = \{A, B, C, D\}$ can be obtained by permuting the four nodes in the Hasse diagram of Figure 9.1 in all 24 possible ways. We also refer to such a permutation as *relabeling the choice alternatives*, or *relabeling* in short.

Figures 9.1-9.3 show all 75 possible strict weak orders on $C = \{A, B, C, D\}$ up to relabeling. For instance, Figure 9.2 shows four different Hasse diagrams, hence four types of strict weak orders on $\{A, B, C, D\}$, none of which are strict linear orders. Here, not every relabeling leads to a distinct Hasse diagram. For example, exchanging A and B in the upper left strict weak order yields an equivalent Hasse diagram that represents the same weak order. By relabeling the nodes, one can find 12 strict weak orders for each of the types shown in the upper left, the upper right and the lower left, and 6 strict weak orders like the one shown in the lower right. Figure 9.3 shows the remaining three types of strict weak orders on $C = \{A, B, C, D\}$. The strict weak order in the top yields four different cases as one relabels the nodes, as does the strict weak order in the center. There is only one strict weak order with the Hasse diagram shown in the bottom display of Figure 9.3, namely $\varnothing$, the complete indifference relation.

A lack of a directed edge between two nodes in the Hasse diagram of a preference relation either means that such a directed edge follows from others by transitivity and is therefore omitted, or it indicates indifference between those two nodes. Because of the orientation of the diagram, omitted directed edges always involve moving from a node above to a node further down via a path of directed edges. Hence, unconnected nodes that are horizontally aligned in the Hasse diagram automatically indicate indifference. (For strict weak orders, horizontal alignment is equivalent to indifference, but this does not hold for all semiorders and interval orders.) In the strict linear order case, there is no indifference between choice options because strict linear orders are complete. Hence all nodes are either explicitly or implicitly connected by a directed edge. This is not the case in the remaining figures. In each of the remaining 51 strict weak orders shown in Figures 9.2-9.3, there are at least some pairs of objects in the indifference relationship.

Figures 9.4 and 9.5 show six more types of Hasse diagrams. These denote the 108 different semiorders on four objects that are not also strict weak orders. What differentiates these from strict weak orders is that they do not satisfy negative transitivity, and hence, indifference is not transitive. For example, on the upper left of Figure 9.4, we have $A\bar{R}C$ and $C\bar{R}B$, but ARB. This is an example like the one we gave in the first paragraph of the chapter, namely, this decision maker is indifferent between A and C, indifferent between C and B, but strictly prefers A over B.

Figure 9.6 shows the Hasse diagram of the 24 interval orders that are not semiorders. These binary relations violate the axiom that $R\,R\,\overline{R^{-1}} \subseteq R$. For example, in the shown diagram, we have $AR\,R\,\overline{R^{-1}}D$ because ARB and BRC and $C\overline{R^{-1}}D$, but $A\bar{R}D$.

We now proceed to a very natural way of probabilizing semiorders and interval orders by means of placing a probability distribution on $\mathcal{SO}$ or on $\mathcal{IO}$.

Definition 9.2. A *random preference model* (aka a *mixture model*) over a collection of binary relations $\mathcal{R}$ is a probability distribution $P : \mathcal{R} \to [0,1]$ that maps each $R \in \mathcal{R}$ into a probability $R \mapsto P_R$, where $P_R \in [0,1]$ and $\sum_{R \in \mathcal{R}} P_R = 1$. A *random preference model of semiorder or interval order preferences* is a random preference model over a collection $\mathcal{SO}$ or $\mathcal{IO}$.

The idea behind "ternary paired comparison probabilities," is that a person who has a choice of a or b will choose a, or express indifference, or choose b, with probabilities $P_{a,b}, 1 - P_{a,b} - P_{b,a}$, and $P_{b,a}$, respectively.

Definition 9.3. A collection $(P_{a,b})_{(a,b) \in \mathcal{C} \times \mathcal{C} \atop a \neq b}$ is called a *system of ternary paired comparison probabilities*, if $\forall a, b \in \mathcal{C}$, with $a \neq b$ we have

$$0 \le P_{a,b} \le 1$$

and

$$P_{a,b} + P_{b,a} \le 1.$$

A system of ternary paired comparison probabilities is *induced by (a random preference model on) semiorders* if there exists a probability distribu-

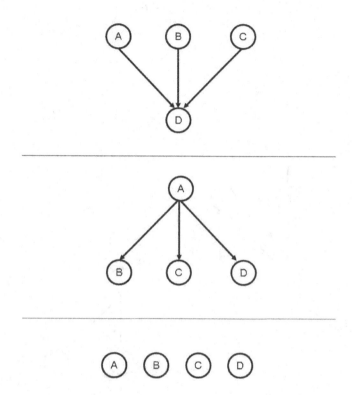

Fig. 9.3. Hasse diagrams of three types of strict weak orders on four objects that are not strict linear orders. After relabeling the choice alternatives, one obtains 4 strict weak orders like the one in the top, 4 strict weak orders like the one in the center. Regardless of relabeling, there is only one strict weak order with the Hasse diagram shown at the bottom.

tion on $\mathcal{SO}$ such that $\forall a, b \in \mathcal{C}, a \neq b$,

$$P_{a,b} = \sum_{\substack{R \in \mathcal{SO} \\ a R b}} P_R. \tag{9.1}$$

A system of ternary paired comparison probabilities is *induced by (a random preference model on) interval orders*, if there exists a probability distribution on $\mathcal{IO}$ such that $\forall a, b \in \mathcal{C}, a \neq b,$

$$P_{a,b} = \sum_{\substack{R \in \mathcal{IO} \\ a R b}} P_R. \tag{9.2}$$

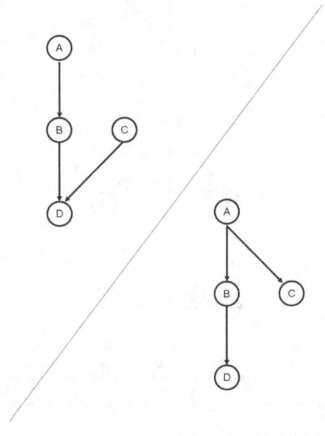

Fig. 9.4. Hasse diagrams of two types of semiorders on four objects that are not strict weak orders. By permuting the nodes, one obtains 24 semiorders of each kind.

9.3. Empirical Illustration and Polytopal Representation

As an illustrating example we use data from a ternary paired comparison pilot study that we ran in the first author's laboratory. In each *trial*, a

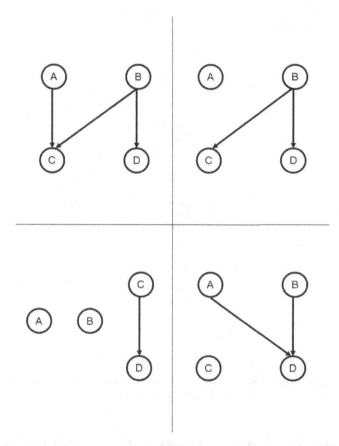

Fig. 9.5.   Hasse diagrams of four types of semiorders on four objects that are not strict weak orders. Permutation of the nodes yields 24 semiorders of the kind shown in the upper left, and 12 semiorders for each of the other kinds.

participant looked at a pair of gambles on a computer display and had three options; they could choose the left gamble, the right gamble, or express indifference between the two gambles. The participant considered each possible pair of gambles 30 times, separated by substantial numbers of decoys. Table 9.3 shows, for one participant, the proportions of times, $Q_{i,j}$, that the participant chose gamble i over gamble j, for all distinct i, j. This particular respondent, although being allowed to express indifference

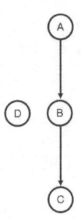

Fig. 9.6.   Hasse diagrams of the only type of interval order on four objects that is not also a semiorder. Relabeling yields 24 interval orders of this kind.

between gambles, in fact never responded with the "indifference" option.[1] Thus, for this particular experimental participant, we have $Q_{i,j} + Q_{j,i} = 1$, even though this was not required by the experimental paradigm. We used the same data in an earlier paper to show that they are consistent with linear and strict weak order preferences (Regenwetter & Davis-Stober, 2008). On the surface it is not obvious how one would generally determine whether or not ternary paired comparison proportions like those in Table 9.3, originated from ternary paired comparison probabilities that are induced by semiorders or interval orders.   Even disregarding the fact that we only observe proportions, it is not straightforward how one would, in general, characterize whether or not a given set of ternary paired comparison probabilities are induced by such a family of binary preference relations.

---

[1] However, the participant did use the "indifference" option among some other pairs of gambles.

Table 9.1. Pilot study using ternary paired comparisons, proportion of choices out of 30 repetitions, reported for one participant.

Gamble pair (i,j)	(A,B)	(A,C)	(A,D)	(B,C)	(B,D)	(C,D)
Prop. i chosen over j : $Q_{i,j}$	$24/30$	$24/30$	$14/30$	$23/30$	$15/30$	$23/30$
Prop. j chosen over i : $Q_{j,i}$	$6/30$	$6/30$	$16/30$	$7/30$	$15/30$	$7/30$
Prop. Indiff. betw. i and j	0	0	0	0	0	0

The key lies in the geometry of the space of ternary paired comparison probabilities that are induced by semiorders or interval orders. We write each $R \in \mathcal{SO}$ (or $R \in \mathcal{IO}$) using $0/1$ coordinates in a suitable high-dimensional space as follows: for each ordered pair (x,y) and each binary relation $R \in \mathcal{SO}$ (or $R \in \mathcal{IO}$), let $R_{xy} = 1$ if xRy, and $R_{xy} = 0$, otherwise. Each binary relation is thereby written as a $0/1$ vector indexed by all ordered pairs of elements in $\mathcal{C}$, i.e., as a point in $[0,1]^{n(n-1)}$ when $|\mathcal{C}| = n$. A probability distribution over $\mathcal{SO}$ or $\mathcal{IO}$ can thus be represented mathematically as a convex combination of such $0/1$ vectors, i.e., as a point in the convex hull of finitely many points in $[0,1]^{n(n-1)}$.

The convex hull of a finite set of points is a convex polytope. Every convex polytope is a bounded intersection of finitely many closed half spaces, each of which can be, in turn, defined by an affine inequality. Thus, characterizing whether or not ternary paired comparison probabilities are induced by binary relations in $\mathcal{SO}$ or $\mathcal{IO}$ is tantamount to finding a 'complete' description of an appropriately selected convex polytope, associated with $\mathcal{SO}$ or $\mathcal{IO}$, via a system of affine inequalities. It can be shown that each affine inequality in a 'minimal and complete' description of a polytope defines a face of maximal dimension, also called a *facet* of the polytope. For the remainder of the chapter, we will focus on cases where $|\mathcal{C}| = 4$.

9.3.1. *Statistical methodology*

The natural empirical sample space for ternary paired comparisons is given by a product of independent trinomials. The basic idea is to assume that empirical data originate from an independent random sample with replacement from the probabilities $(P_{a,b})_{\substack{a,b \in \mathcal{C} \\ a \neq b}}$. Suppose that a respondent evaluated each paired comparison N times. Let $N_{a,b}$ denote the number of times the respondent chose a (i.e., neither 'indifference' nor b) in the paired comparison among a, b. Here, we assume that $N_{a,b}$ and $N_{b,a}$ result from a trinomial sampling process with probabilities $P_{a,b}, P_{b,a}$, and $1 - P_{a,b} - P_{b,a}$ for the three outcome categories. The likelihood of the observed data takes the form

$$\prod_{\substack{a,b \in C \\ a \neq b}} P_{a,b}{}^{N_{a,b}} P_{b,a}{}^{N_{b,a}} \left(1 - P_{a,b} - P_{b,a}\right)^{N - N_{a,b} - N_{b,a}}. \tag{9.3}$$

In our illustration, $N = 30$ denotes the number of (repeated) observations per pair from a single participant in an experiment. We determine the goodness-of-fit of a given model to a given set of data by computing a log-likelihood ratio test statistic. Here, we need to distinguish, for a moment, between unconstrained choice probabilities and choice probabilities that are constrained by a given model. Writing $P_{a,b}$ for the unconstrained probabilities, and writing $\theta_{a,b}$ for the corresponding constrained probabilities, the standard log-likelihood ratio is given by

$$2 \ln \left(\frac{\prod_{\substack{a,b \in C \\ a \neq b}} P_{a,b}{}^{N_{a,b}} P_{b,a}{}^{N_{b,a}} \left(1 - P_{a,b} - P_{b,a}\right)^{N - N_{a,b} - N_{b,a}}}{\prod_{\substack{a,b \in C \\ a \neq b}} \theta_{a,b}{}^{N_{a,b}} \theta_{b,a}{}^{N_{b,a}} \left(1 - \theta_{a,b} - \theta_{b,a}\right)^{N - N_{a,b} - N_{b,a}}} \right). \tag{9.4}$$

If standard assumptions were to hold, this test statistic would asymptotically follow a χ^2 distribution (Bickel & Doksum, 2001). Unfortunately, a fundamental assumption on the asymptotic distribution of the likelihood ratio test is violated. The reason for this problem is that (9.1) and (9.2) imply ordinal constraints on the choice probabilities such as, for example:

$$SO_3 : \ P_{a,b} + P_{b,c} - P_{a,c} \leq 1,$$

$$SO_{26} : \ P_{a,b} + P_{b,c} + P_{c,d} + P_{d,a} + P_{b,d} - 2P_{a,d} - P_{d,b} - P_{b,a} - P_{c,a} \leq 2.$$

If a decision maker's observed choice proportions $(Q_{a,b})_{a \neq b}$ do not satisfy each and every ordinal constraint on the corresponding choice probabilities $(P_{a,b})_{a \neq b}$ for the respective random preference model, say if

$$Q_{a,b} + Q_{b,c} - Q_{a,c} > 1, \ \text{for some } a, b, c \in C,$$

then the maximum likelihood estimate, via the likelihood function in (9.3), must necessarily lie on the boundary of the parameter space. Specifically, in our model, the vector of point estimates will lie on a face of the convex polytope. This violates the standard assumption underlying an asymptotic likelihood ratio test, namely that the hypothesized vector $(P_{a,b})_{a \neq b}$ of true null values of the parameters lies in the strict interior of the parameter space, i.e., in the interior of the polytope. In the next sections, we provide 563 facet-defining inequalities for the polytope that characterizes (9.1) and 191 facet-defining inequalities for the polytope that characterizes (9.2). We also show examples, in (9.5)-(9.15), where the choice proportions violate some of these constraints, demonstrating that the standard assumptions for likelihood testing are indeed violated.

The difficulties we have just discussed can be recast as a problem of *order-constrained inference*. An entire sub-discipline within statistics is devoted to dealing with precisely this type of inferential problem (see e.g., Robertson, Wright, & Dykstra, 1988; Silvapulle & Sen, 2005). A key observation is that, under the order-constrained inference problem, the true asymptotic distribution of the likelihood ratio test critically depends upon the local structure of the parameter space at the maximum likelihood point estimate.

We make use of recent developments in this domain. Suck (1997) discussed statistical testing of probabilistic representations in a slightly different context, and more recently, Myung, Karabatsos, and Iverson (2005) developed a general Bayesian approach to test models that involve order constraints on model parameters. Davis-Stober (2009) expanded groundbreaking results of Iverson and Falmagne (1985) to a broad class of models in which the model parameters of a probabilistic model are constrained by a union of convex polytopes. This technique directly applies to the models we discuss here. The key to analyzing relevant data correctly is to know the proper asymptotic distribution of the log-likelihood ratio test statistic. Davis-Stober (2009) uses recent developments in mathematical statistics and convex geometry to derive this distribution. This distribution is a convex combination of χ^2 distributions with varying degrees of freedom, commonly referred to as a $\bar{\chi}^2$ ("chi-bar squared") distribution.

To summarize briefly: Given a set of choice proportions and a random preference model with the appropriate linear and ordinal constraints on the choice probabilities, we proceed as follows.

(1) We determine the maximum likelihood point estimate of the vector of model parameters, from the data, subject to the constrained model. The maximum likelihood estimate can be obtained via a standard constrained optimization algorithm. For the applications in this chapter, we used the *fmincon* algorithm from the optimization toolbox that is part of the *Matlab*© programming platform.

(2) We calculate the likelihood ratio value using the constrained and unconstrained estimates of the choice probabilities.

(3) We carry out the methodology described in Davis-Stober (2009) to obtain the appropriate $\bar{\chi}^2$ distribution under the null hypotheses that the true choice probabilities $(P_{a,b})_{a \neq b}$ satisfy the random preference model in (9.1) or (9.2).

(4) We compare the calculated likelihood ratio value against the $\bar{\chi}^2$ distribution and draw the appropriate hypothesis testing conclusion.

Another benefit of analyzing random preference models under the likelihood framework in (9.3) is that they avoid certain identifiability issues. Davis-Stober (2009) proves that for general multinomial models (including the above trinomial case) the maximum likelihood estimate is both unique and identifiable within this framework, i.e., there is no worry of local optima.

Note, however, that the number of parameters P_R on the right-hand side of (9.1) and (9.2) exceeds the number of degrees of freedom in the data. For example, when $|\mathcal{C}| = 4$, there are 12 degrees of freedom in the data, as we have illustrated in Table 9.3. On the other hand, we have seen that there are 183 distinct semiorders, and an additional 24 interval orders, so the semiorder mixture model has 182 free parameters and the interval order mixture model has 206 free parameters. Hence, the best fitting probability distribution over semiorders or interval orders is not unique, even if the best fitting marginal trinomial probabilities are unique. The lack of identifiability for the semiorder or interval order probabilities does *not*, however, affect the testability of (9.1) and (9.2) via the likelihood in (9.3). As we will see in the next section, both the semiorder and the interval order polytope make parsimonious predictions about permissible choice probabilities.

By making restrictive predictions that can then be evaluated against data, we offer an approach more similar to standard methods in the natural sciences, and we can avoid some of the weaker methods that are common in the behavioral sciences. Among these weak methods are the frequent reliance on descriptive measures for model fit and model comparison, as well as the omnipresent null hypothesis testing framework, where support for one claim is gathered by statistically rejecting a different claim.

9.3.2. *Testing for semiorder preferences*

We obtained the following proposition by running "PORTA - POlyhedon Representation Transformation Algorithm," a public domain software made available by Christof and Löbel at the Zuse Institut Berlin.[2] Some of the facet-defining inequalities have previously been known (Suck, 1995) to be general facet-defining inequalities for the semiorder polytope (for any size of $|\mathcal{C}|$).

---

[2]PORTA can be downloaded at
http://www2.iwr.uni-heidelberg.de/groups/comopt/software/PORTA/.

Proposition 9.1. *Let* $|\mathcal{C}| = 4$. *Ternary paired comparison probabilities on* $\mathcal{C}$ *are induced by semiorders if and only if the following list of facet-defining inequalities for the semiorder polytope on four objects are satisfied (with quantifiers omitted, and* a, b, c, d *distinct).*

$SO_1 : \; -P_{a,b} \leq 0,$

$SO_2 : \; P_{a,b} + P_{b,a} \leq 1,$

$SO_3 : \; P_{a,b} + P_{b,c} - P_{a,c} \leq 1,$

$SO_4 : \; P_{a,b} + P_{c,d} - P_{a,d} - P_{c,b} \leq 1,$

$SO_5 : \; P_{a,b} + P_{b,c} - P_{a,d} - P_{d,c} \leq 1,$

$SO_6 : \; P_{a,b} + P_{b,c} + P_{c,a} - P_{a,c} - P_{c,b} - P_{b,a} \leq 1,$

$SO_7 : \; P_{a,b} + P_{b,c} + P_{c,d} - P_{a,c} - P_{c,b} - P_{a,d} - P_{b,a} \leq 1,$

$SO_8 : \; P_{a,b} + P_{b,c} + P_{c,d} - P_{a,c} - P_{c,b} - P_{a,d} - P_{b,d} \leq 1,$

$SO_9 : \; P_{a,b} + P_{b,c} + P_{c,d} - P_{d,c} - P_{c,b} - P_{a,d} - P_{b,d} \leq 1,$

$SO_{10} : \; P_{a,b} + P_{b,a} + P_{c,d} - P_{a,d} - P_{b,d} - P_{c,a} - P_{c,b} \leq 1,$

$SO_{11} : \; P_{a,b} + P_{b,c} + P_{c,a} - P_{a,d} - P_{d,c} - P_{c,b} - P_{b,a} \leq 1,$

$SO_{12} : \; P_{a,b} + P_{b,c} + P_{c,a} - P_{a,d} - P_{d,b} - P_{b,a} - P_{c,d} - P_{d,c} \leq 1,$

$SO_{13} : \; P_{a,b} + P_{b,c} + P_{c,a} - P_{a,d} - P_{d,a} - P_{b,d} - P_{d,b} - P_{c,d} - P_{d,c} \leq 1,$

$SO_{14} : \; P_{a,b} + P_{b,c} + P_{c,d} + P_{d,a} - P_{a,d} - P_{d,c} - P_{c,b} - P_{b,a} - P_{b,d} - P_{d,b} \leq 1,$

$SO_{15} : \; P_{a,b} + P_{b,a} + P_{c,d} + P_{d,c} - P_{a,c} - P_{c,a} - P_{a,d} - P_{d,a}$
$$-P_{b,c} - P_{c,b} - P_{b,d} - P_{d,b} \leq 1,$$

$SO_{16} : \; P_{a,b} + P_{b,c} + P_{c,d} - 2P_{a,d} - P_{d,c} - P_{c,b} - P_{b,a} \leq 1,$

$SO_{17} : \; P_{a,b} + P_{b,c} + P_{c,d} + P_{d,a} - P_{a,d} - P_{d,c} - P_{c,b} - 2P_{b,a} - P_{a,c} - P_{d,b} \leq 1,$

$SO_{18} : \; P_{a,b} + P_{b,c} + P_{c,d} + P_{d,a} - 2P_{a,d} - 2P_{d,c} - P_{c,b} - P_{b,a} - P_{c,a} \leq 1,$

$SO_{19} : \; P_{a,b} + P_{b,c} + P_{c,d} + P_{d,a} - 2P_{a,d} - 2P_{d,c} - 2P_{c,b} - 2P_{b,a} \leq 1,$

$SO_{20} : \; P_{a,b} + P_{b,c} + P_{c,d} + P_{d,c} + P_{b,d} - P_{a,c} - P_{a,d} - P_{b,a} \leq 2,$

$SO_{21} : \; P_{a,b} + P_{b,a} + P_{b,d} + P_{a,d} + P_{d,c} - P_{a,c} - P_{b,c} - P_{c,d} \leq 2,$

$SO_{22} : \; P_{a,b} + P_{b,a} + P_{b,c} + P_{c,b} + P_{c,a} - P_{b,d} - P_{d,b} - P_{c,d} - P_{d,a} \leq 2,$

$SO_{23} : \; P_{a,b} + P_{b,c} + P_{c,d} + P_{d,a} + P_{a,c} - P_{a,d} - P_{d,b} - P_{b,d} - P_{d,c} \leq 2,$

$SO_{24} : \; P_{a,b} + P_{b,a} + P_{a,c} + P_{c,a} + P_{b,c} + P_{c,b} - P_{a,d} - P_{d,a}$
$$-P_{b,d} - P_{d,b} - P_{c,d} - P_{d,c} \leq 2,$$

$SO_{25} : \; P_{a,b} + P_{b,c} + P_{c,d} + P_{d,a} + P_{b,d} - 2P_{b,a} - P_{d,b} - P_{a,c} - P_{a,d} \leq 2,$

SO_{26} : $P_{a,b} + P_{b,c} + P_{c,d} + P_{d,a} + P_{b,d} - 2P_{a,d} - P_{d,b} - P_{b,a} - P_{c,a} \leq 2,$

SO_{27} : $2P_{a,b} + P_{b,c} + P_{c,d} + P_{d,a} + P_{c,a} - P_{a,c} - P_{a,d}$
$$-P_{b,a} - P_{b,d} - P_{d,b} - 2P_{c,b} \leq 2,$$

SO_{28} : $2P_{a,b} + P_{b,c} + P_{c,d} + P_{d,a} + P_{b,d} - P_{a,c} - P_{c,a}$
$$-2P_{a,d} - P_{b,a} - P_{c,b} - P_{d,b} \leq 2$$

SO_{29} : $2P_{a,b} + P_{b,c} + P_{c,d} + P_{d,c} + P_{b,d} - 2P_{a,c} - 2P_{a,d} - P_{b,a}$
$$-P_{c,b} - P_{d,b} \leq 2,$$

SO_{30} : $2P_{a,b} + 2P_{b,c} + 2P_{c,d} + P_{d,a} - P_{a,c} - P_{c,a} - 2P_{a,d}$
$$-P_{b,a} - P_{b,d} - P_{d,b} - 2P_{c,b} - P_{d,c} \leq 2,$$

SO_{31} : $2P_{a,b} + 2P_{b,c} + 2P_{c,d} + P_{d,a} + P_{c,a} + P_{d,b} - 2P_{a,c} - P_{a,d}$
$$-P_{b,a} - 2P_{b,d} - 4P_{c,b} - P_{d,c} \leq 3.$$

In the case of $|\mathcal{C}| = 4$ that we consider in this chapter, $\mathcal{SO}$ is a 12-dimensional polytope that has 183 vertices, all of which have 0/1 coordinates. This polytope is characterized by 563 facets, that is, faces of maximal dimension. Specifically, taking into account the omitted quantifiers, there are

12 facets of type SO_1, 6 of type SO_2, 24 of type SO_3, 12 of type SO_4,
24 of type SO_5, 8 of type SO_6, 24 of type SO_7, 24 of type SO_8,
24 of type SO_9, 12 of type SO_{10}, 24 of type SO_{11}, 24 of type SO_{12},
8 of type SO_{13}, 12 of type SO_{14}, 3 of type SO_{15}, 24 of type SO_{16},
24 of type SO_{17}, 24 of type SO_{18}, 6 of type SO_{19}, 12 of type SO_{20},
12 of type SO_{21}, 24 of type SO_{22}, 24 of type SO_{23}, 4 of type SO_{24},
24 of type SO_{25}, 24 of type SO_{26}, 24 of type SO_{27}, 24 of type SO_{28},
24 of type SO_{29}, 24 of type SO_{30}, and 24 facets of type SO_{31}.

Now, we consider the choice proportions $Q_{i,j}$. Of the 563 facet defining inequalities, the observed choice proportions in Table 9.3 violate 11 inequalities and satisfy 552 inequalities. The 11 violations are as follows:

$$Q_{c,d} + Q_{d,a} - Q_{c,a} = 1.1 > 1, \quad (9.5)$$

$$Q_{c,d} + Q_{d,b} - Q_{c,b} = 1.04 > 1, \quad (9.6)$$

$$Q_{b,c} + Q_{c,d} - Q_{b,d} = 1.04 > 1, \quad (9.7)$$

$$Q_{d,b} + Q_{b,c} - Q_{d,c} = 1.04 > 1, \quad (9.8)$$

$$Q_{a,c} + Q_{c,d} - Q_{a,d} = 1.1 > 1, \quad (9.9)$$

$$Q_{d,a} + Q_{a,c} - Q_{d,c} = 1.1 > 1, (9.10)$$

$$Q_{b,c} + Q_{c,d} + Q_{d,b} - Q_{b,d} - Q_{d,c} - Q_{c,b} = 1.08 > 1, (9.11)$$

$$Q_{a,c} + Q_{c,d} + Q_{d,a} - Q_{a,d} - Q_{d,c} - Q_{c,a} = 1.2 > 1, (9.12)$$

$$Q_{a,b} + Q_{b,a} + Q_{c,d} + Q_{d,a} + Q_{d,b} - Q_{c,b} - Q_{c,a} - Q_{d,c} = 2.14 > 2, (9.13)$$

$$Q_{a,b} + Q_{b,a} + Q_{a,c} + Q_{b,c} + Q_{c,d} - Q_{a,d} - Q_{b,d} - Q_{d,c} = 2.14 > 2, (9.14)$$

$$Q_{a,b} + Q_{a,c} + Q_{b,c} + Q_{c,d} + Q_{d,a} - Q_{a,d} - 2Q_{d,c} - Q_{c,a} - Q_{b,d} = 2.04 > 2. (9.15)$$

Inequalities (9.5)-(9.10) are violations of SO_3, which has been called the *triangle inequality* in slightly different contexts. Inequalities (9.11) and (9.12) both are violations of SO_6, (9.13) is a violation of SO_{20}, whereas (9.14) is a violation of SO_{21}, and (9.15) is a violation of SO_{26}.

These violations may or may not be indicative of a statistically significant violation of the semiorder model. Many researchers count the number of violated constraints of a model as a descriptive measure of goodness-of-fit. This descriptive approach is not a good idea, and can be highly counterproductive. As we have shown in previous work for another polytope (Regenwetter, Dana & Davis-Stober, 2010; Regenwetter, Dana, & Davis-Stober, 2011), the number of violated inequality constraints is not monotonically related to goodness-of-fit.

Since the choice proportions $(Q_{a,b})_{a \neq b}$ in our example violate some of the constraints on the choice probabilities $(P_{a,b})_{a \neq b}$, we know that the maximum likelihood estimate does not lie in the strict interior of the parameter space. Hence, we need to rely on order-constrained inference methodology. Following Davis-Stober (2009), we find that the log-likelihood ratio test statistic (at the point estimate) has a $\bar{\chi}^2$ distribution equal to $\frac{1}{2} + \frac{1}{2}\chi_1^2$ (with estimated weights rounded to two significant digits). The point estimate of the log-likelihood ratio is .4964, which gives a p-value of .24 for the goodness-of-fit, indicating a good statistical fit of the semiorder model.

It is important to note that the maximum likelihood point estimate does not necessarily lie in the intersection of the facets whose facet defining inequalities the data vector violates. In this case here, the point estimate lies on a four-dimensional face of the semiorder polytope given by the intersection of the semiorder polytope and the subspace generated by the following (redundant) system of 16 linear equations.

$$P_{a,b} + P_{b,a} = 1, \quad P_{a,c} + P_{c,a} = 1, \quad P_{a,d} + P_{d,a} = 1, (9.16)$$
$$P_{b,c} + P_{c,b} = 1, \quad P_{b,d} + P_{d,b} = 1, \quad P_{c,d} + P_{d,c} = 1, (9.17)$$
$$P_{c,d} - P_{c,a} + P_{d,a} = 1, \quad P_{c,d} - P_{c,b} + P_{d,b} = 1, \quad P_{b,c} - P_{b,d} + P_{c,d} = 1, (9.18)$$
$$P_{b,c} + P_{d,b} - P_{d,c} = 1, \quad P_{a,c} - P_{a,d} + P_{c,d} = 1, \quad P_{a,c} + P_{d,a} - P_{d,c} = 1, (9.19)$$
$$P_{b,c} - P_{b,d} + P_{c,d} - P_{c,b} + P_{d,b} - P_{d,c} = 1, (9.20)$$
$$P_{a,c} - P_{a,d} + P_{c,d} - P_{c,a} + P_{d,a} - P_{d,c} = 1, (9.21)$$
$$P_{a,b} + P_{c,d} + P_{b,a} - P_{c,a} + P_{d,a} - P_{c,b} + P_{d,b} - P_{d,c} = 2, (9.22)$$
$$P_{a,b} + P_{a,c} - P_{a,d} + P_{b,c} - P_{b,d} + P_{c,d} + P_{b,a} - P_{d,c} = 2. (9.23)$$

This face is the face of lowest dimension containing the maximum likelihood point estimate, and it does not equal the intersection of those facets whose facet defining inequalities the choice proportions violate.

This completes our discussion of the semiorder polytope for four choice alternatives. Next, we consider the interval order case, where we expand the set of permissible preference states from 183 semiorders to 207 interval orders.

9.3.3. Testing for interval order preferences

We obtained the following proposition using PORTA. Some of the facet-defining inequalities have previously been known (Mueller & Schulz, 1995; Suck, 1995) to be general facet-defining inequalities for the interval order polytope (for any size of $|\mathcal{C}|$).

Proposition 9.2. Let $|\mathcal{C}| = 4$. Ternary paired comparison probabilities on $\mathcal{C}$ are induced by interval orders if and only if the following list of facet-defining inequalities for the interval order polytope on four objects are satisfied (with quantifiers omitted, and a, b, c, d distinct).

$IO_1: \quad -P_{a,b} \leq 0,$

$IO_2: \quad P_{a,b} + P_{b,a} \leq 1,$

$IO_3: \quad P_{a,b} + P_{b,c} - P_{a,c} \leq 1,$

$IO_4: \quad P_{a,b} + P_{c,d} - P_{a,d} - P_{c,b} \leq 1,$

$IO_5: \quad P_{a,b} + P_{b,c} + P_{c,a} - P_{a,c} - P_{c,b} - P_{b,a} \leq 1,$

$IO_6: \quad P_{a,b} + P_{b,c} + P_{c,d} - P_{a,c} - P_{a,d} - P_{b,d} - P_{c,b} \leq 1,$

$IO_7: \quad P_{a,b} + P_{b,a} + P_{d,c} - P_{a,c} - P_{b,c} - P_{d,a} - P_{d,b} \leq 1,$

$IO_8: \quad P_{a,b} + P_{b,c} + P_{c,d} + P_{d,a} - P_{a,d} - P_{d,c} - P_{c,b} - P_{b,a}$
$$-P_{a,c} - P_{c,a} - P_{b,d} - P_{d,b} \leq 1,$$

$IO_9: \quad P_{a,b} + P_{b,a} + P_{c,d} + P_{d,c} - P_{a,c} - P_{a,d} - P_{b,c} - P_{b,d}$
$$-P_{c,a} - P_{c,b} - P_{d,a} - P_{d,b} \leq 1,$$

$IO_{10}: \quad P_{a,b} + P_{b,a} + P_{b,c} + P_{c,d} + P_{d,b} - P_{a,c} - P_{c,b}$
$$-P_{b,d} - P_{d,a} \leq 2,$$

$IO_{11-12}: \quad -2 \leq 2P_{a,b} + P_{b,c} + P_{c,d} + P_{d,a} - P_{a,c} - P_{c,b} - P_{b,a}$
$$-P_{a,d} - P_{d,b} \leq 2,$$

$IO_{13-14}: \quad -3 \leq 2P_{a,b} + 2P_{b,c} + 2P_{c,d} + 2P_{d,a} - P_{b,a} - P_{a,d} - P_{d,c}$
$$-P_{c,b} - P_{a,c} - P_{c,a} - P_{b,d} - P_{d,b} \leq 3.$$

This is a 12-dimensional polytope with 207 vertices and 191 facets. Taking into account the omitted quantifiers, there are

12 facets of type IO_1, 6 of type IO_2, 24 of type IO_3, 12 of type IO_4,
8 of type IO_5, 24 of type IO_6, 12 of type IO_7, 6 of type IO_8,
3 of type IO_9, 24 of type IO_{10}, 24 of type IO_{11}, 24 of type IO_{12},
6 of type IO_{13}, and 6 of type IO_{14}.

It is interesting to note that this polytope, which has 24 more vertices than the semiorder polytope, has many fewer facets. By moving to a larger polytope, we have actually proceeded to a simpler geometric structure. Hence, there is no straightforward way to predict from an axiomatic characterization of the preference relations that make up the vertices of the polytope whether the resulting system of facet-defining inequalities will be simple or complicated.

Returning to our empirical illustration, of these 191 facet defining inequalities, the observed choice proportions in Table 9.3 violate 8 and satisfy 183. The triangle inequality IO_3 is the same as SO_3, that we have already seen in $\mathcal{SO}$, and hence it is violated 6 times here as well. In addition to the triangle inequality, we have violations (9.11) and (9.12) that we already discussed in the semiorder case and that carry over to the interval order polytope. After rounding, the log-likelihood ratio test statistic point estimate, distribution and p-value match those we have reported in the semiorder model.

The interval order polytope face of lowest dimension that the binary choice point estimate lies in is given by the intersection of the interval order polytope with the subspace generated by (9.16)-(9.21). This means that the face on the interval order polytope lies in a subspace of the subspace that contains the semiorder polytope face (of lowest dimension) that the point estimate lies on in the semiorder polytope. Furthermore, in the interval order case, the face of lowest dimension containing the maximum likelihood point estimate is, in fact, the intersection of the facets whose facet defining inequalities the choice proportions violated. This is not automatically the case, as we have seen in the semiorder case.

9.4. Conclusions

Transitive preferences can be modeled in many different ways. Building directly on our previous work (Regenwetter & Davis-Stober, 2008; Regenwetter, Dana, & Davis-Stober, 2010, 2011), we have considered probabilistic specifications of semiorder and interval order preferences via "mixture," aka "random preference" models.

A well-known property for two-alternative forced choice induced by linear orders (Cohen & Falmagne, 1990; Fishburn & Falmagne, 1989; Gilboa, 1990; Heyer & Niederée, 1992; Koppen, 1991; Suck, 1992; Loomes & Sugden, 1995), the *triangle inequality* resurfaced here for ternary paired comparisons, in the form of SO_3 and IO_3. The triangle inequality can be derived directly from the assumption of a probability distribution over transitive binary relations. However, in the case of the semiorder polytope for four choice alternatives, there are altogether 31 different types of facet-defining inequalities, and for the interval order polytope for four choice alternatives, there are altogether 14 different types of facet-defining inequalities. We leave it for future work to place these into a broader context and study classes of inequalities that are generally facet-defining for any number of choice alternatives.

We have provided an illustration of order-constrained inference (Davis-Stober, 2009) for these two polytopes. Two important features are worth highlighting once more.

(1) The maximum likelihood point estimate obtained from a given set of choice proportions may or may not lie on the intersection of the facets whose facet-defining inequalities the choice proportions violate. We have seen an example of each scenario in our illustration.

(2) It is important to note that the number of violated constraints (here, the number of violated facet-defining inequalities) need not (and should not) be used as a descriptive measure of goodness-of-fit. Order-constrained statistical inference permits a quantitative goodness-of-fit analysis by using chi-bar squared distributions for the asymptotic distribution of the log-likelihood statistic that can be derived from the geometric properties of the probabilistic models under consideration, here the semiorder and interval order polytopes.

We are hopeful that the development of these modeling and testing techniques will open up new avenues for quantitative testing of decision theories in the future.

Acknowledgments

This work was supported by AFOSR grant FA9550-05-1-0356, NIH-NIMH training grant PHS 2 T32 MH014257, and NSF grant SES08-20009 (M. Regenwetter, PI). The human participants research in our reported pilot study was approved under IRB Nr. 05178 (University of Illinois). We are grateful to Lyle Regenwetter for his valuable assistance with preparation of this manuscript. Jean-Paul Doignon, Samuel Fiorini, William Messner, Aleksandr Sinayev, Reinhard Suck, and Christopher Zwilling gave helpful comments on earlier drafts. Any opinions, findings, and conclusions or recommendations expressed in this publication are those of the authors and do not necessarily reflect the views of colleagues, the funding agencies, or of the University of Illinois.

References

Bickel, P. J., & Doksum, K. A. (2001). *Mathematical Statistics: Basic Ideas and Selected Topics*, vol. 1. Upper Saddle River, NJ: Prentice-Hall.

Cohen, M., & Falmagne, J. C. (1990). Random utility representation of binary choice probabilities: A new class of necessary conditions. *Journal of Mathematical Psychology, 34*, 88–94.

Davis-Stober, C. P. (2009). Analysis of multinomial models under inequality constraints: Applications to measurement theory. *Journal of Math-*

ematical Psychology, 53, 1–13.

Fishburn, P. C. (1985). *Interval Orders and Interval Graphs*. NY: John Wiley & Sons.

Fishburn, P. C., & Falmagne, J. C. (1989). Binary choice probabilities and rankings. *Economic Letters, 31*, 113–117.

Gilboa, I. (1990). A necessary but insufficient condition for the stochastic binary choice problem. *Journal of Mathematical Psychology, 34*, 371–392.

Hey, J. D. (2005). Why we should not be silent about noise. *Experimental Economics, 8*, 325–345.

Heyer, D., & Niederée, R. (1992). Generalizing the concept of binary choice systems induced by rankings: One way of probabilizing deterministic measurement structures. *Mathematical Social Sciences, 23*, 31–44.

Iverson, G. J., & Falmagne, J. C. (1985). Statistical issues in measurement. *Mathematical Social Sciences, 10*, 131–153.

Koppen, M. (1991). Random utility representations of binary choice probabilities. In J. P. Doignon and J. C. Falmagne (Eds.), *Mathematical Psychology: Current Developments* (pp. 181–201). New York, NY: Springer.

Loomes, G., & Sugden, R. (1995). Incorporating a stochastic element into decision theories. *European Economic Review, 39*, 641–648.

Luce, R. D. (1956). Semiorders and a theory of utility discrimination. *Econometrica, 26*, 178–191.

Luce, R. D. (1995). Four tensions concerning mathematical modeling in psychology. *Annual Review of Psychology, 46*, 1–26.

Luce, R. D. (1997). Several unresolved conceptual problems of mathematical psychology. *Journal of Mathematical Psychology, 41*, 79–87.

Müller, R., & Schulz, A. (1995). The interval order polytope of a digraph. In E. Balas and E. Clausen (Eds.), *Integer Programming and Combinatorial Optimization* (pp. 50–64). Springer-Verlag.

Myung, J., Karabatsos, G., & Iverson, G. (2005). A Bayesian approach to testing decision making axioms. *Journal of Mathematical Psychology, 49*, 205–225.

Pirlot, M., & Vincke, P. (1997). *Semiorders: Properties, Representations, Applications*. Dordrecht: Kluwer Academic Publishers.

Regenwetter, M., Dana, J., & Davis-Stober, C. P. (2010). Testing transitivity of preferences on two-alternative forced choice data. *Frontiers in Quantitative Psychology and Measurement*. doi: 10.3389/fpsyg.2010.00148.

Regenwetter, M., Dana, J., & Davis-Stober, C. P. (2011). Transitivity of preferences. *Psychological Review, 118*, 42–56.

Regenwetter, M., & Davis-Stober, C. (2008). There are many models of transitive preference: A tutorial review and current perspective. In *Decision Modeling and Behavior in Uncertain and Complex Environments* (pp. 99–124). Springer-Verlag.

Robertson, T., Wright, F., & Dykstra, R. (1988). *Order Restricted Statistical Inference*. NY: John Wiley & Sons.

Silvapulle, M. J., & Sen, P. K. (2005). *Constrained Statistical Inference: Inequality, Order, and Shape Restrictions*. NY: John Wiley & Sons.

Suck, R. (1992). Geometric and combinatorial properties of the polytope of binary choice probabilities. *Mathematical Social Sciences, 23*, 81–102.

Suck, R. (1995). Random utility representations based on semiorders, interval orders, and partial orders. Unpublished manuscript, Universität Osnabrück.

Suck, R. (1997). Probabilistic biclassification and random variable representations. *Journal of Mathematical Psychology, 41*, 57–64.

Chapter 10

Knowledge Spaces Regarded as Set Representations of Skill Structures

Reinhard Suck

University of Osnabrück

In this chapter knowledge space theory is developed from a different point of view. It takes the skills as primitives of the theory and shows how the usual theory can be derived from this setup. The skills are known to be closely related to the basis of the space, they are assumed to be partially ordered. The test items are introduced via a set representation of this partial order. The knowledge space is seen to be determined both by the skill order and the set representation. The interaction between both components is described and viewed from a lattice theoretic perspective. When the lattice structure is regarded, it must be kept in mind that non-isomorphic spaces can have isomorphic lattices.

10.1. Introduction

Knowledge space theory is in some sense an alternative to psychological tests. This is not to suggest that we can base an intelligence test — to give a prominent example of a psychological test — on this theory, at least not at present. However, in situations where the achievement of a student is to be determined and the field of knowledge is sort of hierarchically structured, it is a viable alternative with many advantages. Meanwhile there are numerous successful applications despite the effort it takes to gather data to elicit a knowledge space. Utilizing this theory requires the construction of the space and a method, preferably automated, to determine the knowledge state of the student. This scheme has been implemented in various fields from first grade teaching for young children to statistic education of psychology students and many more. In the sequel we review the basic definitions, give an alternative view, demonstrate the equivalence between the two approaches, outline a few consequences of the latter, discuss lattice theoretic issues of our approach, and, finally, outline consequences for the problem of constructing a space of a given field of knowledge. For a general reference on knowledge spaces the reader is referred to Doignon and Falmagne (1999).

In this chapter we regard mostly finite knowledge spaces. There is an interesting ramification to infinite spaces; in places Doignon et al. (1999) point at this issue. A deeper analysis of the infinite case of the point of view purported in the present work will be published elsewhere.

10.1.1. Notation

We denote the power set of a given set Q by $\mathfrak{P}(Q)$. Subsets of $\mathfrak{P}(Q)$ are called families and denoted by calligraphic letters such as $\mathcal{F}$, $\mathcal{K}$, etc. A function f (or mapping) between two sets A and B is denoted by $f : A \to B$; for $X \subseteq A$ the set $f(X)$ is the set of values $f(x)$ for $x \in X$. It is called *surjective* if $f(A) = B$ and *injective* if $f(x) = f(y)$ implies $x = y$ for all $x, y \in A$. A surjective and injective map is called *bijective* or a *bijection*.

A *partial order* (or *order*, for short) is a pair $(S, \leq)$ where S is a set and $\leq$ a binary relation on S satisfying reflexivity, antisymmetry, and transitivity. In places we will also loosely speak of $\leq$ as a partial order. If $S' \subseteq S$ then the *induced suborder* is the partial order $(S', \leq')$ where $x \leq' y$ iff $x, y \in S'$ and $x \leq y$. It will be convenient to omit the prime in $\leq'$ when dealing with induced suborders.

A partial order $(S, \leq)$ is a *lattice* when for any two elements $x, y \in S$ *infimum* and *supremum* exist. Here infimum is the (unique) greatest lower bound of the two elements and supremum the (unique) least upper bound. Usually, the infimum is denoted by $x \wedge y$ and the supremum by $x \vee y$. Sometimes the notation $(S, \wedge, \vee)$ is used. The partial order is determined by $\wedge$ because defining $x \leq y$ iff $x = x \wedge y$ has the required properties. Clearly, induction yields that for any finite subset infimum and supremum exist. Infinite subsets may or may not have infimum and supremum. When they exist for all infinite subsets then the lattice is called *complete*.

The power set of a given set is a lattice with intersection and union as infimum and supremum, respectively. However, it is a very special one because of the distributivity conditions and existence of complements. The connection between knowledge spaces and lattices will be discussed in Section 10.5.

10.2. Knowledge Spaces and Learning Spaces

We begin by defining a knowledge space. The ingredients are a set Q of items, problems, questions pertinent to a field of knowledge. Some subsets of Q are knowledge states, i.e., subfields of the area which a person can be able to know and the rest not to know.

Definition 10.1.

a) Let $\mathcal{F}$ be a family of subsets of Q. The closure under union of $\mathcal{F}$ is denoted by $\mathcal{F}^*$, i.e.,

$$\mathcal{F}^* := \left\{ \bigcup_{F \in \mathcal{F}'} F; \; \mathcal{F}' \subseteq \mathcal{F} \right\}.$$

b) A knowledge space is a pair $(Q, \mathcal{K})$ where Q is a non-empty set and $\mathcal{K}$ a family of subsets of Q closed under union containing $\varnothing$ and $\mathcal{K}$.

When Q is finite, closure under union according to Definition 10.1a) can be replaced by the simpler

$$K_1 \cup K_2 \in \mathcal{K} \quad \text{for all} \quad K_1, K_2 \in \mathcal{K}.$$

Given a space $\mathcal{K}$. If for some $\mathcal{F} \subseteq \mathcal{K}$ we have $\mathcal{F}^* = \mathcal{K}$, then $\mathcal{F}$ is said to *generate* or to *span* $\mathcal{K}$.

If there is a smallest family $\mathcal{B} \subseteq \mathfrak{P}(Q)$ which generates $\mathcal{K}$, i.e.,

$$\mathcal{B}^* = \mathcal{K}$$

and for all $\mathcal{F}$ which generate $\mathcal{K}$ we have

$$\mathcal{B} \subseteq \mathcal{F}$$

then $\mathcal{B}$ is called the basis of $\mathcal{K}$.

Thus, when a basis exists then for any generating systems $\mathcal{F}$ and $\mathcal{F}'$ of $\mathcal{K}$, the basis must be in $\mathcal{F} \cap \mathcal{F}'$. It can be shown that for finite spaces a basis always exists and is unique. For infinite $\mathcal{K}$ we have to distinguish spaces with basis and without basis.

Recently, knowledge space theorists tend to regard learning spaces or well-graded spaces as the most important subclass.

Definition 10.2. Let Q be a finite set of items. A family $\mathcal{K} \subseteq \mathfrak{P}(Q)$ such that $\varnothing, Q \in \mathcal{K}$ is a learning space if the following two conditions are satisfied:

a) If $K, L \in \mathcal{K}$ and $K \subset L$ then there is a chain of states

$$K \subset K_1 \subset K_2 \subset \cdots \subset K_n = L$$

such that $K_i = K_{i-1} \cup \{q_i\}$ with $q_i \in Q$.

b) If $K, L \in \mathcal{K}$ and $K \subset L$ with $K \cup \{q\} \in \mathcal{K}$ and $q \notin L$ for some $q \in Q$ then $L \cup \{q\} \in \mathcal{K}$.

The meaning of condition a) is that a student can graduate from state K to state L by learning one item at a time. Condition b) in Definition 10.2 says that if it is possible to learn item q from state K directly then it can also be learned directly from any state $L \supset K$ with $q \notin L$. In a learning space the maximal chains in the partial order $(\mathcal{K}, \subseteq)$ have a particular characteristic: They form a gradation, i.e., they are of the form

$$\varnothing \subset \{q_1\} \subset \{q_1, q_2\} \subset \cdots \subset \{q_1, \ldots, q_n\} = Q. \tag{10.1}$$

Each maximal chain has the same structure just with another permutation of the elements of Q.

Property (10.1) has been denoted well-gradedness and according to a theorem of Cosyn and Uzun (2009) the learning space properties of Definition 10.2 are equivalent to (10.1). In this chapter we shall use the terms well-graded space and learning space interchangeably.

It should be noted that Falmagne and Doignon (2011) is in parts a revised version of Doignon and Falmagne (1999). The choice of the new title ("Learning spaces" instead of "Knowledge spaces") stresses the importance which is nowadays attached to this special class of knowledge spaces.

In applications well-gradedness is a useful property for a variety of reasons. Here are a few:

- Well-graded spaces are closely related to other interesting structures — called 'media' — which are recently enjoying much attention in quite separate fields of mathematics and applications, cf. Eppstein, Falmagne and Ovchinnikov (2008).
- Well-graded spaces are pedagogically or didactically called for because they describe the most 'cautious' learning processes (progress is made step by step mastering one piece of knowledge at a time).
- Well-graded spaces are special cases of 'well graded families of relations' (other prominent examples are: linear orders, partial orders, semiorders on a given set, etc., cf. Doignon and Falmagne, 1997, and Eppstein et al., 2008, Chapter 5).

However, their construction poses difficult problems. Put another way, formulating a procedure which goes directly for learning spaces is still an open problem. The development in Section 10.3 is partly motivated by this problem.

In applications the knowledge space is built via a so-called *entailment relation*. After a set of items or questions has been formulated an expert of the field for which the knowledge space is to be developed has to answer numerous questions of the kind:

> Suppose a student of the field has just failed to answer questions $q_1, q_2, \ldots, q_r \in Q$. Is it practically certain that he/she will also fail question q?

Koppen (1993) devised an algorithm by which the number of questions can be considerably reduced because given some answers to such questions many further answers can be inferred and need not be elicited from the expert.

Another approach to construction is to derive the space from students' data. Naturally, in both approaches a lot of thought has to go into the handling of errors. Cosyn and Thiéry (2000) combined the two alternative construction modes.

So far any attempts to modify this procedure to go straightforward for a learning space have failed. It is however questionable that such an attempt can be carried through without modification of the item set. Section 10.6 takes up this issue.

Two knowledge spaces $(Q, \mathcal{K})$ and $(Q', \mathcal{K}')$ are isomorphic when there is a bijection $f : Q \to Q'$ such that

$$K \in \mathcal{K} \Leftrightarrow f(K) \in \mathcal{K}', \tag{10.2}$$

where $f(K) = \{f(q); \ q \in K\}$.

We denote isomorphy by $(Q, \mathcal{K}) \cong (Q', \mathcal{K}')$. Obviously, the partial orders $(\mathcal{K}, \subseteq)$ and $(\mathcal{K}', \subseteq)$ are also isomorphic for isomorphic spaces. In Section 10.5 we will see that the converse is not necessarily true.

10.3. Knowledge Spaces as Set Representations

In this section we shift the focus a bit resulting in an equivalent definition of knowledge spaces. The primitives here are a set Q (as in Section 10.2), but instead of the set $\mathcal{K}$ of knowledge states we assume a partial order $(S, \leq)$ and a function mapping S into the power set of Q, i.e.,

$$\varphi : S \to \mathfrak{P}(Q).$$

This function φ is called a *set representation* of $(S, \leq)$ if it satisfies:

$$s_1 \leq s_2 \quad \text{iff} \quad \varphi(s_1) \subseteq \varphi(s_2). \tag{10.3}$$

Thus, the partial order $(S, \leq)$ is 'represented' as a suborder of the lattice of subsets of Q. These set representations of partial orders are closely related to knowledge spaces. We prove

Theorem 10.1. *Given a partial order $(S, \leq)$, a finite set Q and a function $\varphi : S \to \mathfrak{P}(Q)$ satisfying (10.3). Then $\left(\bigcup_{s \in S} \varphi(s), \{\varphi(s); s \in S\}^*\right)$ is a knowledge space with basis $\{\varphi(s); s \in S\}$ if and only if φ satisfies*

$$|\varphi(s)| > \left|\bigcup_{t<s} \varphi(t)\right|. \tag{10.4}$$

Functions φ which fulfill (10.4) are called 'basic set representations'. Theorem 10.1 gives the reason for this name and why they are important for knowledge space theory.

In general, nothing prevents a set representation to utilize the empty set provided the element represented by $\varnothing$ is a unique minimal element in $(S, \leq)$. However, for a basic set representation we always have $\varphi(s) \supset \varnothing$ because for a minimal element the right-hand side of (10.4), being the number of elements of the union of an empty set of subsets, is 0.

Proof of Theorem 10.1. Denote $\{\varphi(s); s \in S\}$ by $\mathcal{B}$. Clearly, $\left(\bigcup_{s \in S} \varphi(s), \mathcal{B}^*\right)$ is a knowledge space because by definition $\mathcal{B}^*$ is closed under union; it contains $\bigcup_{s \in S} \varphi(s)$ and $\varnothing$, the latter by the same argument as just given on the union of an empty set of subsets. It remains to show that $\mathcal{B}$ is the basis of this knowledge space if and only if φ is basic.

Assume first that $\mathcal{B}$ is the basis of the space. Since φ is a set representation, it must satisfy $|\varphi(s)| \geq \left|\bigcup_{t<s} \varphi(t)\right|$. If it is not basic then there exists $s \in S$ such that

$$|\varphi(s)| = \left|\bigcup_{t<s} \varphi(t)\right|. \tag{10.5}$$

As a set representation φ satisfies

$$\varphi(s) \supseteq \bigcup_{t<s} \varphi(t). \tag{10.6}$$

Equations (10.5) and (10.6) together imply

$$\varphi(s) = \bigcup_{t<s} \varphi(t).$$

Therefore, $\varphi(s)$ is not needed in $\mathcal{B}$ to generate $\mathcal{B}^*$, hence $\mathcal{B}$ is not a basis in contradiction to the assumption.

Conversely, assume that φ is basic. If $\mathcal{B}$ is not the basis then at least one of its elements, B can be omitted from it, meaning it is the union of other sets $B_1, \ldots, B_k \in \mathcal{B}$. Thus,

$$B = \bigcup_{i=1}^{k} B_k.$$

Let $B = \varphi(s)$ and $B_i = \varphi(s_i)$ for some $s, s_1, \ldots, s_k \in S$. Clearly, $B \supseteq B_i$ for $i = 1, \ldots, k$; hence $s \geq s_1, \ldots, s_k$ yielding

$$\varphi(s) = \bigcup_{i=1}^{k} \varphi(s_i) \subseteq \bigcup_{t<s} \varphi(t) \subset \varphi(s),$$

which is absurd. Note that the strict containment $\subset$ in this argument is a consequence of (10.4). Thus, $\mathcal{B}$ is the basis of the knowledge space. $\qquad\square$

To fully understand the concept of a basic set representation, a note of caution seems appropriate at this stage. The essence of Theorem 10.1 is not that $\left(\bigcup_{s \in S} \varphi(s), \{\varphi(s); s \in S\}^*\right)$ is a knowledge space — this is trivial — but rather that its basis is $\{\varphi(s); s \in S\}$.

To clarify the difference look at the example in Figure 10.1. It shows two set representations of the order on the left. The first one is not basic, closing the representing sets under union yields the space left in the second row; its basis consists of the sets $\{1\}$ and $\{2\}$. The second representation is basic and the knowledge space resulting from it is different. The sets $\{1\}$, $\{2\}$, and $\{1, 2, 3\}$ constitute its basis. As a partial order under $\subseteq$ it is isomorphic to the order on the left.

In Section 10.4 we emphasize an important interpretation of the elements of the basis (they correspond to the skills of a field of knowledge) which transfers to the basic set representations.

The most important subclass of knowledge spaces are the well-graded ones (or the learning spaces). They can be characterized in this approach. Theorem 4 in Suck (2003) shows that a space is well-graded if and only if the set representation fulfills:

$$|\varphi(s)| = \left|\bigcup_{t<s} \varphi(t)\right| + 1. \tag{10.7}$$

Thus, the increase in representing elements from all the predecessors of s necessary according to (10.4) is minimal, namely one. This kind of set representations are called 'parsimonious'.

Obviously, any partial order can be represented parsimoniously by the so-called principal ideal representation (i.e., we take $Q = S$ and $\varphi(s) =$

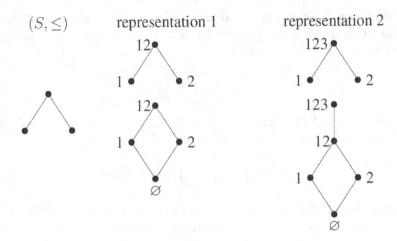

Fig. 10.1. An order with a non-basic and a basic set representation and the ensuing knowledge spaces.

$\{t \leq s\}$). This is a parsimonious representation because $\varphi(s) = \{t < s\} \cup \{s\}$. With this choice of φ the knowledge space constructed according to Theorem 10.1 is closed under intersection or in the usual nomenclature it is an ordinal space.

To see that it is closed under intersection let $s, s' \in S$; we argue

$$\varphi(s) \cap \varphi(s') = \{t \in S; \ t \leq s, s'\} = \bigcup_{t \leq s, s'} \varphi(t).$$

Thus, the intersection of $\varphi(s)$ and $\varphi(s')$ is the union of states $\varphi(t)$, hence in $\mathcal{K}$.

In Suck (2004), Theorem 5, a necessary and sufficient condition is given for those partial orders which allow only one parsimonious set representation. Thus, they can become the basis of only one well-graded knowledge space, namely by the principal ideal representation. We look at this point in relation to the problem of learning space construction in Section 10.6.

10.4. Skills and Knowledge Spaces

In view of Theorem 10.1 we could have introduced knowledge spaces as a triple

$$((S, \leq), Q, \varphi)$$

where $(S, \leq)$ is a partial order, Q a finite set and φ a basic set representation of $(S, \leq)$, i.e., a mapping $\varphi : S \to \mathfrak{P}(Q)$ satisfying (10.3) and (10.4). The set $\mathcal{K} = \{\varphi(s); \; s \in S\}^*$ is the set of knowledge states and $(Q, \mathcal{K})$ is the space if $\bigcup_{s \in S} \varphi(s) = Q$. Its basis is $\{\varphi(s); \; s \in S\}$. The set S is interpreted as the skills which are pertinent to the field of knowledge which is described by the space.

In the classical approach skills enter the picture via *skill maps*. They are defined by Doignon and Falmagne (1999), Chapter 4. We briefly summarize their definition and the investigation of their association with knowledge spaces. Essentially, a skill map is a triple

$$(Q, S, \tau)$$

where Q is an item set, S is a skill set, and τ a map such that

$$\tau : Q \longrightarrow 2^S - \{\varnothing\},$$

with the understanding that $\tau(q)$ is the set of skills assigned to q.

Such a structure *delineates* a knowledge space in the sense that any $T \subseteq S$ defines a knowledge state K by

$$K = \{q \in Q; \tau(q) \cap T \neq \varnothing\}. \tag{10.8}$$

A skill map is called *minimal* if the omission of any element in S delineates a non-isomorphic knowledge space according to (10.8).

In Chapter 4 of Doignon et al. (1999) the following important result is proven:

A minimal skill map corresponds to the basis, i.e., the sets

$$K(s) := \{q \in Q; \; s \in \tau(q)\}$$

form the basis of the knowledge space $\mathcal{K}$ delineated by (Q, S, τ).

In applications skills are utilized to describe the competencies a student can be expected to have or not to have once his or her knowledge state has been determined.

10.4.1. *The competence based structures in the Graz approach*

Apart from the literature already cited, there are more places where skills have been mentioned or utilized in knowledge space contexts. Among others we refer to Düntsch and Gediga (1995), Korossy (1997, 1999), and Heller, Steiner, Hockemeyer and Albert (2006). In these papers the main thrust is different.

In particular, Heller et al. (2006) summarizes what I would call the Graz approach. It is an extension of knowledge space theory, mostly motivated to adapt it to personalized learning, preferably in an automated setting.

They start with three different entities:

(1) a set Q of assessment problems,
(2) a set L of learning objects, and
(3) a set S of competencies which appear relevant to solve the problems in Q and which are taught by the learning objects.

On all three sets a structure is assumed to exist. The structure on Q is essentially a knowledge space, sometimes a bit more general because closure under union need not hold for all pairs of states (but for most — I assume).

The structure on S is again close to a knowledge space, in this case, however, formed with the competencies in S. (The elements in S are often referred to as 'skills' but I want to avoid this name here in order not to confuse it with the Doignon et al. (1999) concept of skills which I use in the present chapter.)

The elements of S and the structure on S are derived in a way which is in some cases fundamentally different from that on Q. A so-called 'concept map' is derived. Rather than defining it mathematically Heller et al. (2006) describe it by an extended example. All the notions of a field of knowledge and their relationships are gathered in it. Next a hierarchical structure is derived from this information. The competence structure is kind of a knowledge space which respects this information.

The description given in Heller et al. (2006) is not mathematically precise. But it can be regarded as a useful heuristic to structure a field of knowledge and to devise a program for teaching this field and for assessing knowledge states of it in students. If a more formal treatment is needed this still has to be provided. Furthermore, its relation to the skill concept from the Doignon/Falmagne approach should be investigated. It remains to be shown to what extent the Graz approach and the one described in this section overlap or diverge.

10.5. Knowledge Spaces as Lattices

Knowledge spaces are lattices. Clearly, they are partial orders with respect to set inclusion. We have to define infimum and supremum. For the latter we can take union of states. However, the infimum requires a more subtle definition because the intersection of two states is not necessarily a state.

If we define for all $K_1, K_2 \in \mathcal{K}$:

$$K_1 \wedge K_2 := \bigcup_{L \in \mathcal{K}, L \subseteq K_i} L$$

then $(\mathcal{K}, \wedge, \cup)$ is a lattice.

Conversely, given a lattice it can be interpreted as a knowledge space. See Rusch and Wille (1996).

With the approach described in Section 10.4 there is however a slight complication when we regard the knowledge space as a lattice. If we start with the partial order of skills, $(S, \leq)$ then there are clearly many sets Q which make for a knowledge space with the given skill order. When the item sets are not in a bijective correspondence which transfers to the states in the sense of (10.2) then the resulting knowledge spaces are not isomorphic whereas the lattices can be isomorphic. In some cases the differences are trivial, for example, when there are two or more items which always occur in the same states. Taking only one of these items and deleting the others from Q will mend this defect. However, there are also more sophisticated cases. We give an example.

Let $S = \{s_1, s_2, s_3, s_4\}$ and $s_1 < s_3$ and $s_2 < s_4$ with all other pairs incomparable. With $Q = \{1, 2, 3\}$ and $Q' = \{1, 2, 3, 4\}$ we define the following set representations φ and φ'.

	φ	φ'
s_1	1	1
s_2	2	2, 4
s_3	1, 3	1, 3, 4
s_4	2, 3	2, 3, 4

Figure 10.2 depicts the order $(S, \leq)$ with the two different set representations and the resulting knowledge spaces.

This phenomenon epitomizes the difference in the normal approach to knowledge spaces described in Section 10.2 and the new one of Section 10.3 and 10.4. In the first the starting point consists of the items Q which are then framed into a knowledge space. In the second we begin with the skills, try to partially order them, and then look for items corresponding to these skills. In the above example the skill structure is the same, however the item sets which are employed to test for these skills happen to be quite different. In the left space item 3 can be known or learned after either item 1 or item 2 and it is possible to know item 1 and 2 but not 3. In the right space of Figure 10.2 things are a bit more intricate. Item 4 can be known together with 2, or with 1 and 3, or with 1 and 2, or with 2 and 3. If a

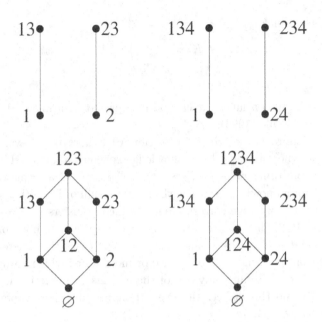

Fig. 10.2. Two set representations of a partial order and the resulting non-isomorphic knowledge spaces with isomorphic lattices.

student does not know item 4, he or she is in state $\varnothing$ or $\{1\}$. When an adviser of such a student wants to teach item 4 next there is no way to do this directly. With a student in state $\varnothing$ item 4 must be taught together with item 2 to aim at the state $\{2, 4\}$. When the student is in state $\{1\}$ the teacher has the choice to teach 4 combined with 2 or with 3; if successful the resulting state is either $\{1, 2, 4\}$ or $\{1, 3, 4\}$. Similar difficulties arise when one wants to teach item 2 or item 3. There is no state in the space from which either of them is learnable separately.

If the skill structure is correctly described by the order of Figure 10.2 it might be advisable to look for items which give the easier space on the left. However, when the usual approach ('items first') is employed and we happen to have chosen items 1,2,3,4 of the space to the right we cannot help ending up with the more complicated space. When we think of the skill structure as the more fundamental property and the items as somewhat arbitrary and exchangeable, then the search for the space on the left of Figure 10.2 should be aimed at. At least we must not wonder that our result is not a learning space when we insist on the other items. Of course, there are more set representations of this skill order possible, for example, the one of Figure 10.7 which yields the only ordinal space with this basis. However, ordinal spaces tend to become quite large. The space of Figure 10.7 has 9 states, the spaces of Figure 10.2 only 6. Clearly this effect increases rapidly when the basis gets larger. For this reason in Section 10.6 we discuss possibilities to save items in the set representations. As a result the number of states will also decrease.

In the example of Figure 10.2 the spaces are easily seen to be not isomorphic because Q and Q' have different cardinality. It is however also possible to construct examples with isomorphic lattices, $|Q| = |Q'|$, and $(Q, \mathcal{K}) \not\cong (Q', \mathcal{K}')$. In Figure 10.3 we give the skill orders and the respective set representations where $Q = Q' = \{a, b, c, 1, 2, 3, 4\}$.

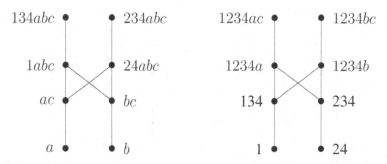

Fig. 10.3. Two set representations of a partial order with $Q = Q'$ but non-isomorphic knowledge spaces with isomorphic lattices.

In the skill order on the left of Figure 10.3 the knowledge to master items a, b, c is a prerequisite for mastering items 1,2,3,4. On the right it is the other way round. The construction of the respective knowledge spaces yields the two spaces of Figure 10.4.

There is no bijection f satisfying (10.2) because, if there were such an f then $f(\{a\})$ and $f(\{b\})$ both are singletons and should both be states; but the space of the right basis has only one state with one element. Nevertheless, the corresponding lattices are isomorphic. Figure 10.5 shows the line diagram of this lattice.

A word of explanation seems called for to reconcile these observations with the Rusch and Wille (1996) paper. There the authors reconstruct knowledge spaces in the framework of formal concept analysis (cf. Ganter & Wille (1996)). The formal contexts from which the spaces are derived via concept analysis are built on a relation which is pretty much the same as the questions asked from an expert in a querying procedure described in Section 10.2. It is known that different contexts can produce the same concept lattice. In concept analysis these contexts are treated equivalent.

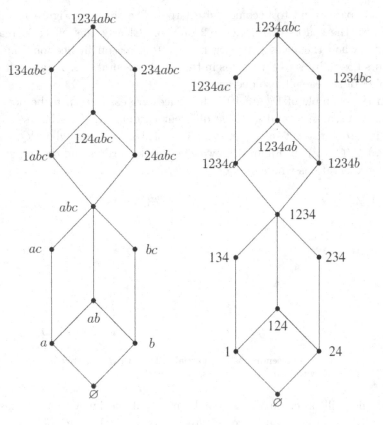

Fig. 10.4.   The knowledge spaces corresponding to the set representations of Figure 10.3.

They can be reduced by certain reduction processes to a standard context. Intuitively, in the example on the right of Figure 10.2 the reduction amounts to introducing a new item 2' for the combination of item 2 and 4 and a new item 3' for 3 and 4. With the item set $\{1, 2', 3'\}$ we obtain a space isomorphic to the one on the left of Figure 10.2. Figure 10.6 shows both knowledge spaces: left, the original, and right, the reduced one. From the viewpoint of knowledge space theory this kind of reduction does not simplify anything; to the contrary, we employ the same number of items, namely four, and combine them in a way which is by no means obvious or simple. Any procedure which truly reduces the number of items is useful and welcome in applications. In this case, however, there is no advantage because the 'reduced' items are chunks of original items.

Fig. 10.5. The lattice of the knowledge spaces corresponding to the bases of Figure 10.3.

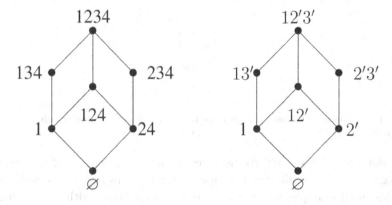

Fig. 10.6. A knowledge space before and after reduction (see text).

10.6. The Construction of Learning Spaces by Set Representations

Going back to the example of Figure 10.2 we note that the space on the left is well-graded, i.e., a learning space, while the other is not. It was already mentioned in Section 10.2 that learning spaces are preferable in many aspects. Clearly, if the construction of the space starts with the item set $Q' = \{1, 2, 3, 4\}$ and $\mathcal{K}'$ is the correct space then no procedure — however sophisticated — can result in a learning space.

This observation might be interpreted as an argument that the construction of well-gradedness must not begin with the items, i.e., the set Q. In this section we investigate how learning spaces are generated in the setting of set representationes. In Section 10.3 we already pointed out that parsimonious mappings of the skill order, i.e., mappings satisfying (10.7) yield them. Given the skill order $(S, \leq)$ it is always possible to utilize the principle ideal representation and end up with an ordinal space. In our standard example the result is shown in Figure 10.7.

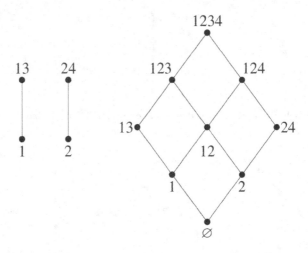

Fig. 10.7.   The ordinal space arising from the principal ideal representation of a skill order.

Comparing this space with the space on the left of Figure 10.2 we see the weakness of this possibility: The space is larger. This fact seems negligible in this small example, however in more realistic settings with spaces on up to a few hundred items and with millions of states the question of how to save items and states becomes serious.

The theory on saturated orders established in Suck (2004) and extended in Dzhafarov (2010) points the way how to proceed towards this aim. We need three more definitions.

Definition 10.3. Let $(S, \leq)$ be a partial order.

a) The induced suborder of a subset $S' \subseteq S$ is a fan if $(S', \leq)$ has a unique maximal element and all other elements are incomparable.

b) Two fans $(S', \leq)$ and $(S'', \leq)$ in $(S, \leq)$ are parallel if the elements s', s'' are incomparable for all $s' \in S'$ and $s'' \in S''$.

c) Two parallel fans $(S', \leq)$ and $(S'', \leq)$ in $(S, \leq)$ with the respective maximal elements m', m'' are skewly topped if there is an element $s \in S$ such that either $s > m''$, s, m' incomparable, and $s > s'$ for all $s' \in S' - \{m'\}$ or $s > m'$, s, m'' are incomparable, and $s > s''$ for all $s'' \in S'' - \{m''\}$.

To illustrate these concepts Figure 10.8 gives an example. The set $\{m', a, b\}$ is a fan as is $\{m'', c\}$. These two fans are parallel and they are skewly topped by s. The orders of Figures 10.2, 10.3, 10.5, and 10.7 also contain parallel fans, but these are not skewly topped.

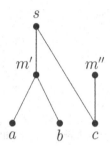

Fig. 10.8.   Two parallel fans skewly topped by s.

In Section 10.2 we briefly described how knowledge spaces are constructed in practice. This procedure can be very time consuming. Furthermore, it seems impossible to modify it in such a way that necessarily a learning space pops out. Our set representation viewpoint can be put to use in this question. Theorem 5 in Suck (2004) shows that a skill order in which all parallel fans are skewly topped allows only one parsimonious set representation, namely the principal ideals, hence only one learning space. In this case $|Q| = |S|$ holds. The proof can be exploited to yield the insight that the only way to reduce the number of items $|Q|$ compared to $|S|$ consists of looking for parallel fans which are not skewly topped and utilize one and the same item in the representation of the two maximal elements of the fans. We applied this technique several times in the examples; in Figure 10.2 with item 3 in the left example and Figure 10.3 with item c in both examples.

In a more general formulation we have to regard any pair of parallel fans. When a particular pair is not skewly topped the expert of the pertinent field of knowledge must rack his or her brain to find an item that distinguishes the maximal skills of both fans from all the preceding skills of both maxima. Note however that parallelism of fans is not transitive and we cannot apply this 'trick' for any pair. To illustrate this issue look at the examples of Figure 10.9. Both orders contain three fans (counting only maximal fans). In the left case four elements suffice to parsimoniously represent it because item 4 helps in each pair. In the example on the right we can only profit from one of the two pairs of parallel fans. In the given representation item

4 is utilized in two fans; so one pair made a reduction possible. We could have done it also with item 5 using the other pair for a reduction; but there is no way to come to terms with one and the same item in both pairs of parallel fans.

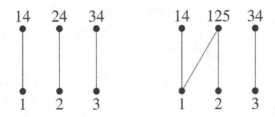

Fig. 10.9.   Two examples of skill orders with different sets of parallel fans.

How can this scheme bring about reductions in general? At present we can only sketch the idea which can possibly be transformed into a general procedure. Let F be a set of fans which are mutually parallel, i.e., each pair of fans in F is parallel. If furthermore no pair of fans of F is skewly topped we call F free. Obviously, only one additional item is needed to represent all the maximal elements of the fans of a free set. This is possible because of the idea already outlined above. Thus, we are advised to partition the set of fans of a partial order in maximal free subsets. Denote these maximal sets by $F_1, F_2, \ldots, F_s$ consisting of $n_1, n_2, \ldots, n_s$ fans, respectively. Then the minimum number of items needed to parsimoniously set represent the skill order $(S, \leq)$ is

$$|S| - n_1 - n_2 - \cdots - n_s. \tag{10.9}$$

This number is identical with the saturation index introduced in Suck (2004); it is a characteristic of the skill order. Theorem 6 in Suck (2004) demonstrates how properties of an order are related to this index. Each knowledge space with a basis isomorphic to $(S, \leq)$ (where we regard the basis as a partial order under set inclusion) needs at least this many items. But apart from saving items in this way we also end up with a learning space when we respect parsimony in our set representation.

The saturation index of the two orders in Figure 10.9 are 4 for the left order and 5 for the other one. The order of Figure 10.3 has a saturation index of $8 - 2 = 6$. In this figure we represented the order with 7 items because of the particular purpose in this section. There is however a representation with 6 items because there is one free set consisting of two fans. Thus, (10.9) yields 6.

In small examples the outlined framework is not difficult to implement. It is however likely to run into complexity problems when the set S becomes large and the order on S is complicated. In particular, it can be surmised that finding the maximal sets of free fans can be hard from a computational viewpoint.

10.7. Discussion

In this final section we will discuss a few issues connected with set representations, namely numeric grading, characterization questions, and problems with a practical handling of this approach.

Classically, psychological tests come up with a number purporting to describe the grade of a student representing his or her ability; or a value for the difficulty of an item, or both. Knowledge space theory gives the knowledge state of a student entailing his or her knowledge and deficiencies. The latter consists of the items which are not in the state determined for this student. In a learning space all learning paths from the state $\varnothing$ to Q have the same length and all states which have the same distance, say r, from $\varnothing$ consist of r items. Hence, it is possible to utilize this number as a grade or gauge it in such a way that a familiar grading system results. This is feasible; it should however be born in mind that the state of a student contains much more information than the number r. Except from very trivial cases (linear learning spaces) students with the same grade possibly know different things and have different skills. On the item side, a difficulty parameter can hardly be introduced. Looking at the example space in Figure 10.2, left, it is tempting to regard items 1 and 2 as equally difficult and 3 as more difficult. However, if a student 'chooses' the path $\varnothing, \{1\}, \{1,3\}, \{1,2,3\}$ then it seems that 2 is the most difficult part of this learning history. In the example on the right it is even more intriguing, but we must not discuss difficulty in this case because the space is not well-graded.

Numerically expressing the ability of a student by the number of items in his or her state (or something related to this value) is reminiscent of the condition that the number of correctly solved items is a 'sufficient statistic' for ability known in item response theory, in particular in the Rasch model. There it is a requirement of the applicability of the model. In the present context it is a property of the space provided it is a learning space.

The other crucial condition of the Rasch model — specific objectivity — does not seem to have an analogue in knowledge space theory. Even in a learning space the difficulty of an item depends on the learning path of a student. If one assumes that all students take the same path — perhaps because of the applied teaching method in a class — then possibly one could rank the items according to difficulty. However this ranking would heavily depend on the population for which this assumption is valid, a far cry from specific objectivity.

If one wants to assign numerical values to the skills instead of the states or the items it is even more futile because they form only a partial order and can thus deviate from linearity in every conceivable way; although there is a kind of hierarchy given by the order. Again in Figure 10.2 the four skills can be divided in two 'basic' and two 'advanced' ones corresponding to the two minimal and the two maximal elements in the partial order. However, acquiring these skills one need not follow the pattern 'basic' first then 'advanced'.

We mention two characterization problems which come up with the theory expounded so far. First, given a skill order $(S, \leq)$, how can the set of all knowledge spaces with bases isomorphic to this skill order be described. This question amounts to somehow summarize all possible basic set representations of $(S, \leq)$. Is there some structure in this set? Which of these spaces are quite similar? How different can they become? The second characterization concerns a subset of the one just described, namely the learning spaces. For this problem we have to consider all parsimonious set representations and ask the same questions as before. It is possible that some of the notions of lattice theory are helpful in this respect and some of these questions are already solved in this context.

For practical application of this theory the separation from skills and items which test for these skills is important. The question arises whether this is in fact possible. This problem must be resolved with the expert of the field for which the space is built. In school curricula skills and competencies are described. They can and should be employed. The partial order is to be derived by questioning the expert. In most cases the task for this expert to partially order the skills will be much easier than answering all the questions described in Section 10.2 in order to work with one of the existing algorithms. Next, items have to be found. To avoid the situation 'one skill one item' which would result in an ordinal knowledge space which would be likely to be too large and not very useful, one has to find items which can, in connection with other skills, test for several skills. We described this situation in Section 10.6. These items might be hard to find. It should however be born in mind that each learning space constructed in the usual way has such items or the space is ordinal. Thus, the existence of these items is not a problem, perhaps finding them might not be trivial.

Connected with the last point is the problem of finding and enumerating the parallel fans and partitioning them in as large as possible 'free subsets' (cf. Section 10.6). When the set of skills is larger than in our standard examples of the previous sections and when the partial order is a bit more complicated, we may run into complexity problems. This issue is a different story, possibly calling for special treatment. The question can be avoided as long as the skill sets are not too large and one does not crave optimality. In this sense one can save a few items if one finds some mutually parallel fans not skewly topped. With some trial and error it may be possible to find so many instances that the number of items is not much larger than the saturation index.

Finally, we want to point out that for theoretical reasons an extension to infinite skill sets, item sets, and knowledge spaces is interesting. Some of the concepts have to be changed. The standard book, Doignon et al. (1999), explores some of these questions in the usual framework. The study of our present approach with set representations has just begun in the infinite case. First results are derived by Dzhafarov (2010).

References

Cosyn, E., & Thiéry, N. (2000). A practical procedure to build a knowledge structure. *Journal of Mathematical Psychology, 44*, 383–407.

Cosyn, E., & Uzun, H. B. (2009). Axioms for learning spaces. *Journal of Mathematical Psychology, 53*, 40–42.

Doignon, J., & Falmagne, J. C. (1997). Well-graded families of relations. *Discrete Mathematics, 173*, 35–44.

Doignon, J., & Falmagne, J. C. (1999). *Knowledge Spaces*. Springer.

Düntsch, I., & Gediga, G. (1995). Skills and knowledge structures. *British Journal of Mathematical and Statistical Psychology, 48*, 9–27.

Dzhafarov, D., (2010). Infinite saturated orders. *Order*, eprint: 1010.2219.

Eppstein, D., Falmagne, J. C., & Ovchinnikov, S. (2008). *Media Theory*. Springer.

Falmagne, J. C., & Doignon, J. (2011). *Learning Spaces*. Springer.

Ganter, B., & Wille, R. (1996). *Formale Begriffsanalyse*. Springer.

Heller, J., Steiner, C., Hockemeyer, C., & Albert, D. (2006). Competence-based knowledge structures for personalised learning. *International Journal on E-Learning, 5*, 75–88.

Koppen, M. (1993). Extracting human expertise for constructing knowledge spaces: An algorithm. *Journal of Mathematical Psychology, 37,* 1–20.

Korossy, K. (1997). Extending the theory of knowledge spaces: A competence-performance approach. *Zeitschrift für Psychologie, 205,* 53–82.

Korossy, K. (1999). Modeling knowledge as competence and performance. In D. Albert and J. Lukas (Eds.), *Knowledge Spaces: Theories, Empirical Research, Applications* (pp. 103–132). Lawrence Erlbaum.

Rusch, A., & Wille, R. (1999). Knowledge spaces and formal concept analysis. In H. H. Bock and W. Polasek (Eds.), *Data Analysis and Information Systems* (pp. 427–436). Springer.

Suck, R. (2003). Parsimonious set representations of orders, a generalization of the interval order concept, and knowledge spaces. *Discrete Applied Mathematics, 127,* 373–386.

Suck, R. (2004). Set representations of orders and a structural equivalent of saturation. *Journal of Mathematical Psychology, 48,* 159–166.

Chapter 11

Experimental Discrimination of the World's Simplest and Most Antipodal Models: The Parallel-Serial Issue

James T. Townsend, Haiyuan Yang, and Devin M. Burns

Indiana University

In general cognitive systems are comprised of more than a single subprocess. The arrangement and linkages of these subprocesses are known as "mental architecture". The simplest non-trivial systems or manner of carrying out multi-tasking on discrete items is found in two diametrically opposite models that have been classically used to describe the architecture of this processing. *Serial* systems allow only one subsystem at a time to operate. The antithesis of serial processing assumes that all subsystems operate simultaneously, or in *parallel*. Mathematical characterizations of these and other architectures have been developed. Models of serial vs. parallel systems, seemingly so distinct, are shockingly hard to tell apart with experimental methodology. Here we review the history of psychological work on the subject, both theoretical and experimental. We also describe three additional properties of systems that must be considered simultaneously (and have often been conflated with architecture): stopping rule, dependency, and capacity. We present formal mathematical descriptions that can allow us to examine in which cases the two will be indistinguishable, and summarize successful paradigms for differentiating the models in the context of various psychological tasks.

11.1. Brief Review

The question of whether people can perform multiple perceptual or mental operations simultaneously, referred to as parallel processing, has intrigued psychologists at least since the late 19th century (Sternberg, 1969; Schweickert, 1978, 1982, 1985; Schweickert & Wang, 1993; Townsend, 1984; Townsend & Ashby, 1983; Townsend & Schweickert, 1989). This innocuous sounding question has turned out to be far more complicated than one might expect, however, since it is not possible (at least in the foreseeable future) to peer inside and count the number of operative channels in our brain at any point in time. A major branch of research resides within Reaction Time (hereafter RT) methodology (e.g., Egeth, 1966; Eriksen & Spencer, 1969; Estes & Taylor, 1964; Garner, 1974; Hick, 1952; Luce, 1986; Neisser, 1967; Posner, 1978; Smith, 1968; Sternberg, 1966; Welford, 1980). The reader is referred to the older tomes, (Welford, 1980; Laming, 1968), as well as newer treatments, (Luce, 1986; Townsend & Ashby, 1983; Zandt, 2000), for general RT surveys and tutorials.

Perhaps the first empirical attempt to answer the parallel-serial question was Sir William Hamilton (Hamilton, 1859). In his informal "experiment" several dice were tossed onto a desk, and he set himself the task of instantaneously assessing the total number of dots showing. From this operation, he attempted to determine the number of objects that could be apprehended simultaneously (in parallel) by a human observer.

Next we find a true pioneer in using the measure of reaction time to explore internal psychological processes. F. C. Donders, a Dutch psychologist, was one of the first to use differences in human reaction time to infer distinct subsystems involved in cognitive processing. The concept of using subtraction to decompose whole reaction times into component parts is still one of the most common tools used for making inferences about mental processes such as learning, memory and attention. Donders (1868) developed the method of subtraction, in which mean reaction times from two different tasks are compared. The second task is designed to require all the stages of the first plus an additional stage. Thus, the difference between mean RTs is taken to be an estimate of the mean duration time of the interpolated stage. The work of Donders was based on the idea that various psychological subprocesses are carried out in a serial fashion, that is, in a series with no overlap in processing time. Such serially arranged psychological systems are now referred to as *Dondersian Systems* (see Figure 11.1).

This method, however, later became the subject of criticism for several reasons. First, the assumption, known as *pure insertion*, that when the task is changed to insert an additional stage, the other stages will remain unchanged is questionable. If this condition is not met, then the difference between RTs cannot be identified as the duration of the inserted stage. Second, if the underlying subprocesses are carried out with any overlap in processing time (as happens in parallel processing), then the method of subtraction will introduce significant error and will typically underestimate the contributions of the subprocess to the overall RT (Taylor, 1976; Townsend, 1984).

Although the method of subtraction was not wholeheartedly accepted,

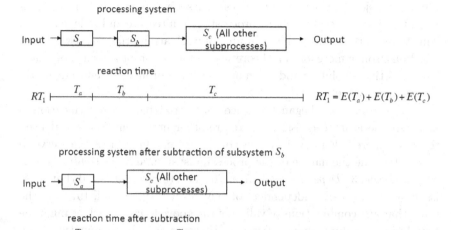

Fig. 11.1. A schematic illustrating the assumptions of Donders's subtractive method.

the attempt to analyze RT into components went on. During the early advent of cognitive science and in particular, the information processing approach, Sternberg (1966) and Egeth (1966) were instrumental in beginning to assay predictions associated with serial vs. parallel systems. Sternberg (1966) examined how reaction times in a short-term memory search experiment changed with an increasing number of items in memory. When Sternberg found that reaction time increased linearly with the number of targets, he inferred that this pattern could only be expected from a serial model. Later work has shown that this pattern is more indicative of capacity limitations than of architecture, but we will get to that later. Interestingly, in a slightly different visual search task, Egeth (1966) found evidence more in favor of parallel processing.

Three years later Sternberg developed a new tool which would subsequently usher in a more potent and universal set of methodologies, called the *additive factors method* (e.g., Sternberg, 1969; see also Ashby & Townsend, 1980; Pachella, 1974; Pieters, 1983; Townsend, 1984). These studies also provide general caveats with regard to factorial methods. As opposed to Donders's method, the additive factors method does not require the postulate that one can insert or take away the contribution of complete stages through experimental changes while leaving the other stages unaltered. Rather, it requires only the weaker assumption of selective influence: that there can be found two (or more) experimental factors that influence disjoint subsets of the stages. These factors can then be manipulated to answer questions about processing architecture and independence. This framework heralded the beginning of work attempting to answer these questions from a more formal, theoretical, information-processing approach. The cognitive revolution and dawn of computing brought a new perspective on cognitive modeling.

Investigators soon began to notice that serial/parallel identification is not a simple isolated question; there are other important facets that need to be considered. Four critical issues or theoretical dimensions that serve to help determine the nature of the processing system are: 1. *Architecture*; 2. *Stopping rule*; 3. *Dependency*; 4. *Capacity*. It is to be emphasized that these issues are all logically independent of one another (Townsend, 1974), in the sense that any combinations of values of the above four dimensions might be found in physically realizable systems. However, certain combinations may be more intuitively or psychologically acceptable than others. Although our focus here is on architecture (serial, parallel, and others), we must provide a brief summary of what these other dimensions are for proper understanding of the models.

Stopping rule: No predictions can be made about processing times until the model designer has a rule for when processing stops. Two main cases of interest are: first-terminating processing and last-terminating (= exhaustive) processing. A first-terminating rule means that processing can be completed as soon as any one of the channels is finished. When paired with a parallel architecture, this produces the familiar race model. This case is often called an OR design because completion of any item from a set of presented items is sufficient to stop processing and ensure a correct response (e.g., Egeth, 1966).

If all items or channels must be processed to ensure a correct response, then exhaustive or AND processing is entailed. For instance, in a target absent trial for a visual search task, every item must be examined to guarantee no targets are present. Figure 11.2 provides schematics of these stopping rules in a serial/parallel system.

Dependency: Models have classically assumed independent information channels, but there are many situations where this presumption does not make sense. There are myriad different ways to allow "cross-talk", or interaction between the channels. In other words, there are a number of types of independence that might be of interest. One type of independence we emphasize is within-stage independence (see Section 11.2 for detailed explanation). Additionally, there are two other types of independence which are

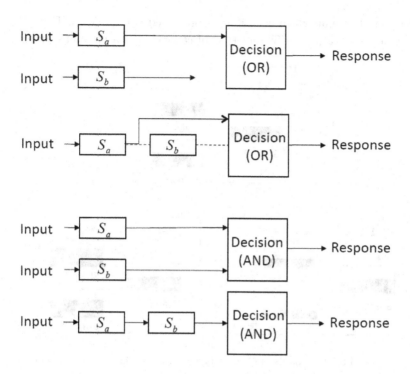

Fig. 11.2. Schematics of different stopping rules and architectures.

related: across-stage independence and independence of total completion time. The former case indicates that knowing the element in position a finishes first and at time t_{a1} tells us nothing about how much longer it will take the element in b to finish. The latter case indicates that the overall time to process a and the overall time to process b are independent. In all cases, we shall investigate the independence in a strict probabilistic sense: two events are independent if and only if the chance that they both happen simultaneously is the product of the chances that each occurs individually, that is, $P(AB) = P(A)P(B)$.

Capacity: This is a generalized notion of the amount of resources available to a given channel. If we say a system has *unlimited capacity* we mean that the overall time to process a single item does not vary with the total number of items undergoing simultaneous processing. The implication is that as a whole, the individual channels are neither slowing down or speeding up as the workload increases or decreases. *Limited capacity*, on the other hand, implies that as the number of items to process goes up, the overall rate at which each item is processed will go down. Figure 11.3 provides schematics of different type of capacities in a parallel system.

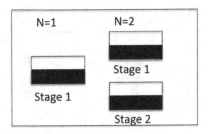

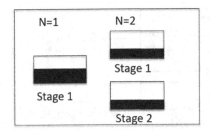
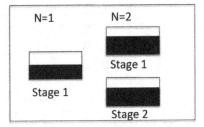

Fig. 11.3.   Schematics of different types of capacity in a parallel system.

To return to the history, we can see that Sternberg's (1966) assumption about the performance of a parallel model implies an unlimited capacity system. A limited capacity parallel system can predict linear increasing reaction times as a function of item number just as does a serial system. Egeth (1966) (see also Townsend & Ashby, 1983; Snodgrass & Townsend, 1980) pointed out that even unlimited capacity parallel processors can yield increasing, but not necessarily linear, mean reaction time functions in a natural way. A number of authors also started using limited capacity parallel processes that could mimic the straight line predictions of standard serial models. Atkinson, Holmgren, and Juola (1969) offered a simple stochastic model which mimicked ordinary serial processing predictions and Townsend (1971, 1972) showed that each type of model could mimic the set size function of the other (see also Murdock, 1971).

Due to these (and other) studies, the phenomenon of linear increasing reaction time curves was no longer considered a fundamental parallel/serial distinction, because it simply indicates first and foremost a limitation in capacity. At the same time, people started paying more attention to analyzing entire distribution functions of reaction time, rather than just the mean. In the next section we will discuss more on serial/parallel mimicking based on the analysis of reaction time distribution functions (Townsend & Ashby, 1983).

Substantial advances in modeling of diverse phenomena using broad classes of serial models have appeared in the work of Treisman and colleagues (e.g., Treisman & Gormican, 1988) and Wolfe and colleagues (e.g., Wolfe, 1994). However, the serial models have not been tested against parallel alternatives.

In the third section we will talk about the empirical side of the parallel/serial issue. There are now several experimental strategies based on mathematical proofs available to help distinguish whether processing is serial or parallel. Most of these methods are based on reaction time but some are based on accuracy.

11.2. Theoretical Treatment

11.2.1. *Stochastic models*

Townsend and others (Townsend, 1972, 1976; Townsend & Ashby, 1983; Townsend & Schweickert, 1989; Schweickert, 1978, 1982, 1989) have developed precise definitions of stochastic parallel and serial processes at the distributional level. For the remainder of this chapter, discussions on natural properties of parallel/serial systems are based on those rigorous definitions.

Many of the results in this chapter will be derived fully only for the case when there are two elements to be processed, but in most cases they could be generalized to any number of elements.

To provide a fine grained analysis of serial and parallel processing, we first introduce some temporal concepts (illustrated in Figure 11.4). Suppose a and b are the two positions of the elements to be processed. Let letter t denote the *intercompletion time* (ICT), which is defined as the interval between successive completions. As can be seen from the figure, the ICT is also the *actual processing time* (duration spent by the system on an element) for elements in a serial system. However, the ICT will not be an actual processing time in parallel processing except for the first element completed. We then denote τ as the *total completion time* on a particular element, which refers to the duration from $t = 0$ until the element is completed. Thus, as is apparent in Figure 11.4, the total completion time of an element in a parallel system is equal to its actual processing time. Also, it should be noticed that the total completion time in either model is always the sum of the number of ICTs preceding the completion of the considered

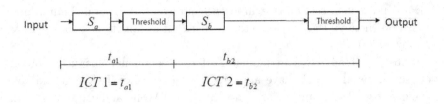

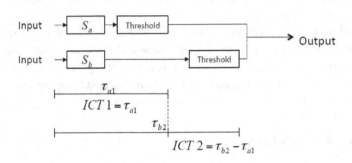

Fig. 11.4.   Illustrations of intercompletion times, actual processing times and total completion times.

element. Finally, the concept of *stage* is needed to decompose the model of processing. By stage we mean the interval between the instants of two successive element completions. Thus "stage k" refers to the interval occupied by the kth intercompletion time. It is easily seen that in a serial model, the length of a stage is equal to the time required for an element to be processed (i.e., the actual completion time), while in a parallel model, the actual completion time of a single element will be equal to the sum of the stage durations up until the element is complete.

Now we are ready to introduce the serial and parallel definitions. Let $f_{ai}(t_{ai})$ be defined as the density function describing the ith intercompletion time when the element in position a is completed ith and processing is serial. Analogous expressions hold for position b. To emphasize the potential difference between the probability densities associate with serial and parallel models, we similarly define $g_{ai}(t_{ai})$ for $i = 1$ or 2 as the density function describing the ith intercompletion time when a is completed ith and processing is parallel. Let $G_{ai}(t_{ai})$ be the parallel distribution function and $G_{ai}(t_{ai}) = 1 - G_{ai}(t_{ai})$ the survivor function for $i = 1$ or 2 on position a. We now state our definitions.

Definition 11.1. A stochastic model of a system for processing positions

a and b is *serial* if and only if

$$f_{a1,b2}(t_{a1}, t_{b2}; \langle a, b \rangle) = p f_{a1}(t_{a1}) f_{b2}(t_{b2} | t_{a1})$$

and

$$f_{b1,a2}(t_{b1}, t_{a2}; \langle b, a \rangle) = (1 - p) f_{b1}(t_{b1}) f_{a2}(t_{a2} | t_{b1}).$$

The quantity p is the probability that a is processed first. $f_{a1,b2}(t_{a1}, t_{b2}; \langle a, b \rangle)$ is an expression of the probability density of the joint occurrence of the completion order $\langle a, b \rangle$, that a consumes t_{a1} time units of processing and that b completes processing t_{b2} time units after a.

This definition is an intuitive description of the behavior of serial models within each stage. For example, the first equation states that with probability p, position a is processed first, in which case the first-stage density is $f_{a1}(t_{ai})$. Now the second-stage density must allow for the length of time taken for a to affect the processing time of b, thus the second-stage density is the conditional density function $f_{b2}(t_{b2} | t_{a1})$.

We can also define the parallel model in the similar way.

Definition 11.2. A stochastic model of a system for processing positions a and b is *within-stage independent and parallel* if and only if

$$g_{a1,b2}(t_{a1}, t_{b2}; \langle a, b \rangle) = g_{a1}(t_{a1}) \bar{G}_{b1}(t_{a1}) g_{b2}(t_{b2} | t_{a1})$$

and

$$g_{b1,a2}(t_{b1}, t_{a2}; \langle b, a \rangle) = g_{b1}(t_{b1}) \bar{G}_{a1}(t_{b1}) g_{a2}(t_{a2} | t_{b1}).$$

Here $g_{a1,b2}(t_{a1}, t_{b2}; \langle a, b \rangle)$ is completely analogous to $f_{a1,b2}(t_{a1}, t_{b2}; \langle a, b \rangle)$ of the serial model. $\bar{G}_{b1}(t_{a1})$ is the survivor function of t_{b1}, that is, $\bar{G}_{b1}(t_{a1}) = p\{T_{b1} > t_{a1}\}$. This survivor function along with $g_{a1}(t_{a1})$ describes the first stage of processing when the completion order is $\langle a, b \rangle$. The product gives the probability density that a is completed first at time t_{a1} and that b is not yet completed by this time. This component of parallel models represents the fact that a and b begin processing simultaneously. That it can be written as a product of functions of the individual elements is a result of our assumption of within-stage independence.

Thus looking at the two definitions, it is easy to notice that the components in the serial and parallel model that describe the second stage of processing are structurally identical. This indicates the fact that on the second stage, whether processing is parallel or serial, only one element remains to be completed, and with only one element left to be processed, serial and parallel models must be equivalent. So what makes the serial model differ from the parallel model? A fundamental difference is in determining processing order. If the system is parallel, from the definition we see that the determination of processing order inherently depends on the rates with which a and b are processed. In contrast, if the system operates serially, then the decision as to whether the processing order will be $\langle a, b \rangle$ or $\langle b, a \rangle$ is made a priori. This difference in how the systems select processing order is a fundamental difference between parallel and serial processing models.

An important aspect in need of clarification is the assumption of within-stage independence. This postulate states that during any single stage (i.e., the time between the completion of two successive elements), the processing of all unfinished elements is independent. This assumption does rule out some interesting parallel models, but it still allows many different kinds of dependency to occur. For example, parallel capacity reallocation models cause a dependency to occur between stages and even on the level of overall completion times, but within-stage independence is still possible.

11.2.2. Serial-parallel equivalence

Now that we have explicitly stated the form of serial and parallel models, we are in a position to ask the question of when there exists a parameter mapping such that the two models are equivalent. As suggested by Townsend and Ashby (1976), given any within-stage independent parallel model, we can always construct a serial model that is completely equivalent to it by setting

$$p = \int_0^\infty g_{a1}(t)\overline{G}_{b1}(t)dt,$$

$$f_{a1}(t_{a1}) = \frac{1}{p}g_{a1}(t_{a1})\bar{G}_{b1}(t_{a1}),$$

$$f_{b1}(t_{b1}) = \frac{1}{1-p}g_{b1}(t_{b1})\bar{G}_{a1}(t_{b1}),$$

$$f_{b2}(t_{b2}|t_{a1}) = g_{b2}(t_{b2}|t_{a1}),$$

$$f_{a2}(t_{a2}|t_{b1}) = g_{a2}(t_{a2}|t_{b1}).$$

Inversely, given a serial model, *if* there exists a within-stage independent parallel model that is completely equivalent to it, it can be found by setting

$$\bar{G}_{a1}(t) = exp[-\int_0^t \frac{pf_{a1}(t')}{p\bar{F}_{a1}(t') + (1-p)\bar{F}_{b1}(t')} dt'],$$

$$\bar{G}_{b1}(t) = exp[-\int_0^t \frac{(1-p)f_{b1}(t')}{p\bar{F}_{a1}(t') + (1-p)\bar{F}_{b1}(t')} dt'],$$

$$g_{b2}(t_{b2}|t_{a1}) = f_{b2}(t_{b2}|t_{a1}),$$

$$g_{a2}(t_{a2}|t_{b1}) = f_{a2}(t_{a2}|t_{b1}).$$

It turns out that either of the parallel survivor functions for stage 1 (indexed by the capital "G's") may not be true survivor functions in that one or more may not approach 0 as t becomes large, as it must for item completion to occur.

Hence, these results prove that serial processes are more general than within-stage parallel processes, at least in the sense that the class of parallel models defined above is contained within the class of serial models. This should not be altogether unexpected, since from the definitions of the models we can see that serial models have one more parameter than parallel models for manipulating processing order. This result was employed in an experimental test of serial vs. within-stage independent parallel memory retrieval (Ross & Anderson, 1981).

Thus the previous discussion has illustrated that many serial and parallel models can mimic each other in the sense of distributional equivalence, which demonstrated the complexity of this serial/parallel issue for any empirical reaction time data set, we can never easily draw the conclusion that the processing is serial or parallel, because either model may give rise to the same RT distribution. However, it is critical to understand that when we say serial models are more general than parallel models, it does not mean that a parallel machine of this type works in real time like a serial machine, only that the mathematical description of the parallel class of machine is contained within the mathematical description of the serial class for a particular paradigm. Armed with the foregoing mathematics and intuition, we now exhibit an even more general mathematical equivalence function between the classes of parallel and serial models.

11.2.3. *Further results on serial-parallel equivalence*

Notice that up to now, within-stage independence plays an important role in the analysis of serial-parallel equivalence problem. To loosen this condition, we need to consider the above stochastic models in a more general way. As suggested in Townsend and Ashby (1983), following up on Vorberg (1977), if we treat the problem from the point of view of foundational probability theory, we find that the two general classes of serial and parallel model are equivalent to each other.

Formally, our goal is to establish that for any model $m_s \in M_s$ (the class of serial models), there exists a model $m_p \in M_p$ (the class of parallel models) such that m_p gives the same probability measure on all the corresponding possible events. Then the class M_s is said to imply the class M_p. If the implication goes both ways, then we can say that these two classes are equivalent to each other. To achieve this goal, we need to employ the concepts of the *Borel field* and *Borel functions* from *measure theory*. We note that the real line R with its usual topology is a locally compact *Hausdorff space*, hence we can define a Borel field on it. To be specific, regarding our present case, the sets in Borel field B will associate with intervals of time in which elements might complete processing, and the Borel function J establishes the relation between the two Borel fields, thus associating the probability spaces. We may now state the formal definitions of parallel and serial models in terms of measure spaces.

Definition 11.3. A model of a parallel system is defined by $m_p = \langle W_p, B_p, P_p \rangle$, where B_p is the Borel field associated with W_p, our space, and P_p is some particular probability measure on B_p. For every point w belonging to W_p, the probability density function assigned by m_p is given by

$$g_p(w) = g(\tau_1, ..., \tau_n; a_1, ..., a_n)$$

with

$$w \in W_p = \{(\tau_1, ..., \tau_n; a_1, ..., a_n) | \tau_i \in R^+,$$

$$1 \le i \le n; \tau_1 \le \tau_2 \le ... \le \tau_n, (a_1, ..., a_n) \in permutation(n)\}$$

where τ_i is the total completion time (also the actual processing time in parallel case) at which the ith element to be finished is completed, and a_i is the serial position of the element completed ith.

Definition 11.4. A model of a serial system is defined by $m_s = \langle W_s, B_s, P_s \rangle$, where B_s is the Borel field on W_s. Thus for every point w belonging to W_s, the probability density function assigned by M_s is

$$f_s(w) = p(a_1, ...a_n)f(t_1, ...t_n | a_1, ...a_n)$$

with

$$w \in W_s = \{(t_1, ..., t_n; a_1, ..., a_n) | t_i \in R^+,$$

$$1 \leq i \leq n; (a_1, ..., a_n) \in permutation(n)\}$$

where t_i is the intercompletion time (also the actual processing time in serial case). $p(a_1, ..., a_n)$ gives the probability of the order shown.

As indicated in Townsend and Ashby (1983), it is always possible to provide a mapping between the parallel and serial spaces that is one-to-one and onto (i.e., the Borel function J), to show that for any model of one type, there exists a model of the other type which is equivalent. Thus, from the point of view of pure mathematics, serial models and parallel models could be actually totally equivalent at the distributional level, once we loosen the condition of within-stage independence.

To be sure, these mappings obscure vital distinctions that may exist between the two types of processing. Another issue that needs to be emphasized is that even when two theories or models make predictions at some level that are equivalent, to the extent that the two theoretical structures are not truly identical, there must exist aspects of the models or theories that relate to different possibilities in the real world. Here, to get a deeper understanding of the mechanisms of serial and parallel processing that can give us guidance in terms of experimental strategies, we list several divergent properties of serial and parallel systems.

Fundamental Distinctions between Serial and Parallel Systems

(i) In the case of serial systems, at any point in time, at most one element can be in a state of partial completion, whereas parallel systems can have any number of partially completed elements.

(ii) In the case of serial systems, only one element can change its processing state at a certain point in time, whereas in parallel systems, any element can increase its processing status at any point in time.

(iii) In the case of serial systems, the time taken to process an element cannot depend on the identity of any element that is not completed until after the given element is finished, whereas in parallel systems such dependence can exist.

(iv) In the case of serial systems, it is possible to build a system so that a processing order of the elements is preselected and then to make the processing times depend on the particular preordained order on any given trial, whereas a parallel system cannot be built so that a processing order can be preselected.

11.2.4. *The principle of correspondent change*

We now introduce an important principle of mathematical models and theory testing as the summary of this section. The name we give to this principle is the *principle of correspondent change* (Townsend & Ashby, 1983, Chapter 15). This principle is rather obvious and employed by scientists on a day-to-day basis in an implicit way, but not always fully appreciated.

 (i) For any given empirical milieu and for any given class of models, there will exist a set of subclasses where models are indistinguishable within their subclass and in that specific milieu.
(ii) Empirical changes in the environment or stimulating situation should be reflected in a nonvacuous theory or model by corresponding changes or invariances in the model or theory. Such changes or invariances should be predictable and in consonance in the correct model.

It is not surprising that two theories or models may make predictions at some level that are equivalent. However, the second point of this principle points out that we do not need to be too upset about this fact, because to the extent that the two theoretical structures are not truly identical, there must exist aspects of the models or theories that relate to different possibilities in the real world. Let us return to the context of serial and parallel model discrimination. If we examine the matter from a static view, it has been demonstrated that for any parallel model, there must exist a serial model which under certain conditions yields the same processing time distribution. However, if we examine the matter from a dynamic view, then due to the structural differences inherent in the models, there must exist one or more factors such that by controlling those factors, the outcome of the two different models will change in different ways. To make the discussion a little more concrete, suppose there are two positions a and b, and suppose we are trying to decide whether the processing behavior is serial or parallel. Each model will certainly possess structure, which will be in the form of functions of parameters. Let the experimenter manipulate the complexity of the element placed in b while leaving the complexity of the element in a

the same. Then surely the structure associated with position a should be invariant whereas the structure connected with position b should vary in an appropriate fashion. We definitely allow that interactions or limitations in capacity affect the rate in position a, but again they should be reflected in regular changes which show different patterns in different models, which then allow us to distinguish between the two models.

The principle of correspondent change could be considered in a more abstract way and could be applied to general model discrimination problems rather than just serial and parallel issues. If we denote δ as the variable indicating the experiment conditions, and denote $F(\delta)$ and $G(\delta)$ as the observed outcomes of the experiment conditions under two different models, then it is highly possible that model equivalence occurs in some sense. This would mean functions F and G have the same range, that is, for any δ' , there exists a δ'' to make $G(\delta'') = F(\delta')$. However, the signature of F and G must not be the same since the structures of models are essentially different. Thus, if we are lucky enough, by moving the value of δ, we can tell the difference of the two models since the direction of the changes of the outcome are different in the different models.

The principle of correspondent change establishes the connection between a static view and a more dynamic view, and provides guidance for experimental design. As stated before, the principle of correspondent change has been employed by scientists for many years, although in an implicit form much of the time. In the next section we will introduce some experimental paradigms aiming to distinguish serial and parallel processing. Note that, for most experimental paradigms, the principle of correspondent change is a necessary condition.

11.3. Experimental Paradigms

In this section, we discuss several experimental paradigms which aim to distinguish serial versus parallel processing.

11.3.1. *Tests using capacity*

In general, as discussed extensively above, limited capacity parallel models can mathematically and intuitively perfectly mimic serial models, so we are generally averse to architectural tests based on that notion. However, it has been long known, that unlimited or super capacity parallel models can make predictions that reasonable serial models simply cannot duplicate (e.g., Townsend, 1972, 1974).

Lately, methodologies associated with capacity distinctions have been proposed that might be able to distinguish serial models from parallel processes (Townsend & Nozawa, 1995; Townsend & Wenger, 2004). Thornton and Gilden (2007) is such an example. Their technology includes a multiple target search (MTS) method and utilizes computational models of serial and parallel processing. Although they do not deal directly with our capacity measure or explicitly consider it, both the information of reaction time and accuracy are used in their method, allowing them to consider both the capacity qua efficiency issue as well as speed-accuracy trade-offs.

They employed a random walk model to simulate the multiple-target visual search. To address the capacity issue in the parallel model, they proposed employing a limitation parameter τ that attenuates the drift rate of each random walk as a function of set size. With regard to the serial model, they introduced two parameters to relax decision criteria as a function of both set size and accumulation time, which allows the serial model to incorporate a rational decision strategy consistent with subjects psychological motivations.

In this way, whether serial or parallel processing was performed in a task could be assessed by comparing goodness of fit metrics for the two models. The authors provide 29 sets of data that cover different difficulty levels of visual search. The results of applying their algorithm to each of the 29 tasks demonstrated that the majority of tasks were performed in parallel, while some difficult searches (e.g., rotation and mirror inversion tasks) were more likely to use serial processing.

Little research has been done in using random walk simulations to model either interactions between channels or the capacity limitation issue in parallel processing. However, Thornton and Gilden (2007) do provide a method of model selection in which at least the capacity issue is considered.

In order to account for the influence of the other three previously mentioned processing characteristics on architecture, Wenger and Townsend (2006) undertook perhaps the first study to simultaneously consider all four properties in a factorial design without having to resort to averaged data. In order to compare serial and parallel models in their task, they employed linear dynamic systems (see Townsend & Wenger, 2004). Unlike Thornton and Gilden (2007), however, this approach did not rely on quantitative model fits of parameterized models, but rather looked for distinguishing qualitative behaviors.

In this study, both meaningful (face-like) and meaningless (scrambled face) stimuli were tested, and in both cases the number of targets and distractors were manipulated. Rather than finding what had been previously assumed in these cases, that the former would be processed in parallel and the latter in serial fashion, it was demonstrated that parallel processing was used in both cases. The primary difference between the two was found via the capacity measure, with much higher capacity for meaningful stimuli.

These findings argue against the theory that there exist separate, distinct processing systems for gestalt-like stimuli, and instead that processing may differ only in the degree of efficiency. Other interesting findings concerning stopping rule and independence are put forth, but are less germane to the current discussion.

11.3.2. *The parallel-serial tester*

PST can be viewed as a descendant of Snodgrass (1972) pattern matching paradigm, combined with mathematical proof (Townsend, 1976) to distinguish parallel vs. serial models. One of the nice properties is that the tester functions at the level of the mean reaction time, which obviates dilemmas with the exact form of the distribution. On the other hand, when it is desirable to employ specific classes of distributions, the sensitivity of the test may be enhanced.

The PST paradigm requires an observer to search through a list of two items for one or more targets. The paradigm consists of the three experimental conditions. Condition I is composed of two types of trials, and requires response R1 if the target is on the left, and response R2 if the target is on the right. Conditions II and III are composed of four types of trials each. Condition II requires that both comparison items match the target in order for response R1, whereas in condition III only one comparison is required to match the target in order for R1 to be the correct response. This is outlined in Table 11.1.

The underlying assumptions of PST are: the processing is self - terminating, the distribution on processing time is somehow different when two matching items are being compared than it is when two mismatching items are compared, accuracy is high enough and does not covary in an important way with reaction time. PST is a distribution-free theory. There are many examples of experimental applications in the work of Townsend and Snodgrass (Townsend & Snodgrass, 1974; Snodgrass & Townsend, 1980; Snodgrass, 1972). PST is founded on the principle that all (and any) items being processing in a parallel system can be in partial states of completion at any point in time, whereas only one item can be in such a state in a serial system.

Table 11.1.   Parallel-serial Testing paradigm

Target = A		
Condition	Comparison items	Response
C I	AB	R2
	BA	R1
C II	AA	R1
	AB	R2
	BA	R2
	BB	R2
C III	AA	R1
	AB	R1
	BA	R1
	BB	R2

The results from Snodgrass and Townsend (1980) suggest that PST results depend on the complexity of the matching required of the subject. With more complex patterns and processing requirements, subjects appear to be forced to resort to serial processing, whereas in the simpler versions evidence showed that subjects can operate in parallel.

11.3.3.   *Tests by time delimitation*

Eriksen and Spencer (1969), Shiffrin and Garner (1972), and Travers (1973) compared the effect of presenting one symbol or part of the display for t msec followed by another symbol for t msec and so on until the entire k symbols had been shown after $k * t$ msec, with the effect of displaying the entire array simultaneously for t msec. They found that accuracy in the latter condition was roughly equal to that in the first condition, which supported parallel processing as opposed to serial processing.

Townsend and his colleagues (Townsend, 1981; Townsend & Ashby, 1983; Townsend & Fial, 1968) developed a counterpart to that paradigm, in which the first condition is identical to the above. In the latter condition, k successive time intervals, each of duration t/k are permitted for processing each item. Hence if processing is serial, performance should be about equal in the two cases, whereas if processing was parallel, the sequential type of presentation should seriously degrade performance. The experimental results were in favor of parallel processing (Townsend, 1981).

Although a potentially useful tool in studying mental architecture such as serial vs. parallel processing, it appears that some degree of model mimicking still might occur (see, e.g., the discussion in Townsend, 1981; and Townsend & Ashby, 1983).

11.3.4. The second response paradigm

From the previous section and the principle of correspondence change, we see that parallel models can predict that if processing is stopped at an arbitrary point in time, any number of items may be in a state of partial processing. For instance, if the cognitive system is processing features in parallel on several items, then cessation of processing can leave each item with some features completed. In contrast, in serial processing, one item is completed at a time, with the succeeding item not being started until the last is finished (Townsend, 1974). Therefore, if processing is sharply terminated, at most one item should be in a state of partial processing. This will not ordinarily show up in the overall accuracy results of a typical experiment.

In order to exploit this distinction, Townsend and Evans (1983) developed a technique based on a second response on each item to be processed, which was later extended by Zandt (1988). It was demonstrated that the pattern of accuracy on the second responses differed for serial and parallel models. Null hypotheses for serial processing within several levels of constraints on responding in the serial models were introduced and the results applied to a pilot experiment. Within the study, the data passed the tests for the most lenient serial hypothesis but ran into trouble with the more restrictive criteria. A potential vulnerability of this strategy is that in some applications, the second response might be based more on the first response than on the cognitive or perceptual processing associated with the first response.

The next section reviews the theoretical and methodological legacy of the additive factors method (Sternberg, 1969) mentioned earlier. It forms by far the most broadly and deeply theoretically developed theory-driven methodology for serial-parallel assessment. It is also the most widely employed and its popularity seems to be growing.

11.3.5. The method of factorial interactions with selective influence of cognitive subprocesses

The concept of *selective influence* is an important notion in perceptual and cognitive processes, which is also one of the most widely used assumptions in approaches to the identification of processing architecture. This notion was first employed by Sternberg (1969) in his additive factors method, in which he stated that there exist manipulations that can separately influence the processing associated with each factor, say X and Y. Thus in a

2×2 factorial design, we can write the Mean Interaction Contrast (MIC) as

$$MIC = [RT(X + \Delta X, Y + \Delta Y) - RT(X + \Delta X, Y)]$$
$$-[RT(X, Y + \Delta Y) - RT(X, Y)]$$

Sternberg (1969) suggested that based on selective influence, serial models with independent processing times would exhibit $MIC = 0$. The theory underlying selective influence was subsequently endowed with a rigorous mathematical foundation (see, e.g., Schweickert, 1978; Schweickert & Townsend, 1989; Townsend, 1984; Townsend & Ashby, 1983; Townsend & Nozawa, 1988; Townsend & Schweickert, 1985, 1989; Townsend & Thomas, 1994) also contributed to this question, and found that the sign of the MIC could be an indicator of not only processing architecture but also stopping rules. The results could be summarized as follows.

If processing is serial, then regardless of whether a self-terminating or exhaustive stopping rule is used, the MIC always equals zero. If processing is parallel, then the MIC value will depend on the stopping rule. That is, parallel exhaustive processing will lead to an underadditivity of the mean RT, thus a negative MIC, whereas parallel self-terminating processing will lead to an overadditivity of the mean RT, thus a positive MIC. Figure 11.5 illustrates the MIC prediction for parallel and serial processing under different types of stopping rules.

Thus we can see that the MIC provides us with important clues to the mental architecture and stopping rules. However, it still has limitations. For example, the MIC cannot distinguish between serial, self-terminating processing and serial, exhaustive processing, and also cannot distinguish between parallel and coactive processing (Miller, 1982; Townsend & Nozawa, 1995). Due to these limitations, it was determined that ordering the means through selective influence is not sufficient to make discriminative predictions for varying architectures and stopping rules. It is necessary that a selective influence variable exert influence on distributions at a higher level. (Check Townsend, 1990a; Townsend, 1990b for the stochastic dominance hierarchy.)

11.3.6. *Systems factorial technology (SFT)*

Townsend and Nozawa (1995) developed the Systems Factorial Technology (SFT) which simultaneously tested parallel vs. serial processing for various stopping rules at the level of distribution functions as well as measuring

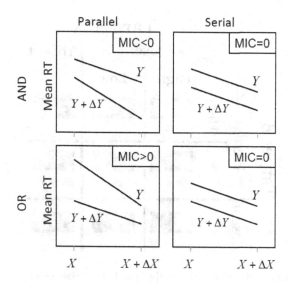

Fig. 11.5. Predictions for mean reaction times and mean interaction contrasts as a function of architecture and stopping rule.

capacity at a distributional level.

SFT includes an experimental paradigm (DFP) which requires the manipulation of two factors: The activation of sub-channel (target presence/absence) and the speed of processing (different salience levels). To use SFT to predict architecture and stopping rule, we assume that salience manipulation could produce selective influence at the distributional level. See Figure 11.6 for double factorial design.

These data are analyzed using Survivor Interaction Contrast (SIC),

$$SIC(t) = [S_{ll}(t) - S_{lh}(t)] - [S_{hl}(t) - S_{hh}(t)],$$

where the subscript denotes the salience level of each process.

Figure 11.7 displays the predictions for the variety of models.

Observe that parallel-processing SICs reveal total positivity in the case of OR conditions but total negativity in the case of AND conditions. Furthermore, OR parallel and coactive parallel processing now are distinguished by their respective SICs: The contrast for OR parallel processing is consistently positive, whereas the contrast for the coactive model possesses a small negative "blip" at the earliest times, before going positive. Since the MIC must be positive in coactivation, it can be shown that the positive portion of the SIC always has to exceed the negative portion.

The advantage associated with use of both the SIC and the MIC goes

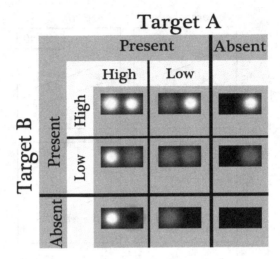

Fig. 11.6.   Illustration of the Double Factorial Design.

beyond the ability to distinguish coactive from parallel processing. It is also intriguing that the OR and the AND serial stopping rules are now experimentally distinguishable, since in the OR case $SIC = 0$ always, but in the AND case there is a large negative portion of SIC, followed by an equally large positive portion. Thus, both the architecture and the stopping rule are experimentally determinable by the factorial tests carried out at the distributional level. The general applicability of the distributional approach has benefited from theoretical extensions by Schweickert and colleagues (e.g., Schweickert, Giorgini, & Dzhafarov, 2000) to general feed-forward architectures, which contain parallel and serial subsystems, and from advances in methods of estimating entire RT distributions (see Zandt, 2000, 2002).

SFT can also help us predict workload capacity. Townsend and Nozawa (1995) proposed a measure called the capacity coefficient. For OR processes, the capacity coefficient is computed as the ratio between the integrated hazard function of the double target condition and the sum of the integrated hazard functions of the single target conditions, based on RT data.

$$C_{OR} = \frac{H_{AB}(t)}{H_A(t) + H_B(t)}$$

Here A and B signify the two processes. The hazard function is the probability density function over the survivor function, $h(t) = (f(t))/(S(t))$,

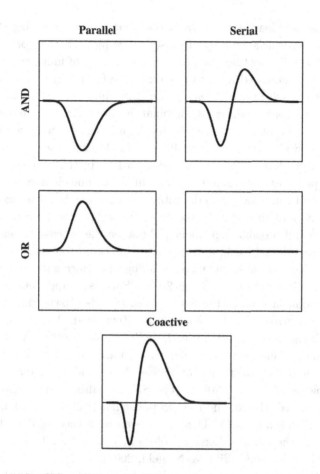

Fig. 11.7. SIC predictions for different information processing models.

$H(t)$ is the integral of the hazard function from zero to t. Townsend and Wenger (2004) developed a comparable capacity coefficient for the AND task.

$$C_{AND} = \frac{K_A(t) + K_B(t)}{K_{AB}(t)}$$

Analogous to the integrated hazard function, $K(t)$ is defined as the integral of $k(t)$ from zero to t, in which $k(t) = (f(t))/(F(t))$.

The interpretation of the two capacity coefficients for OR and AND conditions is the same: $C(t)$ should be equal to 1 in a system with *unlimited capacity*, i.e., the processing in a given channel is not affected by the increase of the workload. $C(t)$ above one indicates that processing efficiency of individual channels actually increases as we increase the workload, and $C(t)$ below 1 indicates that the processing efficiency of individual channels decreases with increased workload. We call the former case super capacity and the latter case *limited capacity*. The SIC function can distinguish independent models with different combinations of architecture and stopping rules. If, however, at least one model has high levels of interaction between channels, the SIC signatures of the different models can resemble each other (Eidels, Houpt, Altieri, Pei, & Townsend, 2010). In this case the capacity coefficient provides information about architecture and channel dependence indirectly, and can allow us to disambiguate the models. Thus, examining both the SIC and the capacity coefficient provides a decisive test for the architecture and possible dependencies between the processing channels.

Beyond its original use in a simple detection task (Townsend & Nozawa, 1995), SFT has extended our understanding of performance in a wide variety of cognitive tasks (e.g., Sung, 2008). Successful applications include classic phenomena such as the Stroop task (Eidels, Townsend, & Algom, 2010), Navon arrows (Blaha, Johnson, & Townsend, 2007), and various Gestalt principles (Eidels, Townsend, & Pomerantz, 2008). Another area in which the SIC has been particularly informative is in search tasks, both visual search (Fific, Townsend, & Eidels, 2008) and short term memory search (Townsend & Fific, 2004). The SIC has also been successful with more complicated stimuli, and faces in particular (Fific & Townsend, 2010; Wenger & Townsend, 2001). There have even been successful applications of the SIC in the clinical domain (Johnson, Blaha, Houpt, & Townsend, 2010; see also Townsend, Fific, & Neufeld, 2007).

Recent theoretical work has opened up the possibility of using the SIC in a wider range of tasks. Eidels et al. (2010) have extended the predictions of the SIC to a wide class of parallel models with processing dependencies. On another front, Houpt and Townsend (2010) have developed a tool for making statistical inferences with the SIC. Fific, Nosofsky, and Townsend (2008) have applied SFT to the domain of multidimensional classification. In their experiment they contrasted SIC signatures for stimuli that varied along highly separable dimensions to the case where they varied along integral dimensions. In the separable case, some subjects were found to be using serial processing with an exhaustive stopping rule, while other subjects were clearly parallel, but also used an exhaustive stopping rule. In the experiment with integral dimensioned stimuli, however, they found clear evidence that subjects were using a coactive architecture, rather than serial or parallel. They assert that this strong distinction in architecture could be used as another method for distinguishing between separable and integral stimulus dimensions (Garner, 1974).

11.4. Discussion

Psychology, and particularly cognitive psychology, aim to uncover the mysteries of the human mind. To do so as a science, it must perforce evolve rigorous theories and methods which are quite analogous, but not homologous, to the systems identification theories found primarily in electrical engineering and computer science curricula. The mind, apparently being run by a machine with 10^{14} or so primary parts (the neurons) and with a miasma of little understood graded chemical and potential fields thrown into the bargain, is an astronomically more complicated system than any discrete or continuous system open to such strategies. Hence, the question must arise not as to whether it is possible to build a computer program, not to mention verbal models that imitate certain aspects of cognition–such tasks are common homework problems in cognitive psychology or cognitive science–but whether said models or theories that are *unique* are achievable. Would not the assiduous theorist wish, in her fondest dreams, to assert that such a model is both *necessary* and *sufficient* to explain a certain phenomenon?

The parallel vs. serial processing issue stands as a kind of icon of cognitive psychology from this point of view. Why? Because one could hardly conceive of an issue of mental function where the opposing concepts are so antipodal as these two modes of processing. Not only of entirely opposite character, but so doggone elementary and straightforward almost to the point of seeming insipidity. And yet, not: 1. Parallel and serial processes form the basic components of well-defined systems of almost arbitrary complexity (a special interesting case is found in systems based on connected graphs; see Schweickert, 1978; Schweickert & Townsend, 1989). 2. Perhaps somewhat astoundingly, given the expected complexity of mental activity, either one or the other type of process has actually been identified to a very high probability employing the approach of Systems Factorial Technology. A few examples were cited earlier.

So, one might have forecast a rather early and decisive method of identification of serial vs. parallel processing. Alas, that has been far from the case as outlined in this essay. First raised in an experimental fashion approximately 140 years ago, and resurrected in the 1960s, the issue has proven itself to be a rather sticky wicket to say the least. The intransigence of this 'simple' issue, primarily due to mutual model mimicking in a strong mathematical sense, should give pause to the most confirmed Pollyannish psychologist or cognitive scientist.

Yet there is clearly room for the industrious optimist, in the fact that a number of years of arduous mathematical effort combined with appraisal of theory-driven methodologies in the laboratory crucible have led to a number of highly promising strategies for assessing whether processing is serial, parallel, or possibly hybrid. The most general and, so far, explored of these, has been those approaches based on systems factorial principles. It is now possible to provide quite persuasive evidence of serial or parallel processing, which was not the case a few decades ago.

There may be an overall message for psychology and cognitive science here. Perhaps in the domain of 'taking the lid off the human black box', as opposed to other venues, and in paraphrasing Ronald Reagan: "There ain't no free lunch, my friend."

Acknowledgments

This work was supported by NIH-NIMH MH 057717-07 and AFOSR FA9550-07-1-0078.

References

Ashby, F. G., & Townsend, J. T. (1980). Decomposing the reaction time distribution: Pure insertion and selective influence revisited. *Journal of Mathematical Psychology, 21*, 93–123.

Atkinson, R. C., Holmgren, J. E., & Juola, J. F. (1969). Processing time as influenced by the number of elements in a visual display. *Attention, Perception, & Psychophysics, 6*, 321-326.

Blaha, L. M., Johnson, S. A., & Townsend, J. T. (2007). An information processing inverstigation of hierarchical form perception: Evidence for parallel processing. *Visual Cognition: Object Perception, Attention and Memory 2006 Conference Report, 15*, 73-77.

Donders, F. C. (1868). Over de snelheid van psychische processen. Onderzoekingen gedaan in het Physiologisch Laboratorium der Uterchtsche Hoogeschool, 1868-1869. *Tweeds Reeks, 2*, 92–120.

Egeth, H. E. (1966). Parallel versus serial processes in multidimensional stimulus discrimination. *Attention, Perception, & Psychophysics, 1*,

245–252.

Eidels, A., Houpt, J. W., Altieri, N., Pei, L., & Townsend, J. T. (2010). Nice guys finish fast, bad guys finish last: Facilitatory vs. inhibitory interaction in parallel systems. *Journal of Mathematical Psychology, 55*, 176–190.

Eidels, A., Townsend, J. T., & Algom, D. (2010). Comparing perception of Stroop stimuli in focused versus divided attention paradigms: Evidence for dramatic processing differences. *Cognition, 114*, 129–150.

Eidels, A., Townsend, J. T., & Pomerantz, J. R. (2008). Where similarity beats redundancy: The importance of context, higher order similarity, and response assignment. *Journal of Experimental Psychology: Human Perception and Performance, 34*, 1441–1463.

Eriksen, C. W., & Spencer, T. R. (1969). Rate of information processing in visual perception: Some results and methodological considerations. *Journal of Experimental Psychology Monographs, 79*, 1–16.

Estes, W., & Taylor, H. (1964). A detection method and probabilistic models for assessing information processing from brief visual displays. *Proceedings of the National Academy of Science, 52*, 446–454.

Fific, M., Nosofsky, R. M., & Townsend, J. T. (2008). Information-processing architectures in multidimensional classification: A validation test of the systems factorial technology. *Journal of Experimental Psychology: Human Perception and Performance, 34*, 356–375.

Fific, M., & Townsend, J. T. (2010). Information-processing alternatives to holistic perception: Identifying the mechanisms of secondary-level holism within a categorization paradigm. *Journal of Experimental Psychology: Learning, 36*, 1290–1313.

Fific, M., Townsend, J. T., & Eidels, A. (2008). Studying visual search using systems factorial methodology with target–distractor similarity as the factor. *Attention, Perception, & Psychophysics, 70*, 583–603.

Garnes, W. R. (1974). *The Processing of Information and Structure*. NY: John Wiley & Sons.

Hamilton, J. (1859). *Lectures on Metaphysics and Logic*. Edinburgh: Blackwood.

Hick, W. E. (1952). On the rate of gain of information. *Quarterly Journal of Experimental Psychology, 4*, 11–36.

Houpt, J. W., & Townsend, J. T. (2010). The statistical properties of the survivor interaction contrast. *Journal of Mathematical Psychology, 54*, 446–453.

Johnson, S. A., Blaha, L. M., Houpt, J. W., & Townsend, J. T. (2010). Systems factorial technology provides new insights on global–local information processing in autism spectrum disorders. *Journal of Mathematical Psychology, 54*, 53–72.

Krueger, L. E. (1978). A theory of perceptual matching. *Psychological Review, 85*, 278–304.

Laming, D. R. J. (1968). *Information Theory of Choice-Reaction Times.* London: Academic Press.

Luce, P. A. (1986). A computational analysis of uniqueness points in auditory word recognition. *Attention, Perception, & Psychophysics, 39*, 155–158.

Miller, J. O. (1982). Divided attention: Evidence for coactivation with redundant signals. *Cognitive Psychology, 14*, 247–279.

Murdock, B. B. (1971). A parallel-processing model for scanning. *Attention, Perception, & Psychophysics, 10*, 289–291.

Neisser, U. (1967). *Cognitive Psychology.* NY: Appleton-Century-Crofts.

Pachella, R. G. (1974). The interpretation of reaction time in information processing research. In B. Kantowitz (Ed.), *Human Information Processing.* Hillsdale, NJ: Erlbaum Press.

Pieters, J. P. (1983). Sternberg's additive factor method and underlying psychological processes: Some theoretical considerations. *Psychological Bulletin, 93*, 411–426.

Posner, M. I. (1978). *Chronometric Explorations of Mind.* Hillsdale, NJ: Erlbaum Press.

Ross, B. H., & Anderson, J. R. (1981). A test of parallel versus serial processing applied to memory retrieval. *Journal of Mathematical Psychology, 24*, 183–223.

Schweickert, R. (1978). A critical path generalization of the additive factor method: Analysis of a Stroop task. *Journal of Mathematical Psychology, 18*, 105–139.

Schweickert, R. (1982). The bias of an estimate of coupled slack in stochastic PERT networks. *Journal of Mathematical Psychology, 26*, 1–12.

Schweickert, R. (1985). Separable effects of factors on speed and accuracy: Memory scanning, lexical decision, and choice tasks. *Psychological Bulletin, 97*, 530–546.

Schweickert, R. (1989). Separable effects on factors on activation functions in discrete and continuous models. *Psychological Bulletin, 106*, 318–328.

Schweickert, R., Giorgini, M., & Dzhafarov, E. N. (2000). Selective influence and response time cumulative distribution functions in serial-parallel task networks. *Journal of Mathematical Psychology, 44,* 504–535.

Schweickert, R., & Townsend, J. T. (1989). A trichotomy: Interactions of factors prolonging sequential and concurrent mental processes in stochastic discrete mental (PERT) networks. *Journal of Mathematical Psychology, 33,* 328–347.

Schweickert, R., & Wang, Z. (1993). Effects on response time of factors selectively influencing processing in acyclic task networks with OR gates. *British Journal of Mathematical and Statistical Psychology, 46,* 1–30.

Shiffin, R. M., & Garner, G. T. (1972). Visual processing capacity and attentional control. *Journal of Experimental Psychology, 93,* 72–82.

Smith, J. E. K. (1968). Models of confusion. Paper presented to the Psychonomic Society, St. Louis, MO.

Snodgrass, J. G. (1972). Reaction times for comparisons of successively presented visual patterns: Evidence for serial self-terminating search. *Perception and Psychophysics, 12,* 364–372.

Snodgrass, J. G., & Townsend, J. T. (1980). Comparing parallel and serial models: Theory and implementation. *Journal of Experimental Psychology: Human Perception and Performance, 6,* 330–354.

Sternberg, S. (1966). High-speed scanning in human memory. *Science, 153,* 652–654.

Sternberg, S. (1969). The discovery of processing stages: Extensions of Donders' method. *Acta Psychologica, 30,* 276–315.

Sung, K. (2008). Serial and parallel attentive visual searches: Evidence from cumulative distribution functions of response times. *Journal of Experimental Psychology: Human Perception and Performance, 34,* 1372–1388.

Taylor, D. A. (1976). Stage analysis of reaction time. *Psychological Bulletin, 83,* 161–191.

Thorton, T. L., & Gilden, D. L. (2007). Parallel and serial processes in visual search. *Psychological Review, 114,* 71–103.

Townsend, J. T. (1971). A note on the identifiability of parallel and serial processes. *Attention, Perception, & Psychophysics, 10,* 161–163.

Townsend, J. T. (1972). Some results concerning the identifiability of parallel and serial processes. *British Journal of Mathematical and Statistical Psychology, 25,* 168–199.

Townsend, J. T. (1974). Issues and models concerning the processing of a finite number of inputs. In B. Kantowitz (Ed.), *Human Information Processing.* Hillsdale, NJ: Erlbaum Press.

Townsend, J. T. (1976). Serial and within-stage independent parallel model equivalence on the minimum completion time. *Journal of Mathematical Psychology, 14*, 219–238.

Townsend, J. T. (1981). Some characteristics of visual whole report behavior. *Acta Psychologica, 47*, 149–173.

Townsend, J. T. (1984). Uncovering mental processes with factorial experiments. *Journal of Mathematical Psychology, 28*, 363–400.

Townsend, J. T. (1990a). Truth and consequences of ordinal differences in statistical distributions: Toward a theory of hierarchical inference. *Psychological Bulletin, 108*, 551–567.

Townsend, J. T. (1990b). Serial vs. parallel processing: Sometimes they look like Tweedledum and Tweedledee but they can (and should) be distinguished. *Psychological Science, 1*, 46–54.

Townsend, J. T., & Ashby, F. G. (1976). Toward a theory of letter recognition: Testing contemporary mathematical models. Paper presented to Midwestern Psychological Association, Chicago, IL.

Townsend, J. T., & Ashby, F. G. (1983). *The Stochastic Modeling of Elementary Psychological Processes*. Cambridge, MA: Cambridge University Press.

Townsend, J. T., & Evans, R. (1983). A systems approach to parallel-serial testability and visual processing. *Advances in Psychology, 11*, 166–191.

Townsend, J. T., & Fial, R. (1968). Spatiotemporal characteristics of multi-symbol perception. Paper presented to the First Annual Mathematical Psychology Meetings, Stanford, CA.

Townsend, J. T., & Fific, M. (2004). Parallel versus serial processing and individual differences in high-speed search in human memory. *Attention, Perception, & Psychophysics, 66*, 953–962.

Townsend, J. T., Fific, M., & Neufeld, R. W. J. (2007). Assessment of mental architecture in clinical/cognitive research. In T. A. Treat, R. R. Bootzin, and T. B. Baker (Eds.), *Psychological Clinical Science: Papers in Honor of Richard M. Mcfall*. Mahwah, NJ: Lawrence Erlbaum Associates.

Townsend, J. T., & Nozawa, G. (1988). Strong evidence for parallel processing with simple dot stimuli. Paper presented to the Twenty-Ninth Annual Meeting of Psychonomics Society, Chicago, IL.

Townsend, J. T., & Nozawa, G. (1995). Spatio-temporal properties of elementary perception: An investigation of parallel, serial, and coactive theories. *Journal of Mathematical Psychology, 39*, 321–359.

Townsend, J. T., & Schweickert, R. (1985). Interactive effects of factors prolonging processes in latent mental networks. In G. d'Ydewalle (Ed.), *Cognition, Information Processing, and Motivation: Proceedings of the XXIII International Congress of Psychology*, vol. 3 (pp. 255-276). Amsterdam: North Holland.

Townsend, J. T., & Schweickert, R. (1989). Toward the trichotomy method of reaction times: Laying the foundation of stochastic mental networks. *Journal of Mathematical Psychology, 33*, 309–327.

Townsend, J. T., & Snodgrass, J. G. (1974). A serial vs. parallel testing paradigm when same and different comparison rates differ. Paper presented to the Psychonomic Society, Bosten, MA.

Townsend, J. T., & Thomas, R. D. (1994). Stochastic dependencies in parallel and serial models: Effects on systems factorial interactions. *Journal of Mathematical Psychology, 38*, 1–34.

Townsend, J. T., & Wenger, M. J. (2004). A theory of interactive parallel processing: New capacity measures and predictions for a response time inequality series. *Psychological Review, 111*, 1003–1035.

Travers, J. R. (1973). The effects of forced serial processing on identification of words and random letter strings. *Cognitive Psychology, 5*, 109–137.

Treisman, A., & Gormican, S. (1988). Feature analysis in early version: Evidence from search asymmetries. *Psychological Review, 95*, 15–48.

Van Zandt, T. (1988). Testing serial and parallel processing hypothesis in visual whole report experiments. Master's thesis, Purdue University, West Lafayette, IN.

Van Zandt, T. (2000). How to fit a response time distribution. *Psychonomic Bulletin & Review, 7*, 424–465.

Van Zandt, T. (2002). Analysis of response time distributions. In J. T. Wixted and H. Pashler (Eds.), *Stevens' Handbook of Experimental Psychology, Vol. 4: Methodology in Experimental Psychology* (3rd ed.). NY: John Wiley & Sons.

Vorberg, D. (1977). On the equivalence of parallel and serial models of information processing. Paper presented to the Tenth Annual Mathematical Psychological Meeting, Los Angeles, CA.

Welford, A. T. (1980). *Reaction Times.* New York: Academic Press.

Wenger, M. J., & Townsend, J. T. (2001). Faces as gestalt stimuli: Process characteristics. In M. J. Wenger and J. T. Townsend (Eds.), *Computational, Geometric, and Process Perspectives of Facial Cognition* (pp. 229–284). Mahwah, NJ: Erlbaum.

Wenger, M. J., & Townsend, J. T. (2006). On the costs and benefits of faces and words: Process characteristics of feature search in highly meaningful stimuli. *Journal of Experimental Psychology: Human Perception and Performance, 33*, 755-779.

Wolfe, J. M. (1994). Visual search in continuous, naturalistic stimuli. *Vision Research, 34*, 1187-1195.

Chapter 12

Model Selection with Informative Normalized Maximum Likelihood: Data Prior and Model Prior

Jun Zhang

University of Michigan

Normalized maximal likelihood (NML) is a probability distribution over the sample space associated with a parametric model class. At each sample point, the value of an NML distribution is obtained by taking the value of the maximum likelihood estimator (ML value) corresponding to that sample point (data), and then normalizing across the sample space to give rise to a probability measure. In this chapter the minimax problem that leads to the above (non-informatve) NML as its solution is revisited by introducing an arbitrary weighting function over sample space. The solution then becomes "informative NML", which involves both a prior distribution over the sample space ("data prior") and a prior distribution over a model class ("model prior"), with the obvious Bayesian interpretations of the ML values. This approach avoids the so-called "infinity problem" of the non-informative NML, namely the unboundedness of the logarithm of the normalization factor (which serves as an index for model complexity), while at the same time providing a notion of consistency between modeler's prior beliefs about models and about data.

12.1. Introduction

Model selection has, in the last decade, undergone rapid growth for evaluating models of cognitive processes, ever since its introduction to the mathematical/cognitive psychology community (Myung, Forster, & Browne, 2000). The term "model selection" refers to the task of selecting, among several competing alternatives, the "best" statistical model given experimental data. To avoid ambiguity, "best" here has a now-standard operational definition – the commonly accepted criterion is that models must not only show reasonable goodness-of-fit in accounting for existing data, but also demonstrate some kind of simplicity so that it would not capture sampling noise in the data. This criteria, emphasizing generalization as opposed to fitting as the goal of modeling, embodies Occam's Razor, the principle of offering parsimonious explanation of data with fewest assumptions. Though mathematical implementations may differ, resulting in the various methods such as AIC, BIC, MDL, etc., each invariably boils down to balancing two aspects of model evaluation, one measuring its goodness-of-fit over existing data and the other measuring its complexity or capability for generalization.

The Minimum Description Length (MDL) Principle (Rissanen, 1978, 1983, 1996, 2001) is an information theoretic approach to inductive inference with roots in algorithmic coding theory. It has become one of the most popular means for model selection (Grünwald, Myung, & Pitt, 2005; Grünwald, 2007). Under this approach, data are viewed as codes to be compressed by the model. The goal of model selection is to identify the model, from a set of candidate models, that permits the shortest description length (code) of the data. The state-of-the-art of MDL approach to model selection has evolved into using the so-called Normalized Maximum Likelihood, or NML for short (Rissanen, 1996, 2001), as the criterion for model selection. In this chapter, this framework is revisited, and then modified by formally introducing the notion of "data prior". This turns the (non-informative) NML framework into the "informative" NML framework, which carries Bayesian interpretations. Informative NML subsumes (the traditional, non-informative) NML for the case of data prior being uniform, much in the same way that Bayesian inference subsumes maximal likelihood inference for the case of prior over hypotheses (parameters) being uniform.

12.2. A Revisit to NML

12.2.1. *Construction of normalized maximal likelihood*

Denote the set of probability distributions f over some sample space $\mathcal{X}$ as[1]

$$\mathcal{B} = \{f : \mathcal{X} \to [0,1], \ f > 0, \ \sum_{x \in \mathcal{X}} f(x) = 1\} \ .$$

We will use the term "model class", denoted by $\mathcal{M}_\gamma$ with a structural index γ, to specifically refer to a parametric family $\mathcal{M}_\gamma$ of probability distribu-

---

[1]We assume, for ease of exposition, that sample space $\mathcal{X}$ is discrete and hence use the summation notation $\sum_{x \in \mathcal{X}} \{\cdot\}$. When $\mathcal{X}$ is uncountable, f is taken to be the probability density function with the summation sign replaced by $\int_{\mathcal{X}} \{\cdot\} d\mu$ where $\mu(dx) = d\mu$ is the background measure on $\mathcal{X}$.

tions all of functional form

$$\mathcal{M}_\gamma = \{f(\cdot|\theta) \in \mathcal{B}, \ \forall \theta \in \Theta \subseteq \Re^m\} \ ;$$

in other words, for any fixed θ,

$$f(x|\theta) > 0 \ , \quad \sum_{x \in \mathcal{X}} f(x|\theta) = 1 \ .$$

The NML distribution $p^*(x)$ computed from the entire model class is, by definition,

$$p^*(x) = \frac{f(x|\hat{\theta}(x))}{\sum_{y \in \mathcal{X}} f(y|\hat{\theta}(y))}, \tag{12.1}$$

where $\hat{\theta}(\cdot)$ denotes the maximum likelihood estimator

$$\hat{\theta}(x) = \mathrm{argmax}_\theta f(x|\theta) \ . \tag{12.2}$$

Note that, in general $p^*(x)$ itself may not be a member of the family $\mathcal{M}_\gamma$ of the distributions in question,

$$p^*(\cdot) \notin \mathcal{M}_\gamma,$$

because it is obtained by a) selecting *one* parameter $\hat{\theta}(x)$ (and hence one distribution in $\mathcal{M}_\gamma$) for *each point* x of the sample space $\mathcal{X}$ (i.e., for each data point), then b) using the corresponding value of the distribution function $f(x|\hat{\theta}(x))$, and finally c) normalizing across all possible data points $x \in \mathcal{X}$. The NML distribution is a *universal* distribution, in the sense of being generated from the family $\mathcal{M}_\gamma$ (i.e., an entire class) of probability distributions; it (generally) does not, however, correspond to any individual distribution within that family. See Figure 12.1.

12.2.2. *Code length, universal distribution, and complexity measure*

In algorithmic coding theory, the negative logarithm of a distribution corresponds to the "code length". Under this interpretation, $p^*(x)$ is identified as the length of an ideal code for a model class

$$\text{ideal code length} = -\log p^*(x) = -\log f(x|\hat{\theta}(x)) + \log \sum_{y \in \mathcal{X}} f(y|\hat{\theta}(y)).$$
$$\tag{12.3}$$

For arguments of such coding scheme being "ideal" in the context of model selection, see Myung, Navarro, and Pitt (2006). It suffices to point out that as a criterion for model selection, the two terms in (12.3) describe on

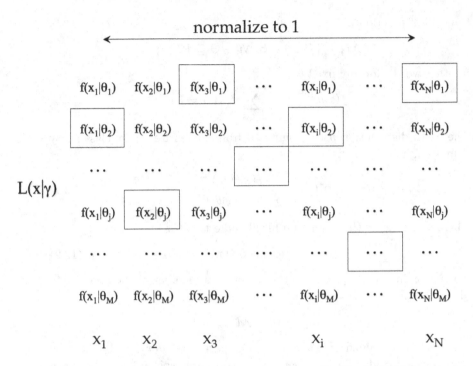

Fig. 12.1. Schematic diagram of normalized maximum likelihood (NML) for a model class M_γ whose likelihood functions $f(\cdot|\theta)$ are parameterized by θ. Each row represents the probability density (mass) indexed by a particular θ_j value as parameter, so each row sums to 1. On the bottom, x represents all possible data, with each data point x_i "selecting" (across the corresponding column) a particular $\hat{\theta}$ with the largest likelihood value, indicated by a box. The ML function $f(x|\hat{\theta}(x))$, which is also denoted $L(x|\gamma) \equiv L_\gamma(x)$, is a map from x to the largest likelihood value shown in the box. Their sum, denoted $\hat{C}_\gamma$, may not equal 1. Normalizing $f(x|\hat{\theta}(x))$ by $\hat{C}_\gamma$ gives the NML function.

the one hand the goodness-of-fit of a model with its best-fitting parameter (first term) and on the other the complexity of a model class (second term). Therefore, the general philosophy of NML falls in the same spirit of properly balancing two opposing tensions in model construction, namely, better approximation versus lower complexity, to achieve the goal of best generalizability.

Note that in (12.1) the probability that the universal distribution p^* assigns to the observed data x is proportional to the maximized likelihood $f(x|\hat{\theta}(x))$, and the normalizing constant

$$C_\gamma = \sum_{y \in \mathcal{X}} f(y|\hat{\theta}(y)) \tag{12.4}$$

is the sum of maximum likelihoods of all potential data that could be observed in an experiment. It is for this reason that p^* is often called the normalized maximum likelihood (NML) distribution associated with model class $\mathcal{M}_\gamma$. The NML distribution is specified once the functional (i.e., parametric) form of the model class is given. It is determined prior to an experiment, that is, prior to any specific data point x being given. Model complexity, as represented by the second term of (12.3), is operationalized as the logarithm of the *sum of all best-fits* a model class can provide collectively. This complexity measure therefore formalizes the intuition that the model that fits almost every data pattern very well would be much more complex than a model that provides a relatively good fit to a small set of data patterns but does poorly otherwise.

The NML distribution p^* is derived as a solution to a *minimax problem*: Find the distribution that minimizes the worst-case average regret (Rissanen, 2001):

$$p^* \longleftarrow \inf_{q \in \mathcal{B}} \sup_{g \in \mathcal{B}} \mathrm{E}_g \left\{ \log \frac{f(y|\hat{\theta}(y))}{q(y)} \right\} \tag{12.5}$$

where p, q ranges over the entire $\mathcal{B}$, the set of all probability distributions, and $\mathrm{E}_g\{\cdot\}$ denotes the taking of expectation

$$\mathrm{E}_g\{F(y)\} = \sum_{y \in \mathcal{X}} g(y) F(y) .$$

The solution, p^*, is *not* constrained to be in the set $\mathcal{M}_\gamma$. The basic idea of this minimax approach to model selection is to identify a single probability distribution that is "universally" representative of an entire parametric family of distributions and that mimics the behavior of any member of that family in the sense formulated in (12.5) (Barron, Rissanen, & Yu, 1998; Hansen & Yu, 2001). Since its computation does not invoke or even assume the existence of a true, data-generating distribution, the NML distribution is said to be "agnostic" from the truth distribution (Myung, Navarro, & Pitts, 2006), though such claim about "agnosticity" is the subject of some debate (Karabatsos & Walker, 2006; Grünwald & Navarro, 2009; Karabatsos & Walker, 2009). The debate is centered around whether the Bayesian approach under a non-informative Dirichlet process prior can be viewed as identical to that of maximal likelihood estimator, and whether the choice of a particular form of penalty function is *a priori* motivated.

12.2.3. NML and Bayesianism with non-informative prior

Under asymptotic expansion, the negative logarithm of the NML distribution can be shown (Rissanen, 1996) to be:

$$-\log p^*(x) = -\log f(x|\hat{\theta}(x)) + \frac{k}{2}\log\left(\frac{n}{2\pi}\right) + \log\int_{\Theta}\sqrt{\det \mathrm{I}(\theta)}\,d\theta + o(1)$$

$$(12.6)$$

where n denotes the sample size, k is the number of model parameters, and $\mathrm{I}(\theta)$ is the Fisher information matrix

$$\mathrm{I}(\theta) = \sum_{x \in \mathcal{X}} f(x|\theta)\,\frac{\partial \log f(x|\theta)}{\partial \theta^i}\frac{\partial \log f(x|\theta)}{\partial \theta^j}\,.$$

The expression (12.6) was called the "Fisher information approximation (FIA) to the NML criterion" (Pitt, Myung, & Zhang, 2002). The first two terms are known as the Bayesian Information Criterion (BIC; Schwartz, 1978). The third term of (12.6) involving the Fisher information also appeared from a formulation of Bayesian parametric model selection (Balasubramanian, 1997). This hints at the deeper connection between NML approach and Bayesian approach to model selection. We elaborate here.

In Bayesian model selection, the goal is to choose, among a set of candidate models, the one with the largest value of the marginal likelihood for observed data x, defined as

$$p_{\mathrm{Bayes}}(x) = \int_{\Theta} f(x|\theta)\pi(\theta)d\theta \qquad (12.7)$$

where $\pi(\theta)$ is a prior on the parameter space. A specific choice is the Jeffrey's prior $\pi_J(\theta)$, which is non-informative

$$\pi_J(\theta) = \frac{\sqrt{\det I(\theta)}}{\int_{\Theta}\sqrt{\det I(\theta)}d\theta}\,.$$

An analysis by Balasubramanian (1997) shows that if $\pi_J(\theta)$ is used in (12.7), then an asymptotic expansion of $-\log p_{\mathrm{Bayes}}(x)$ yields an expression with the same three leading terms as in (12.6). In other words, for large n, Bayesian model selection with (the non-informative) Jeffrey's prior and NML become virtually indistinguishable. This observation parallels the findings by Takeuchi and Amari (2005) that the asymptotic expressions of various estimators, including MDL, projected Bayes estimator, bias-corrected MLE, each of which indexes a point (value of θ) in the model manifold, were related to the choice of priors; this in turn has an information geometric interpretation (Matsuzoe, Takeuchi, & Amari, 2006).

Note that in the NML approach, data is assumed to be drawn from the sample space according to a uniform distribution: the summation $\sum_{x \in \mathcal{X}}$ treats every data x with the same weight. In algorithmic coding applications, this is not a problem because here the data are the symbols under transmission which can be pre-defined to occur equally likely by the encoder and the decoder. In model selection applications where data will most likely be generated from a non-uniform distribution, care must be taken to calculate such quantities like (12.4). If the summation is taken over the stream of data that follow each other (i.e., as the data generation process is being realized), then the multiplicity in any sample value x will be naturally taken into account. On the other hand, if the summation is taken a priori (i.e., the data generation process is being assumed), then proper weighting of the data stream is called for. In this contribution, we explore a generalization of the NML formulation about model complexity measure by explicitly considering the modeler's prior belief about data and prior belief of the model classes ("prior" in comparison with data collecting and model fitting).

12.3. NML with Informative Priors

Recall that the normalizing constant in (12.1) is obtained by first finding the maximum likelihood value for each sample point and then summing all such maximum likelihood values across the sample space. An implicit assumption behind this definition of model complexity is that every sample point is equally likely to occur a priori (i.e., before data collection). In terms of the Bayesian language, this amounts to assuming no prior information about possible data patterns. In this sense, NML may be viewed as a "non-informative" MDL method.

In practice, however, it is common that information about the possible patterns of data is available prior to data collection. For example, in a memory retention experiment, one can expect that the proportion of words recalled is likely to be a decreasing function of time rather than an increasing function, that retention performance will be in general worse under free recall than under cued recall, that the rate in which information is forgotten or lost in memory will be greater for uncommon, low frequency words than for common, high frequency words, etc. Such prior information implies that not all data patterns are equally likely. It would be advantageous to incorporate such information in the model selection process. The exposition below explores the possibility of developing an "informative" version of NML.

12.3.1. *Universal distribution with data-weighting*

Recall that in point estimation, a given data point $x \in \mathcal{X}$ selects, within the entire model class $\mathcal{M}_\gamma$, a particular distribution with parameter $\hat{\theta}$:

$$x \to \hat{\theta} \rightsquigarrow f(\cdot|\hat{\theta}) \in \mathcal{M}_\gamma .$$

Here $\hat{\theta} : \mathcal{X} \to \Theta$ is some estimating function, for example, the MLE as given by (12.2). The $\rightsquigarrow$ sign is taken to mean "selects". The expression $f(y|\hat{\theta}(x))$, when viewed as a function of y for any fixed x, is a probability distribution that belongs to the family $\mathcal{M}_\gamma$ (i.e., is one of its elements). Evaluated at $y = x$, we denote $f(x|\hat{\theta}(x)) \equiv L_\gamma(x)$, viewed now as a function of the data x explicitly (recall that γ is the index for model class $\mathcal{M}_\gamma$). Note that $L_\gamma(x)$ is not a probability distribution; $\sum_x L_\gamma(x) \neq 1$ in general. The NML distribution $p^*(x)$, which is the normalized version of $L_\gamma(x)$, is derived as the solution of the minimax problem (12.5), over the yet-to-be determined distribution $q(x)$, with regret given as $\log(L_\gamma(x)/q(x))$. Now, instead of using this regret function, we use $\log(s(x)L_\gamma(x)/q(x))$ and consider a more general minimax problem

$$\inf_{q \in \mathcal{B}} \sup_{g \in \mathcal{B}} \mathrm{E}_g \left\{ \log \frac{s(y)\, L_\gamma(y)}{q(y)} \right\} , \qquad (12.8)$$

where $s(x)$ is any positively-valued function of x.

Proposition 12.1. *The solution to the minimax problem (12.8) is given by $q(\cdot) = p(\cdot|\gamma)$ where*

$$p(x|\gamma) \equiv \frac{s(x)\, L_\gamma(x)}{\hat{C}_\gamma} = \frac{s(x)\, L_\gamma(x)}{\sum_{y \in \mathcal{X}} s(y)\, L_\gamma(y)} ; \qquad (12.9)$$

the minimaximizing bound is $\log \hat{C}_\gamma$ where

$$\hat{C}_\gamma = \sum_{y \in \mathcal{X}} s(y)\, L_\gamma(y) . \qquad (12.10)$$

Proof. Our proof follows that of Rissenan (2001) with only slight modifications. First, noting the elementary relation

$$\inf_{q \in \mathcal{B}} \sup_{g \in \mathcal{B}} G(g,q) \geq \sup_{g \in \mathcal{B}} \inf_{q \in \mathcal{B}} G(g,q)$$

for any functional $G(g, p)$. Applying this to (12.8), the quantity $\{\cdot\}$ under minimaximizing,

$$\mathrm{E}_g \left\{ \log \frac{s(y) L_\gamma(y)}{q(y)} \right\} = \mathrm{E}_g \left\{ \log \frac{g(y)}{q(y)} \right\} - \mathrm{E}_g \left\{ \log \frac{g(y)}{s(y) L_\gamma(y)} \right\}$$

$$= \mathrm{E}_g \left\{ \log \frac{g(y)}{q(y)} \right\} - \mathrm{E}_g \left\{ \log \frac{g(y)}{p(x|\gamma)} \right\} + \log \hat{C}_\gamma = D(g\|q) - D(g\|p) + \log \hat{C}_\gamma \ ,$$

where $D(\cdot\|\cdot)$ is the non-negative Kullback-Leibler divergence

$$D(g\|q) = \mathrm{E}_g \left\{ \log \frac{g(y)}{q(y)} \right\} = \sum_{y \in \mathcal{X}} g(y) \log \frac{g(y)}{q(y)} \ .$$

Therefore

$$\inf_{q \in \mathcal{B}} \sup_{g \in \mathcal{B}} \mathrm{E}_g \left\{ \log \frac{s(y) \, L_\gamma(y)}{q(y)} \right\} \geq \sup_{g \in \mathcal{B}} \inf_{q \in \mathcal{B}} \left(D(g\|q) - D(g\|p) + \log \hat{C}_\gamma \right)$$

$$= \sup_{g \in \mathcal{B}} \left(-D(g\|p) + \log \hat{C}_\gamma \right) = \log \hat{C}_\gamma$$

where the infimum (over q) in the last-but-one step is achieved for $q = g$ and the supremum (over g) in the last step is achieved for $g = p$. Therefore, the solution to (12.8) is achieved when $q = p(\cdot|\gamma)$. $\qquad\square$

Remark 12.1. The distribution $p^*(x)$, that is, non-informative NML (12.1), is known (Shtarkov, 1987) also to be the solution of the following slightly different minimax problem:

$$\inf_{q \in \mathcal{B}} \sup_{y \in \mathcal{X}} \log \frac{f(y|\hat{\theta}(y))}{q(y)} \ .$$

We can modify the above to yield a minimax problem (with given $s(y)$)

$$\inf_{q \in \mathcal{B}} \sup_{y \in \mathcal{X}} \log \frac{s(y) f(y|\hat{\theta}(y))}{q(y)} \ ,$$

and show that (12.9) is also its solution. The proof of this statement follows readily from the proof of Proposition 12.1.

We call (12.9) the informative NML distribution, which depends on an arbitrary positively-valued function $s(\cdot)$. Clearly, for all densities g,

$$\mathrm{E}_g \left\{ \log \frac{s(y) \, L_\gamma(y)}{p(y|\gamma)} \right\} = \mathrm{E}_g \log \hat{C}_\gamma = \log \hat{C}_\gamma$$

is constant. When $s(y) = const$, then

$$\hat{C}_\gamma \rightsquigarrow const \sum_y L_\gamma(y) = const \, C_\gamma$$

with

$$p(x|\gamma) \rightsquigarrow p^* = \frac{L_\gamma(x)}{\sum_y L_\gamma(y)} \ ,$$

both reducing to the (non-informative) NML solution derived by Rissanen (2001). The difference between $p(x|\gamma)$ and p^* is, essentially, the use of $s(x)L_\gamma(x)$ in place of $L_\gamma(x)$, that is, the maximal likelihood value $L_\gamma(x)$ at a data point x is weighted by a non-uniform, data-dependent factor $s(x)$. The data-dependency of the universal distribution (which in general still lies outside the manifold of the model class) qualifies it for the term "informative" NML (just as the parameter-dependency of a prior distribution in the Bayesian formulation qualifies it as an "informative prior").

Note that the function $s(x)$ in Proposition 12.1 can be any positively-valued function defined on $\mathcal{X}$. And the choice of $s(x)$ would affect the complexity measure $\hat{C}_\gamma$, which is also always positive.

12.3.2. *Prior over data and prior over model class*

The maximal likelihood values $L_\gamma(x)$ from model class γ over data x are a series of positive values; normalization over x gives the (non-informative) NML distribution $p^*(x)$ in Rissanen's (2001) analysis. Here, it is presupposed that the modeler has a prior belief π_γ about the plausibility of various model classes γ (with $\pi_\gamma > 0, \sum_\gamma \pi_\gamma = 1$), and a prior belief $\pi(x)$ about the credibility of the data x (with $\pi(x) > 0, \sum_x \pi(x) = 1$). These two types of prior beliefs may not be "compatible", in some sense yet to be specified more accurately below.

Let us take

$$s(x) = \frac{\pi(x)}{\sum_\gamma \pi_\gamma L_\gamma(x)} \ . \tag{12.11}$$

The meaning of such $s(x)$ will be elaborated later — it is related to, but not identical with, the so-called "luckiness prior" (Grünwald, 2007).

Note that (12.9) can be re-written as

$$p(x|\gamma) = \frac{p(\gamma|x)\, \pi(x)}{\sum_{y \in \mathcal{X}} p(\gamma|y)\, \pi(y)} \ , \tag{12.12}$$

where $p(\gamma|x)$ is defined by

$$p(\gamma|x) \equiv \frac{\pi_\gamma L_\gamma(x)}{\sum_\gamma \pi_\gamma L_\gamma(x)} \ .$$

Since the denominator of the right-hand side of the above expression involves a summation over γ (and not x), we can then obtain

$$p(\gamma|x) = \frac{\pi_\gamma \, p(x|\gamma)}{\sum_\gamma \pi_\gamma \, p(x|\gamma)} \; . \tag{12.13}$$

The two equations (12.12) and (12.13) clearly have Bayesian interpretations: when $\pi(x)$ is taken to be the modeler's initial belief about the data *prior to* modeling, the solution to the minimax problem, now in the form of (12.12), can be viewed as the *a posterior* description of the data from the perspective of the model class $\mathcal{M}_\gamma$, with $p(\gamma|x)$ as likelihood functions *about the various model classes*. Likewise, when π_γ is taken to be the modeler's initial belief about the model class $\mathcal{M}_\gamma$ *prior to* an experiment, $p(\gamma|x)$ as given by (12.13) can be viewed as the *a posterior* belief about the various model classes after experimentally obtaining and fitting data x, whereas the informative NML solution $p(x|\gamma)$ serves as the likelihood functions *about the data*. So, informative NML has *two* interpretations, a) as the posterior of the data given model, in (12.12), or b) as the likelihood function of the model given data, in (12.13).

The above two interpretations correspond to two ways (see Figure 12.2) the maximum likelihood values $L_\gamma(x)$ can be normalized: a) across data x to become the probability distribution over data $p(x|\gamma)$; and b) across model class $\mathcal{M}_\gamma$ to become the probability distribution over model class $p(\gamma|x)$. This demonstrates a duality between data and model from the modeler's perspective.

12.3.3. *Model complexity measure*

Let us now address the model complexity measure associated with the informative NML approach. Substituting (12.11) into the expression for the complexity measure $\hat{C}_\gamma$, we have

$$\pi_\gamma \hat{C}_\gamma = \sum_x p(\gamma|x)\pi(x) \; . \tag{12.14}$$

Explicitly written out

$$\hat{C}_\gamma = \sum_x \frac{\pi(x) L_\gamma(x)}{\sum_\gamma \pi_\gamma L_\gamma(x)} \; .$$

From (12.14), we obtain

$$\sum_\gamma \pi_\gamma \hat{C}_\gamma = 1 \; .$$

data prior $\pi(x)$

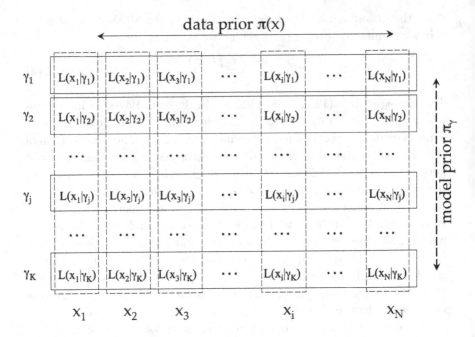

Fig. 12.2.   Illustration of data prior, model prior, and the ML values $L(x|\gamma) \equiv L_\gamma(x)$ for data points $x_1, x_2, \cdots, x_N$ across various model classes $M_{\gamma_1}, M_{\gamma_2}, \cdots, M_{\gamma_k}$. When model prior π_γ is given, ML values (viewed as columns) are used as the likelihood function of a particular data point for different model classes, in order to derive posterior estimates of model classes. When data prior $\pi(x)$ is given, ML values (viewed as rows) are used as the likelihood function of a particular model class for different data points, in order to derive posterior estimates of data (informative NML solution). Data prior and model prior can be made to be compatible (see Proposition 12.2).

This indicates that the new model complexity measure proposed here, $\hat{C}_\gamma$, is normalized after weighted by π_γ. The fact that $\pi_\gamma > 0$ implies that

$$\hat{C}_\gamma < \infty \ .$$

This solves a long-standing problem, the so-called "infinity problem" (Grünwald, Myung, & Pitt, 2005) associated with C_γ in the non-informative NML.

Recall that the non-informative NML follows the "two-part code" idea of MDL, that is, one part that codes the description of the hypothesis space (the functional form of the model class), the other part that codes the description of the data as encoded with respect to these hypotheses (the maximum likelihood value of the MLE). As such, the original minimax problem (12.5) has a clear interpretation of the "ideal code" from algorithmic coding perspective, with C_γ as the complexity measure of the model class. Here the normalization factor $\hat{C}_\gamma$ associated with the informative NML solution (12.9) has an analogous interpretation. The only difference is that the complexity measure now is dependent on the prior belief of the data $\pi(x)$ and the prior belief of the model classes π_γ, in addition to its dependency on the best-fits provided by each model class for all potential data.

The $s(x)$ factor introduced in the minimax problem given in (12.8) is related (but with important differences, see next subsection) to the "luckiness prior" introduced by Grünwald (2007). In the current setting, with $s(x)$ taking the specific form of (12.11), we have the following interpretation: for the occurrence of any data point x, the denominator $\sum_\gamma \pi_\gamma L_\gamma(x)$ gives expected occurrence of x from the modeler's prior knowledge about all models he/she builds, whereas the numerator $s(x)$ gives the modeler's prior knowledge about the data occurrence from a known data-acquisition procedure. Since the knowledge of the modeler/experimenter about model building and experimentation may come from different sources, the "luckiness" of acquiring data x as resulting from an experiment thus can be operationalized as the ratio of these two probabilities associated with different types of uncertainty about data.

Note that if and only if

$$\pi(x) = \text{const} \sum_\gamma \pi_\gamma L_\gamma(x), \qquad (12.15)$$

the luckiness factor $s(x) = \text{const}$; this is the case when the informative NML solution (12.9) reduces to the non-informative NML solution (12.1), both in this formulation, and in the approach reviewed by Grünwald (2007). We say that the modeler's prior belief over data and prior belief over model class are mutually *compatible* when (12.15) is satisfied (over all possible data values x). It is easy to see that luckiness is a constant (i.e., same across all data points) if and only if model prior and data prior are compatible.

Proposition 12.2. *The following three statements are equivalent:*

(a) Luckiness $s(x)$ is constant;
(b) Model prior π_γ and $\pi(x)$ are compatible;
(c) Informative NML is identical with non-informative NML.

12.3.4. *Data prior versus "luckiness prior"*

The data-dependent factor $s(x)$ introduced here, while in the same spirit of the so-called "luckiness prior" as in Grünward (2007, pp. 308-312), carries subtle differences. In Grünward's case, the corresponding minimax problem is

$$\inf_{q \in \mathcal{B}} \sup_{g \in \mathcal{B}} E_g \left\{ \log p(y|\hat{\theta}(y)) - \log q(y) - a(\hat{\theta}(y)) \right\}$$

and the extra factor $a(\hat{\theta}(y))$ is a function of the maximum likelihood estimator $\hat{\theta}(\cdot)$. In the present case, the minimax problem is

$$\inf_{q \in \mathcal{B}} \sup_{g \in \mathcal{B}} E_g \left\{ \log p(y|\hat{\theta}(y)) - \log q(y) + \log s(y) \right\},$$

with $s(y)$ a function defined on the sample space directly (and not through "pull-back"). However, both approaches to informative NML afford Bayesian interpretations. The approached described in Grünwald (2007) will lead to the *luckiness-tilted Jeffreys' prior* (p.313, ibid.),

$$\pi_{J,a}(\theta) = \frac{\sqrt{\det I(\theta)}\, e^{-a(\theta)}}{\int_{\Theta} \sqrt{\det I(\theta)}\, e^{-a(\theta)} d\theta},$$

which has the information geometric interpretation as an invariant volume form under a generalized conjugate connection on the manifold of probability density functions (Takeuchi & Amari, 2005; Zhang & Hasto, 2006; Zhang, 2007). The approach adopted in this chapter gives rise to a dual interpretation between model and data. Just as the maximum likelihood principle can be used to select the parameter (among all "competing" parameters) of a certain model class, the NML principle has been used to select a model class out of a set of competing models. Just as there is a Bayesian counterpart to the ML principle for parameter selection, what is proposed here is the Bayesian counterpart to NML, i.e., the use of maximum $p(\gamma|x)$ value (with fixed x, i.e., the given data) for model selection (among all possible model classes). The same, old debate and argument surrounding ML and Bayes can be brought back here — we are back to square one. Except that we are now operating at a higher level of explanatory hierarchy, namely, at the level of model classes (whereby each class is represented by a universal distribution through its maximum likelihood values after proper normalization); yet the duality between model and data still manifests itself.

12.4. General Discussions

To summarize this chapter, from the maximum likelihood function $f(x|\hat{\theta}(x)) \equiv L_\gamma(x)$ (where $\hat{\theta}$ is the MLE for the model class $\mathcal{M}_\gamma$), one can *either* construct the (non-informative) NML as a universal distribution of the model class γ through normalizing with respect to x, as Rissanen (2001) did, *or* derive the posterior distribution for model selection (12.13) through normalizing with respect to γ, as is done here. This has significant implications for model selection. In the former case, model selection is through the comparison of NML values for various model classes. Because the NML solution (12.1) is a probability distribution (in fact, universal distribution representing the particular model class) with total mass 1, then necessarily no single model can dominate (i.e., be the preferred choice) across all data! In other words, for any data x where model class γ_1 is preferred to model class γ_2, there exists some other data x' where model class γ_2 outperforms model class γ_1. Here, in our situation, we use an (informative) universal distribution which is interpreted as the likelihood function, with respect to a prior belief about all model classes — model selection is through computing the Bayes factor which combines the two data scenario. The dominance or superiority of one model class over another in accounting for all data is permitted under the current method.

12.5. Conclusion

Normalized maximal likelihood is a probability distribution over the sample space associated with a parametric model class. At each sample point, the value of an NML distribution is obtained by taking the likelihood value of the maximum likelihood estimator (ML value) corresponding to that sample point (data), and then normalizing across the sample space to give rise to the unit probability measure. Here, the minimax problem that leads to the above (non-informative) NML as its solution is revisited by our introducing an arbitrary weighting function over sample space. The solution then becomes "informative NML", which involves both a prior distribution over the sample space ("data prior") and a prior distribution over model class ("model prior"), with obvious Bayesian interpretations of the ML values. This approach avoids the so-called "infinity problem" of the non-informative NML, namely the unboundedness of the logarithm of the normalization factor (which serves as an index for model complexity), while at the same time providing a notion of consistency between the modeler's prior beliefs about models and data.

Acknowledgments

This chapter is based on preliminary results grown out of a discussion between the author and Jay Myung during 2005, which was reported to the 38th Annual Meeting of the Society for Mathematical Psychology held at the University of Memphis, TN (Zhang & Myung, 2005). The work has since been greatly expanded – the author benefited from subsequent discussions on this topic with Richard Shiffrin, who encouraged the development of the notion of a "data prior", and with Woojae Kim and Jay Myung, who caught an error in an earlier, circulating draft. The views of this chapter (and any mistakes therein) represent solely those of the author, and may be different from those commentators.

References

Balasubramanian, V. (1997). Statistical inference, Occam's razor and statistical mechanics on the space of probability distributions. *Neural Computation, 9*, 349–368.

Barron, A., Rissanen, J., & Yu, B. (1998). The minimum description length principle in coding and modeling, *IEEE Transactions on Information Theory, 44*, 2743–2760.

Grünwald, P., Myung, I. J., & Pitt, M. A. (2005). *Advances in Minimum Description Length: Theory and Applications.* Cambridge, MA: MIT Press.

Grünwald, P. (2007). *The Minimum Description Length Principle.* Cambridge, MA: MIT Press.

Grünwald, P., & Navarro, D. J. (2009). NML, Bayes and true distributions: A comment on Karabatsos and Walker (2006). *Journal of Mathematical Psychology, 53*, 43–51.

Hansen, M. H., & Yu, B. (2001). Model selection and the principle of minimum description length. *Journal of the American Statistical Association, 96*, 746–774.

Karabatsos, G., & Walker, S. G. (2006). On the normalized maximum likelihood and Bayesian decision theory. *Journal of Mathematical Psychology, 50*, 517–520.

Karabatsos, G., & Walker, S. G. (2009). Rejoinder on the normalized maximum likelihood and Bayesian decision theory: Reply to Grünwald and Navarro (2009). *Journal of Mathematical Psychology, 53*, 52.

Matsuzoe, H., Takeuchi, J., & Amari, S. (2006). Equiaffine structures on statistical manifolds and Bayesian statistics. *Differential Geometry and*

Its Applications, 24, 567–578.

Myung, I. J. (2000) The importance of complexity in model selection. *Journal of Mathematical Psychology, 44*, 190–204.

Myung, I. J., Forster, M. R., & Browne, M. W. (2000). Guest editors' introduction: Special issue on model selection. *Journal of Mathematical Psychology, 44*, 1–2.

Myung, J. I., Navarro, D. J., & Pitt, M. A. (2006). Model selection by normalized maximum likelihood. *Journal of Mathematical Psychology, 50*, 167–179.

Pitt, M. A., Myung, I. J., & Zhang, S. (2002). Toward a method of selecting among computational models of cognition. *Psychological Review, 109*, 472–491.

Rissanen, J. (1978). Modeling by the shortest data description. *Automata, 14*, 465–471.

Rissanen, J. (1983). A universal prior for integers and estimation by minimum description length. *Annals of Statistics, 11*, 416–431.

Rissanen, J. (1986). Stochastic complexity and modeling. *Annals of Statistics, 14*, 1080–1100.

Rissanen, J. (1996). Fisher information and stochastic complexity. *IEEE Transactions on Information Theory, 42*, 40–47.

Rissanen, J. (2000). MDL denoising. *IEEE Transactions on Information Theory, 46*, 2537–2543.

Rissanen, J. (2001). Strong optimality of the normalized ML models as universal codes and information in data. *IEEE: Information Theory, 47*, 1712–1717.

Schwarz, G. (1978). Estimating the dimension of a model. *The Annals of Statistics, 6*, 461-464.

Takeuchi, J., & Amari, S. (2005). α-Parallel prior and its properties. *IEEE Transaction on Information Theory, 51*, 1011–1023.

Zhang, J. (2007). A note on curvature of -connections on a statistical manifold. *Annals of Institute of Statistical Mathematics, 59*, 161–170.

Zhang, J., & Hasto, P. (2006). Statistical manifold as an affine space: A functional equation approach. *Journal of Mathematical Psychology, 50*, 60–65.

Zhang, J., & Myung, J. (2005). Informative normalized maximal likelihood and model complexity. Talk presented to the 38th Annual Meeting of the Society for Mathematical Psychology, University of Memphis, TN.

Subject Index